Lecture Notes in Computer Science   11608

Commenced Publication in 1973
Founding and Former Series Editors:
Gerhard Goos, Juris Hartmanis, and Jan van Leeuwen

More information about this series at http://www.springer.com/series/7409

Elisabeth Métais · Farid Meziane ·
Sunil Vadera · Vijayan Sugumaran ·
Mohamad Saraee (Eds.)

Natural Language Processing and Information Systems

24th International Conference on Applications
of Natural Language to Information Systems, NLDB 2019
Salford, UK, June 26–28, 2019
Proceedings

 Springer

Editors
Elisabeth Métais
Conservatoire National des Arts et Métiers
Paris, France

Sunil Vadera
University of Salford
Salford, UK

Mohamad Saraee
CSE
University of Salford
Salford, UK

Farid Meziane
University of Salford
Salford, UK

Vijayan Sugumaran
Oakland University
Rochester, MI, USA

ISSN 0302-9743   ISSN 1611-3349   (electronic)
Lecture Notes in Computer Science
ISBN 978-3-030-23280-1   ISBN 978-3-030-23281-8   (eBook)
https://doi.org/10.1007/978-3-030-23281-8

LNCS Sublibrary: SL3 – Information Systems and Applications, incl. Internet/Web, and HCI

This Springer imprint is published by the registered company Springer Nature Switzerland AG
The registered company address is: Gewerbestrasse 11, 6330 Cham, Switzerland

Preface

This volume of *Lecture Notes in Computer Science* (LNCS 11608) contains the papers presented at the 24th International Conference on Application of Natural Language to Information Systems, held at MediacityUK, University of Salford on the, during June 26–28, 2019 (NLDB 2019). Since its foundation in 1995, the NLDB conference has attracted state-of-the-art research and followed closely the developments of the application of natural language to databases and information systems in the wider meaning of the term.

The NLDB conference is now a well-established conference that attracts participants from all over the world. The conference evolved from the early years when most of the submitted papers where in the areas of natural language, databases and information systems to encompass more recent developments in the data and language engineering fields. The content of the current proceedings reflects these advancements. The conference also supports submissions on studies related to language that have not been well supported in the early years such as Arabic, Tamil, Hindi, and Farsi.

We received 75 papers and each paper was sent to at least three reviewers and reviewed by at least two. The conference co-chairs and Program Committee co-chairs had a final consultation meeting to look at all the reviews and make the final decisions on the papers to be accepted. We accepted 21 papers (28%) as long/regular papers and 16 short papers.

We would like to thank all the reviewers for their time, effort, and for completing their assignments on time albeit under tight deadlines. Many thanks to the authors for their contributions.

May 2019

Elisabeth Métais
Farid Meziane
Vijayan Sugumaran
Sunil Vadera
Mohamad Saraee

Organization

Conference Chairs

Elisabeth Métais	Conservatoire National des Arts et Metiers, Paris, France
Farid Meziane	University of Salford, UK
Sunil Vadera	University of Salford, UK

Program Committee Chairs

Mohamed Saraee	University of Salford, UK
Vijay Sugumaran	Oakland University Rochester, USA

Program Committee

Akoka Jacky	CNAM & TEM, France
Hidir Aras	FIZ Karlsruhe, Germany
Faten Atigui	CNAM, France
Imran Sarwar Bajwa	The Islamia University of Bahawalpur, Pakistan
Mithun Balakrishna	Lymba Corporation, USA
Pierpaolo Basile	University of Bali, Italy
Nicolas Béchet	IRISA, France
Sana Belguith	University of Salford, UK
Sandra Bringay	LIRMM, France
Paul Buitelaar	National University of Ireland Galway, Ireland
Raja Chiky	ISEP, France
Kostadin Cholakov	HU Berlin, Germany
Philipp Cimiano	Universität Bielefeld, Germany
Isabelle Comyn-Wattiau	CNAM, France
Flavius Frasincar	Erasmus University Rotterdam, The Netherlands
André Freitas	University of Passau, Germany/Insight, Ireland
Debasis Ganguly	Dublin City University, Ireland
Ahmed Guessoum	USTHB, Algiers, Algeria
Yaakov Hacohen-Kerner	Jerusalem College of Technology, Israel
Udo Hahn	Jena University, Denmark
Siegfried Handschuh	University of St.Gallen, Switzerland
Michael Herweg	IBM, Germany
Helmut Horacek	Saarland University, Germany
Ashwin Ittoo	HEC, University of Liege, Belgium
Paul Johannesson	Stockholm University, Sweden
Epaminondas Kapetanios	University of Westminster, UK
Zoubida Kedad	UVSQ, France

Eric Kergosien	GERiiCO, University of Lille, France
Christian Kop	University of Klagenfurt, Austria
Valia Kordoni	Saarland University, Germany
Elena Kornyshova	CNAM, France
Leila Kosseim	Concordia University, Canada
Mathieu Lafourcade	LIRMM, France
Els Lefever	Ghent University, Belgium
Jochen Leidner	Thomson Reuters, USA
Nguyen Le Minh	Japan Advanced Institute of Science and Technology, Japan
Cédric Lopez	VISEO – Objet Direct, France
D. Manjula	Anna University, Chennai, India
Heinrich C. Mayr	Alpen-Adria-Universität Klagenfurt, Austria
John McCrae	CITEC, Universität Bielefeld, Germany
Farid Meziane	Salford University, UK
Elisabeth Métais	CNAM, France
Marie-Jean Meurs	UQAM, Montreal, Canada
Luisa Mich	University of Trento, Italy
Andres Montoyo	Universidad de Alicante, Spain
Andrea Moro	Università di Roma La Sapienza, Italy
Rafael Muñoz	Universidad de Alicante, Spain
Yulong Pei	Eindhoven University of Technology, The Netherlands
Davide Picca	UNIL, Switzerland
Behrang Qasemizadeh	Heinrich-Heine-Universität Düsseldorf, Germany
Paolo Rosso	NLEL València, Spain
Mohamed Saraee	University of Salford, UK
Bahar Sateli	Concordia University, Canada
Khaled Shaalan	The British University in Dubai, UAE
Max Silberztein	Université de Franche-Comté, France
Kamel Smaili	University of Lorraine, France
Veda Storey	Georgia State University, USA
Vijayan Sugumaran	Oakland University Rochester, USA
Bernhard Thalheim	Kiel University, Germany
Krishnaprasad Thirunarayan	Wright State University, USA
Geetha T. V.	Anna University, India
Christina Unger	CITEC, Universität Bielefeld, Germany
L. Alfonso Ureña-López	University of Jaén, Spain
Sunil Vadera	University of Salford, UK
Panos Vassiliadis	University of Ioannina, Greece
Tonio Wandmacher	IRT SystemX, Saclay, France
Feiyu Xu	DFKI Saarbrücken, Germany
Wlodek Zadrozny	UNCC, USA

Contents

Short Papers

Full Papers

Deep Neural Network Models for Paraphrased Text Classification in the Arabic Language

Adnen Mahmoud[1,2](✉) and Mounir Zrigui[1]

[1] Algebra, Numbers Theory and Nonlinear Analyzes Laboratory LATNAL,
University of Monastir, Monastir, Tunisia
mahmoud.adnen@gmail.com, mounir.zrigui@fsm.rnu.tn
[2] Higher Institute of Computer Science and Communication Techniques,
Hammam Sousse, University of Sousse, Sousse, Tunisia

Abstract. Paraphrase is the act of reusing original texts without proper citation of the source. Different obfuscation operations can be employed such as addition/deletion of words, synonym substitutions, lexical changes, active to passive switching, etc. This phenomenon dramatically increased because of the progressive advancement of the web and the automatic text editing tools. Recently, deep leaning methods have gained competitive results than traditional methods for Natural Language Processing (NLP). In this context, we consider the problem of Arabic paraphrase detection. We present different deep neural networks like Convolutional Neural Network (CNN) and Long Short Term Memory (LSTM). Our aim is to study the effective of each one in extracting the proper features of sentences without the knowledge of semantic and syntactic structure of Arabic language. For the experiments, we propose an automatic corpus construction seeing the lack of Arabic resources publicly available. Evaluations reveal that LSTM model achieved the higher rate of semantic similarity and outperformed significantly other state-of-the-art methods.

Keywords: Paraphrase detection · Deep learning · Word embedding ·
Convolutional neural network · Long short term memory ·
Arabic corpus construction

1 Introduction

The technological advancement of the Web and text editing tools, it became very easy to find and re-use any kind of information. This increased dramatically the paraphrase practice, which is difficult to detect. It means including other person's text as your own without proper citation. The texts must be semantically the same but rephrased using different obfuscation operations such as addition/deletion of words, synonym substitutions, lexical changes, active to passive switching, etc. [1]. As a result, automatic detection of text reuse is a fundamental issue in Natural Language Processing (NLP). It has attracted the attention of the research community due to the wide variety of applications associated with it (e.g. information retrieval, question answering, essay grading, text summarization, etc.). Often, neural networks have provided powerful learning methods for analyzing semantic textual similarity through feed forward and

© Springer Nature Switzerland AG 2019
E. Métais et al. (Eds.): NLDB 2019, LNCS 11608, pp. 3–16, 2019.
https://doi.org/10.1007/978-3-030-23281-8_1

recurrent neural networks architectures. In this paper, we consider the problem of Arabic paraphrase detection. Thus, we study different models of deep neural networks and demonstrate the best one for extracting high-level features and capturing long-range dependencies between words. The rest of this paper is organized as follows: Sect. 2 briefly describes a literature review on paraphrase detection systems based on semantic textual similarity approaches. Then, Sect. 3 details our proposed methodology. Subsequently, Sect. 4 presents the experimental dataset and evaluation results. Finally, Sect. 5 concludes the paper.

2   Literature Review

Natural Language Processing (NLP) has long been one of the holy grails of computer science [2]. Although processing language and comprehending the contextual meaning is an extremely complex task, paraphrase detection is a sensitive field of research for specific language. Therefore, various methods have been produced for estimating textual similarity between documents.

Lexical based methods compared documents if they contained the same characters or words, like: Prayogo et al. [3] studied the structure of Bayesian network for Indonesian paraphrase identification using lexical features (Levenshtein distance, term frequency based on cosine similarity, and Long Common Substring (LCS)). Subsequently, they calculated similarity in the semantic tree WordNet by applying Wu and Palmer and Shorest path methods. Similarly, Ali et al. [4] detected plagiarism in Urdu documents based on a distance measuring method, structural alignment algorithm, and vector space model. The performance of this system is evaluated using Support Vector Machine (SVM) and Naïve Bayes classifiers. The experimental results demonstrated that the detection result is improved using cosine similarity with the Term Frequency-Inverse Document Frequency (TF-IDF) technique and the simple Jaccard measure. These approaches were very accurate on detecting verbatim cases of plagiarism (i.e. copy-paste), but they were useless to detect complex cases of plagiarism, such as paraphrase, where texts showed significant differences in wording and phrasing. However, Bag of Words (BoW) models regarded texts as a set of words without taking into consideration their order. Few works have been proposed, distinguish: Al-Shenak et al. [5] enhanced a method for Arabic question answering. They used Latent Semantic Analysis (LSA) for modeling term and document to the same concept space and SVM for classification. Similarly, Kurniawan et al. [6] detected plagiarism of writing and image on Facebook's social media using LSA method and Smith-Waterman algorithm. In contrast, Latent Dirichlet Allocation (LDA) method computed text similarity according to the topic distribution as shown in the system of Aljoha et al. [7]. They recognized external plagiarism by combining LDA and Part of Speech Tags (POS) techniques. Thus, semantic information is added even if the part-of-speech features alone could be used satisfactorily. We distinguish also knowledge-based methods that stored and queried structured information using lexical knowledge databases (e.g. Wikipedia, WordNet, DBpedia, etc.). They aimed to measure the semantic overlap, distinguish: Al-Shamry et al. [8] tested whether the research entered under the specialization of computer science or not, where only such research would subject to semantic plagiarism detection using WordNet. Furthermore, Ghanam et al. [9] detected Arabic plagiarism using WordNet combined with TF-IDF and feature-based semantic similarity methods.

In recent decades, neural networks models have been employed for analyzing semantic information without depending on any external knowledge resource: Mikolov et al. [10] proposed word2vec model for generating word embedding. It had two architectures: The first was a Continuous Bag-of-Words (CBOW) model for predicting the current word from the context of words. Nagoudi et al. [11] used it for detecting semantic similarity in Arabic sentences. The second was a Skip-gram model for predicting the context of words from the current word. Mahmoud et al. [12] combined it with TF-IDF method for representing the most descriptive sentence and identifying paraphrase in Arabic documents. Thereafter, global embedding (GloVe) is introduced in [13] combining Skip-gram and word-word co-occurrences. It was more useful to train word embedding and compare semantic differences between sentences in different languages (e.g. Spanish [14], Arabic [15], English [16], etc.). In addition, Kenter et al. [17] compared different word embeddings like word2vec, FastText, and GloVe. Contrariwise, Niraula et al. [18] showed that words relatedness and similarity could be measured by combining word embeddings models (i.e. LSA, LDA, word2vec and GloVe). Similarly, Al-Smadi et al. [19] extracted lexical, syntactic, and semantic features to train Maximum Entropy (MaxEnt) and Support Vector Regression (SVR) classifiers. Their main advantage was to complement, better represent the coverage of semantic aspects of words and overcome their limitation in analyzing the specificities of any language. Therefore, deep learning methods have yielded competitive results than traditional text classification models and have brought gains to NLP. They achieved good results in extracting the proper features of sentences without the knowledge of semantic and syntactic structure of a language, via feed forward neural networks like Convolutional Neural Network (CNN). It was useful in different systems and achieved good results for text classification and semantic similarity analysis as demonstrated in the studies of Kim [20], Hu et al. [21], Mahmoud et al. [22], Lazemi et al. [23], and Salem et al. [24]. The drawback of these methods that were limited to process each word of sentence as a single feature by ignoring their order of occurrence. Therefore, recurrent neural networks are introduced to make sequential data process and learn sentence from the context of words considering their previous information. However, these models were susceptible to explode the problem of gradient. That is why; Long Short Term Memory (LSTM) and Gated Recurrent Units (GRU) were efficient. They treated long sequences using a gating mechanism to create a memory control of values proposed over the time. They were successful for analyzing semantic similarity between long sentences as demonstrated in the studies of Duong et al. [25] and Reddy et al. [26].

Throughout the state of the art, little attention has been considered for Arabic paraphrase detection due to the following reasons: Arabic is the official language of the Arab world [27]. It counts more than 445 million speakers and ranked the 8th in the number of pages that circulate on the Internet [28]. Arabic is known to be a morphological rich language because of the existence of diacritics and stacked letters above or below the base line [29, 30]. This results the existence of more than one meaning and category to which the word belongs. Among the Arabic language specificities that produce its processing complexity, we cite [31, 32]: non-vocalization, homograph, agglutination, derivation, no concatenation, and phrase types (verbal, nominal and prepositional). In general, an Arabic word can present ambiguity and can be interpreted with different meanings as the word "ذهب" means he went or gold. In addition, the word "وجهة" means destination or a side [33].

3 Proposed Model

In this section, the methodology behind paraphrase detection method is briefly described. We propose a deep learning approach to learn sentences representations and estimate the degree of semantic relatedness. It is decomposed into the following phases: Text pre-processing to eliminate irrelevant data, features extraction to represent the most discriminant information, and finally similarity computation to determine how much suspect and source documents convey the same meaning.

3.1 Pre-processing

Pre-processing is fundamental in Arabic NLP systems for storing texts into machine-readable formats and facilitating further processing (e.g. parsing or text mining). We eliminate the less useful parts of the text through removing diacritics, extra white spaces, titles numeration, special characters, duplicated letters and non-Arabic words. Then, the exploration of words in the sentence is done. We apply tokenization operation dividing the text into tokens.

3.2 Global Word Embedding

While count based matrix factorization and contextual Skip-gram models have produced the data sparsity problem with large-scale data, words embeddings are generated with the unsupervised global vector representation (GloVe). Let w_i and $\breve{w}_j$ are the vectors of words i and j; b_i and $\breve{b}_j$ are the scalar biases of the main word i and the context of word j; V is the vocabulary size and f (x) is the weighting function for rare and frequent co-occurrences. Training is performed on aggregating word co-occurrence statistics and producing representations with linear substructures of the word vector space using an objective function J defined as follows in Eq. (1):

$$ J = \sum_{i,j=1}^{V} f(X_{ij})(w_i^T \breve{w}_j + b_i + \breve{b}_j - \log(X_{ij}))^2 \qquad (1) $$

More formally, let $S = [w_1, w_2, \ldots, w_n]$ be a sentence of length n, where w_i is the i-th word of the sentence represented by its word embedding x_i. It is a row vector of K dimension in a matrix $X = [x_1, x_2, \ldots, x_n]$ of size N × K.

3.3   Similarity Computation

Neural network models have provided powerful learning sentence representations for many natural language applications. There are two major types of neural networks architectures: feed-forward networks and recurrent networks. While feed forward neural networks are able to extract local patterns, recurrent neural networks are able to capture long-range dependency in the data by abandoning the Markov assumption. This is the aim of the proposed paraphrase detection model. We intend to study the capability of Convolutional Neural Network (CNN) and Long Short-Term Memory (LSTM) models and compare their effectiveness. Our goal is to encode the semantic and syntactic properties of Arabic sentences and compute thereafter efficiently the semantic similarity.

3.3.1   Convolutional Neural Network

Convolutional Neural Network (CNN) is a feed forward architecture. It is advantageous in features engineering through the independence from prior knowledge and human efforts. The proposed model consists of three layers as shown in Fig. 1.

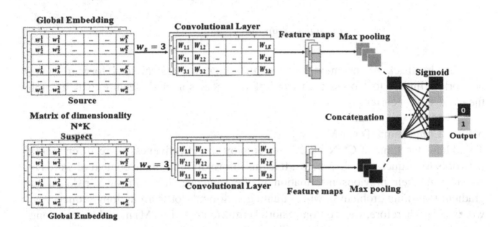

Fig. 1.   Proposed GloVe-CNN model

Convolutional Layer. We extract high-level features from the input matrix X. Given different window sizes w_s, a filter is seen as a weight F of size 64 moved in each sentence of words $n - w_s + 1$. The core of this layer is obtained from the application of convolutional operator on F and X according to a non-linear function. The output of this layer is computed as follows in Eq. (2) [34]:

$$Y_i = \sum_{j=1}^{K} \sum_{h=1}^{w} F_{ijh} \times X_{n-w_s+1,j} + b_i \qquad (2)$$

Where: $i = 1, 2, \ldots, n$; the term of bias $b = [b_1, \ldots, b_n]$. After applying the convolutions, we introduce the non-linearity through an activation function using the Rectified Linear Unit (ReLU) function defined in Eq. (3):

$$f(x) = \max\{0, x\} \tag{3}$$

Pooling Layer. The most relevant and common features are extracted by applying the max pooling operation. More precisely, this layer produces the reduced feature maps $P = [p_1, p_2 \ldots, p_n]$, where:

$$p_i = max_{1 \leq i \leq n-w_s+1} Y_i \tag{4}$$

All vectors are concatenated to enhance the generalization ability of the model. These are fed into the fully connected layer to perform the classification.

Fully Connected Layer. This layer with a dropout improves the performance of the model and reduce the over fitting problem. It generates the output score in the range of [0, 1] using sigmoid function defined as follows in Eq. (5):

$$Output = \text{Sigmoid}(x) = \frac{e^x}{(1 + e^x)} \tag{5}$$

The drawback of this model is the following: a classical feed forward neural network is limited to process each word of sentence as a single feature, which ignore their order of occurrence.

3.3.2   Long Short-Term Memory

To address the limit of CNN model, we employ the effectiveness of recurrent neural networks for sequential data process. It learns information from the context of words by considering their previous information in the sentence. In contrast, it risks of the gradient vanishing problem in which training is difficult for learning long sequences of words [35]. Therefore, we use Long Short-Term Memory (LSTM) model to learn long-term dependencies as represented in Fig. 2. It consists of a hidden unit composed of a memory cell for storing one or multiple values controlled by an input gate (is written to) that decides how many values enter the unit, an output gate (read from) decides how many values output from the unit, and a forget gate (delete from) decides whether value remains in the unit. Each of gates receives all of the current and past inputs to the cell and combine them according to a unique set of weights. Then, each gate pass the output of this combination to an activation sigmoid function at time t:

$$y_t = \frac{1}{1 + e^{(\sum_i w_i y_i^{t-1})}} \tag{6}$$

Finally, this internal state is updated by another activation function and multiplied by the output gate to generate the output of the memory o_t, at time t is the following:

$$o_t = y_t a S_t \qquad (7)$$

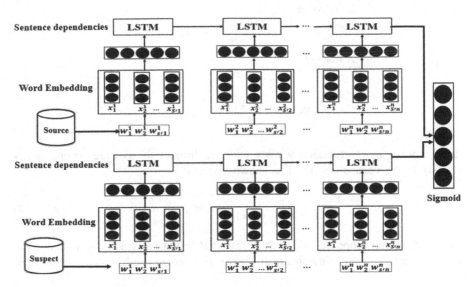

Fig. 2. Proposed GloVe-LSTM model

4 Experiments

4.1 Datasets

The lack of publicly available resources make hard in experimenting the performance of Arabic paraphrase detection methods [36]. As a solution, we develop automatically a corpus consisting of the following process: Open Source Arabic Corpora (OSAC) [37] is used as an original corpus containing 22,429 documents of 10 categories (e.g. economics, history, sports, etc.) as shown in Table 1.

An automatic development of an Arabic paraphrased corpus is proposed. Indeed, the degree of paraphrase D is fixed arbitrary in the range of [0.45, …, 0.75] using random uniform function. This rate is used thereafter to count the number of words to replace R from the OSAC source corpus of N words as follows in Eq. (8):

$$R = N \times D \qquad (8)$$

Paraphrased corpus needs an analogy reasoning in which words with similar meanings tend to have similar contextual representations. To do this, we use Skip-gram model for capturing various degrees of similarity among words and offering efficient representations with low-dimensional vectors. It predicts the context of the middle

Table 1. Open Source Arabic Corpora (OSAC)

Categories	Number of documents
Economics	3102
History	3233
Education & Family	3602
Religious and Fatwas	3171
Sports	2419
Health	2296
Astronomy	557
Low	944
Stories	726
Cooking Recipes	2373

original word w_i in a surrounding window c for maximizing the average log probability according to the vocabulary size T as shown in Eq. (9):

$$\frac{1}{T}\sum_{i=1}^{T}\sum_{-c \le j \le c, j \neq 0} \log p(w_{i+j}|w_i) \tag{9}$$

POS technique proposed by the Stanford NLP group is used for annotating sentences by their syntactic (grammatical) classes (e.g. noun, verb, adjective, adverbs, etc.).

Finally, a random shuffle function replaces the source words according to an index chosen, according to the following constraint: To preserve syntactic and semantic properties of the original sentences, source words should be replaced from the vocabulary model with their most similar ones that have the same grammatical classes.

Experiments are carried out on 15701 documents for training (4710 paraphrases and 10991 different documents) and 6728 documents for the test (2019 paraphrases and 4709 different documents).

4.2 Word Embedding

Different neural network models are employed in our study. That is why; we transform word tokens into fixed vectors by looking up the pre-trained local and global word embeddings. For the paraphrased corpus development, each word from the OSAC corpus is mapped to its pre-trained 300-dimensional word vector. It is produced by the Skip-gram model trained on more than 2.3 billion words from various datasets (Arabic Corpora Resource (AraCorpus), King Saud University Corpus of Classical Arabic (KSUCCA) and Arabic papers from Wikipedia.). Table 2 summarizes the parameters of the word2vec model.

For the paraphrase detection, we use the GloVe model trained on the same resources used for word2vec. Table 3 presents its configuration in details.

Table 2. Configuration of word2vec model

Parameters	Values
Vocabulary size	2.3 billion
Vector dimension	300
Window size	3
Min_count	≤ 5
Workers	8
Iterations number	7

Table 3. Configuration of GloVe model

Parameters	Values
Co-occurrence matrix size	1.119.436 *1.119.436
Embedding size	300
Context size	3
Minimum occurrence	25
Learning rate	0.05
Batch size	512
Numbers of epochs	20

4.3 Paraphrased Corpus Analysis

To validate the effectiveness of the proposed corpus, let a sentence S of N words, we calculate an average of all cosine similarities of its word embedding's $\{w_1, \ldots, w_n\}$ as defined in Eq. (10). The objective is to identify the degree of relatedness between the source and the resulted paraphrased sentences and determine the impact of the proposed annotations. After several experiments, the combination of Skip-gram and POS is reported to be good in capturing syntactic and semantic properties of words with the

Table 4. Example of a paraphrased sentences construction

Source	ادى التلميذ احسن عمل 'The student did the best job'		
Parameters	Paraphrased sentences	Average (word2vec)	Average (word2vec-POS)
(250, 5)	ادى الطالب احسن عمل 'The student did the best job'	0.75	0.78
(300, 3)	ابتكر الطالب افضل شغل 'The student created the best work'	0.81	0.85
(300, 7)	ابتكار الطالبات افضل الشغل 'Innovation of students best job'	0.70	0.74
(450, 3)	ابتكر الطالب اجمل شغل 'The student created the most beautiful work'	0.75	0.80
(450, 6)	يفعل الطفل عمل 'The child does work'	0.60	0.66

following configuration: 3 as a window size (three words after and before the target one) and 300 as a vector dimension. Table 4 illustrates an example of a paraphrased sentence construction:

$$\text{Average} = \sum\nolimits_{i=1}^{n} \frac{w_i}{n} \tag{10}$$

4.4   Results and Discussion

CNN Results. The convolutional layer has a filter size of 64, a kernel width of 3, a ReLU as an activation function, a max-pooling layer of size 4, and a fully connected layer with sigmoid function for the classification. We study the effect of window size to the accuracy of CNN models with GloVe embedding. Figure 3 shows the test accuracy curves with different window sizes. The x-axis is the window sizes and the y-axis is the accuracy ratio of the models on the test set. It is clear that w = 3 is the most appropriate window size, which gives the best accuracy of 79.5% for training CNN model. The overall experimental results show its benefit in capturing high-level contextual features within sentences.

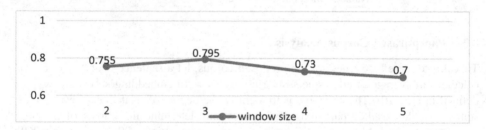

Fig. 3. Accuracy of CNN models with different window sizes

LSTM Results. We tune the parameters of LSTM model as follows: The number of hidden units in all models is fixed at 256, and a dropout of probability 0.5. Compared to CNN models, we investigate the performance of LSTM model for sentence modeling and similarity computation. After 100 training iterations, it achieves an accuracy of 83% and outperforms CNN models as shown in Fig. 4. Although CNN models are efficient in extracting invariant features, this experimental result demonstrates that LSTM model is better in analyzing long-sequence dependencies of words and conserving the semantic of sentences.

Discussion. Overall experiments demonstrate that CNN and LSTM models are similar and successful by sharing parameters between neurons. In addition, they are different for the following reasons: CNN models are efficient in representing features with a fixed number of computation steps and the output depends only on the current input. In contrast, LSTM models are more advantageous in sequential modeling sentences and sharing parameters across the temporal dimension. Table 5 and Fig. 5 represent a

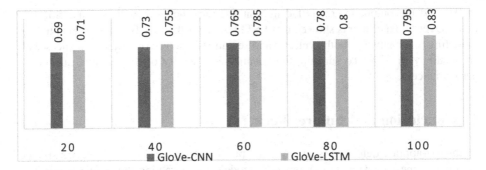

Fig. 4. Accuracy of CNN and LSTM models with different word embeddings

Table 5. Overall comparisons

Systems	Datasets	Language	Models	Accuracy
Proposed Models	OSAC corpus	Arabic	GloVe-CNN	79.500
			GloVe-LSTM	83.000
Salem et al. [24]	10 long texts	Arabic	CNN-clustering	82.970
Le-Hong et al. [34]	UIUC question types	English	GloVe-CNN	83.000
			GloVe-LSTM	76.800
Duong et al. [25]	MSRP	English	Word2vec-Dense Softmax	78.600
			GloVe- Dense Softmax	75.238

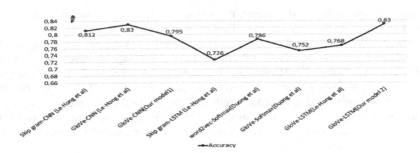

Fig. 5. Comparative study regarding accuracy

comparison with other methods in the literature: Seeing the diversity specificities of Arabic language and the complexity of their processing, the proposed GloVe and CNN based model achieved an accuracy of 79.5%. Using the same parameters, Le-Hong et al. [34] demonstrated that it was more efficient for English text classification with an accuracy of 83%. However, Salem et al. [24] clustered segments of a small Arabic dataset and found which one had a different stylometry in comparison to the other using CNN model as a classifier. Contrariwise, Duong et al. [25] achieved 78.6% of accuracy for English paraphrase detection. They encoded pre-trained word vectors by an LSTM model and the output matrix is fed into an attention network with a dense Softmax layer. Overall experiments showed that LSTM is better than CNN model for capturing, modeling and analyzing both syntactic and semantic structure of Arabic sentences with an accuracy of 83%. To sum up, the quantity and quality of data greatly affects the performance.

5   Conclusion and Future Work

Arabic paraphrase detection is an important problem to solve. In this paper, we studied the effectiveness of deep neural networks models for extracting the proper features of sentences including CNN and LSTM. For experiments, we proposed an automatic approach for Arabic paraphrased corpus construction combining Skip-gram model and POS technique. It preserved both semantic and syntactic properties of sentences. Results showed that Global-embedding GloVe with LSTM model significantly outperformed the state-of-the-art methods with an accuracy of 83%. We note that LSTM model was efficient in capturing long sentence dependencies while CNN was capable in extracting discriminant contextual features from texts. Despite the promising results, we will deal with additional stylistic feature sets (e.g. quantitative, function word, vocabulary richness, etc.). Moreover, we will use other deep neural networks models to capture more local context relations between sentences and improve the prediction of semantic similarity in less time like Gated Recurrent Unit (GRU), etc.

References

1. Sameen, S., Sharjeel, M., Muhammad, R., Nawab, A., Rayson, P., Muneer, I.: Measuring short text reuse for the Urdu language. IEEE Access **6**, 7412–7421 (2018)
2. Kumar, V., Verma, A., Mittal, N., Gromov, S.V.: Anatomy of pre-processing of big data for monolingual corpora paraphrase extraction: Source language sentence selection. In: Abraham, A., Dutta, P., Mandal, J., Bhattacharya, A., Dutta, S. (eds.) Emerging Technologies in Data Mining and Information Security. AISC, vol. 814, pp. 495–505. Springer, Singapore (2019). https://doi.org/10.1007/978-981-13-1501-5_43
3. Prayogo, A.H., Syahrul, M., Adiwijaya, M.: On the structure of Bayesian network for Indonesian text document paraphrase identification. In: International Conference on Data and Information Science (ICoDIS), pp. 1–15 (2018)
4. Ali, W., Ahmed, T., Rehman, Z., Rehman, A.U., Slaman, M.: Detection of plagiarism in Urdu text documents. In: 14th International Conference on Emerging Technologies (ICET), Islamabad, pp. 1–6 (2018)
5. Al-Shenak, M., Nahar, K., Halwani, H.: AQAS: Arabic question answering system based on SVM, SVD, and LSI. J. Theor. Appl. Inf. Technol. **97**(2), 681–691 (2019)
6. Kurniawan, M., Surendro, K.: Similarity measurement algorithms of writing and image for plagiarism on Facebook's social media. In: 1st International Conference on Engineering and Applied Technology (ICEAT), pp. 1–10 (2018)
7. Aljohani, N., Alowibdi, J., Daud, A., Khan, J., Nasir, J., Abbasi, R.: Latent Dirichlet Allocation and POS Tags based method for external plagiarism detection: LDA and POS tags based plagiarism detection. Int. J. Semant. Web Inf. Syst. (IJSWIS) **14**(3), 53–69 (2018)

8. Al-Shamery, E., Gheni, H.: Plagiarism detection using semantic analysis. Indian J. Sci. Technol. **9**(1), 1–8 (2016)
9. Ghanem, B., Arafeh, L., Rosso, P., Sánchez-Vega, F.: HYPLAG: hybrid Arabic text plagiarism detection system. In: Silberztein, M., Atigui, F., Kornyshova, E., Métais, E., Meziane, F. (eds.) NLDB 2018. LNCS, vol. 10859, pp. 315–323. Springer, Cham (2018). https://doi.org/10.1007/978-3-319-91947-8_33
10. Mikolov, T., Chen, K., Corrado, G., Dean, J.: Efficient estimation of word representations in vector space. arXiv preprint arXiv:1301.3781 (2013)
11. Nagoudi, E., Schwab, D.: Semantic similarity of Arabic sentences with word embeddings. In: Third Arabic Natural Language Processing Workshop (WANLP), Valencia, Spain, pp. 18–24 (2017)
12. Mahmoud, A., Zrigui, M.: Semantic similarity analysis for paraphrase identification in Arabic texts. In: 31st Pacific Asia Conference on Language, Information and Computation, (PACLIC), Philippine, pp. 274–281 (2017)
13. Pennington, J., Socher, R., Manning, C.: GloVe: global vectors for word representation. In: Conference on Empirical Methods in Natural Language Processing (EMNLP), Qatar, pp. 1532–1543 (2014)
14. Rodríguez, I.: Text similarity by using GloVe word vector representations. Ph.D. (2017)
15. Alkhatlan, A., Kalita, J., Alhaddad, A.: Word sense disambiguation for Arabic exploiting Arabic WordNet and word embedding. Procedia Comput. Sci. **142**, 50–60 (2018)
16. Cer, D., Diab, M., Agirrec, E., Lopez-Gazpi, I., Specia, L.: SemEval-2017 Task 1: semantic textual similarity multilingual and cross-lingual focused evaluation, pp. 1–14. arXiv:1708.00055 (2017)
17. Kenter, T., Rijke, M.: Short text similarity with word embeddings. In: 24th ACM International Conference on Information Knowledge Management (CIKM), pp. 1411–1420 (2015)
18. Niraula, N., Gautam, D., Banjadae, R., Maharjan, N., Rus, V.: Combining word representations for measuring word relatedness and similarity. In: 28th International Florida Artificial Intelligence Research Society Conference (FLAIRS), Florida (2015)
19. AL-Smadi, M., Jaradat, Z., AL-Ayyoub, M., Jararweh, Y.: Paraphrase identification and semantic text similarity analysis in Arabic news tweets using lexical, syntactic, and semantic features. ACM Digit. Libr. **53**(3), 640–652 (2016)
20. Kim, Y.: Convolutional neural networks for sentence classification. In: Conference on Empirical Methods in Natural Language Processing (EMNLP), Doha, pp. 1746–1751 (2014)
21. Hu, B., Lu, Z., Li, H., Chen, Q.: Convolutional neural network architectures for matching natural language sentences. In: Annual Conference on Neural Information Processing Systems Montreal (NIPS), Canada, pp. 2042–2050 (2014)
22. Mahmoud, A., Zrigui, A., Zrigui, M.: A text semantic similarity approach for Arabic paraphrase detection. In: International Conference on Computational Linguistics and Intelligent Text Processing (CICLing), Budapest, pp. 338–349 (2017)
23. Lazemi, S., Ebrahimpour-komleh, H., Noroozi, N.: Persian plagirisim detection using CNN s. In: 2018 8th International Conference on Computer and Knowledge Engineering (ICCKE), pp. 171–175. IEEE (2018)
24. Salem, A., Almarimi, A., Andrejkova, G.: Text dissimilarities predictions using convolutional neural networks and clustering. In: World Symposium on Digital Intelligence for Systems and Machines (DISA), pp. 343–347 (2018)
25. Duong, P., Nguyen, H., Duong, H., Ngo, K., Ngo, D.: A hybrid approach to paraphrase detection. In: 5th NAFOSTED Conference on Information and Computer Science (NICS), pp. 366–371 (2017)

26. Aravinda Reddy, D., Anand Kumar, M., Soman, K.P.: LSTM based paraphrase identification using combined word embedding features. In: Wang, J., Reddy, G.R.M., Prasad, V. K., Reddy, V.S. (eds.) Soft Computing and Signal Processing. AISC, vol. 898, pp. 385–394. Springer, Singapore (2019). https://doi.org/10.1007/978-981-13-3393-4_40
27. Abdellaoui, H., Zrigui, M.: Using tweets and emojis to build TEAD: an Arabic dataset for sentiment analysis. Computacion y Sistemas **22**(3), 777–786 (2018)
28. Terbeh, N., Maraoui, M., Zrigui, M.: Arabic discourse analysis: a naïve algorithm for defective pronunciation correction. Computación y Sistemas **23**(1), 153–168 (2019)
29. Mansouri, S., Charhad, M., Zrigui, M.: A heuristic approach to detect and localize text in Arabic news video. Computacion y Sistemas **23**(1), 75–82 (2018)
30. Aouichat, A., Hadj Ameur, M.S., Geussoum, A.: Arabic question classification using support vector machines and convolutional neural networks. In: Silberztein, M., Atigui, F., Kornyshova, E., Métais, E., Meziane, F. (eds.) NLDB 2018. LNCS, vol. 10859, pp. 113–125. Springer, Cham (2018). https://doi.org/10.1007/978-3-319-91947-8_12
31. Ghezaiel Hammouda, N., Torjmen, R., Haddar, K.: Transducer cascade to parse Arabic corpora. In: Silberztein, M., Atigui, F., Kornyshova, E., Métais, E., Meziane, F. (eds.) NLDB 2018. LNCS, vol. 10859, pp. 230–237. Springer, Cham (2018). https://doi.org/10.1007/978-3-319-91947-8_22
32. Batita, M., Zrigui, M.: Derivational relations in Arabic Wordnet. In: 9th Global WordNet Conference (GWC), Singapore, pp. 137–144 (2018)
33. Chouigui, A., Khiroun, O.B., Elayeb, B.: A TF-IDF and co-occurrence based approach for events extraction from arabic news corpus. In: Silberztein, M., Atigui, F., Kornyshova, E., Métais, E., Meziane, F. (eds.) NLDB 2018. LNCS, vol. 10859, pp. 272–280. Springer, Cham (2018). https://doi.org/10.1007/978-3-319-91947-8_27
34. Le-Hong, P., Le, A.: A comparative study of neural network models for sentence classification. In: 2018 5th NAFOSTED Conference on Information and Computer Science (NICS), pp. 360–365. IEEE (2018)
35. Bsir, B., Zrigui, M.: Bidirectional LSTM for author gender identification. In: 10th International Conference on Computational Collective Intelligence (ICCCI), pp. 393–402 (2018)
36. Mahmoud, A., Zrigui, M.: Artificial method for building monolingual plagiarized Arabic corpus. Computacion y Systemas **22**(3), 767–776 (2018)
37. Saad, M., Ashour, W.: OSAC: Open source Arabic corpora. In: 6th International Conference on Electrical and Computer Systems (EECS), pp. 1–6 (2010)

Model Answer Generation for Word-Type Questions in Elementary Mathematics

Sakthithasan Rajpirathap$^{(\boxtimes)}$ and Surangika Ranathunga$^{(\boxtimes)}$

Department of Computer Science and Engineering, Faculty of Engineering,
University of Moratuwa, Katubedda 10400, Sri Lanka
{rajpirathaps,surangika}@cse.mrt.ac.lk

Abstract. There are several categories of word-type questions at elementary level Mathematics. These include addition, subtraction, multiplication, division and ratio. Addition and subtraction problems can be further divided based on their textual information. Those types are change type (join-separate type), compare type, and whole-part type. This paper presents a set of ensemble classifiers to automatically generate model answers for these three types of addition and subtraction problems. Currently, questions with one unknown variable are considered. In addition to the existing data sets, a new data set is created for the training and the evaluation purpose. Our results outperform the existing statistical approaches.

Keywords: Elementary Mathematics · Math Word problems ·
Question classification · Answer generation

1 Introduction

Managing assessments through computers can be considered as an alternative for manual assessment. For Mathematics questions expressed only using Mathematics formulae such as those referring to Algebra, Computer Aided Assessment systems such as SymPy can be used [1]. However, in automated assessment of student answers, for Math Word problems (a mathematical problem expressed using natural language), a model answer should be provided in advance. Many systems that focus on assessment of this type of Mathematics questions have assumed that a teacher manually provides the model answer for each question for assessment to be carried out [2]. Although there is research to automate the answer generation process as well [3–5], answer generation for Math Word problems (MWPs) is still an open issue.

In this research, we focus on simple elementary level Mathematics questions that are expressed as Math Word problems (MWPs). These mostly contain addition, subtraction, division, multiplication and ratio calculations, and geometry based questions. Addition and subtraction problems can be further divided into sub-types by considering the textual information in those questions.

© Springer Nature Switzerland AG 2019
E. Métais et al. (Eds.): NLDB 2019, LNCS 11608, pp. 17–28, 2019.
https://doi.org/10.1007/978-3-030-23281-8_2

These categories are change type (join-separate type), compare type, and whole-part type [12].

Template/rule based, graph based, ontology based, statistical, and deep learning based approaches have been used for answer generation for Math Word problems. Template/Rule based approaches rely on rules/templates for input sentences [8,9]. They use stored knowledge in the form of rules and templates. The presence of rules and constraints affects the flexibility to adapt to new question types. Ontology-based approaches rely on ontology relationships that exist within the entities. In this case, domain-experts are needed to derive the domain knowledge to write logic to create an ontology map. Statistical approaches rely on traditional Machine Learning approaches [3]. In some hybrid systems, rule logics are used in the initial stage of the system's process, and statistical models in the final stage [5], while some other approaches did vice versa [3]. In recent research, these hybrid systems have produced more promising results. However, manually defining rules or templates for simple elementary Mathematics word type questions is an issue. In the recent past, deep learning approaches have also been used [10], however, they require a large number of training samples to perform at an acceptable level. Graph based approaches [11] use a graph representation of word problems, where entities and relevant quantities are plotted in the graphs and equations are generated from the graph based on the question types.

This paper presents a statistical system to automatically generate model answers for the aforementioned three types of addition and subtraction questions. We experimented with multiple classifiers such as Random forest, Gaussian NB, decision tree, Support Vector Machines (SVM), Perceptron, and their ensembles. We also introduce new features that perform better than those used by existing research [3,5]. For evaluation, relevant questions were extracted from Add-Sub dataset [13], ARIS dataset [11] and Roy et al.'s [3] dataset. SingleOP dataset [14] was taken only for the purpose of testing. In addition, 782 new questions were collected from O/L Mathematics teachers in Sri Lanka to create a dataset of 1713 questions. An accuracy of 94.7% was achieved for 10-fold cross validation on the combined data set, while an accuracy of 88.7% was achieved for the SingleOP test dataset evaluated in decision tree ensemble classifier using a combination of the newly defined features along with those borrowed from previous research [3] and [5]. (The source is added to link https://github.com/rajpirathap/ModelAnsGenProj)

2 Question Types

This research focuses on "change", "compare" as well as "whole-part" type of elementary Mathematics questions.

Change Type: In the "change" type of questions, a particular numerical value or a quantitative value of an entity is changed over time throughout sentences of a question. This formula can be explained as below,

Start value+ change value = summation or result
Or
Start value - change value = difference or summation or result

Here the "start value" is the initial value of a quantity of an entity. "Change value" represents the change of a quantity value of the entity/variable. As a result, the answer is derived as a sum or difference value of a quantity in a question.

Q1 below is an example of a Change type question,

Q1: Pete had 3 apples. Ann gave Pete 5 more apples, how many apples does Pete have now?

Comparison Type: In this type of questions, there is a comparison of two numerical values of entities. The comparison is implicitly captured by identifying the difference between the quantities or numerical values in a question. Simply the formula can be stated as below:

Initial value + or - difference value = second value

The quantity value of "Initial value" is compared with the quantity value of "second value". "difference value" denotes the quantity difference between the "initial value" and "second value". In most cases, this type of questions has keywords such as "more", "less", and "than" in its context. For example consider the question Q2,

Q2: Joe has 3 balloons. His sister Connie has 5 balloons. How many more balloons does Connie have than Joe?

By considering the above question, the formula can be expressed as

first value (5) - second value (3) = difference (unknown)

Whole-Part Type: In this type of questions, value of the same kind of variable/entity is divided into parts and the "whole" quantity is expressed by considering the "part" values. It can be simply expressed as,

part value + part value = whole value

In some cases, the whole value and one of the part values are mentioned in the context of the question, the remaining part value has to be derived. Some other questions mention about the quantity of "parts" and the "whole" quantity value has to be derived. For example, consider the question Q3. Here, boys and girls are parts of child entity/variable.

Q3: There are 6 boys and 8 girls in the volleyball team. How many children are in the team?

In the above question, the formula can be derived as

part value + part value = whole value.

3   Related Work

Template/rule based, graph based, ontology based, statistical, and deep learning based approaches have been used in previous research for answer generation for MWPs.

Template/Rule based approaches have been used to solve question types such as real and natural number arithmetic, 2D and 3D geometry, pre-calculus, and calculus related mathematical questions [8,9]. The main drawback of these approaches is the presence of manually prepared templates/rules and constraints, which affects the flexibility to adapt to a new question type.

Ontology based systems have commonly been designed for some selected domain(s). For example, the fuzzy logic ontology model presented by Morton and Qu [15] is aimed at the domains of investment, distance, and projectile domains. Also they mostly considered addition type of linear equation based questions. The main limitation of this approach is the need for a domain expert for new question type integration.

We refer to systems that employ traditional classification algorithms as statistical approaches. Roy et al.'s [3] proposed the most recent statistical approach for word type problems. This approach limits itself by allowing questions with only one arithmetic operation (among addition, multiplication, subtraction and division) at a time with two or three operand candidates. This system employs a system of cascading classifiers to identify the operations to solve questions. These cascading classifiers are dependent on each other for their continued work, meaning that errors of classifiers in the early stages get propagated to latter stages. Amnueypornsakul and Bhat [5] also used a multi-stage classifier with a rule-based learner to generate answers for change and whole-part type linear equation related MWPs.

Liang et al. [7] used a tag-based statistical framework to perform understanding and reasoning in solving MWPs. In this approach, logical inferences is used to identify the proper tags which can help to identify desired operands and filter out irrelevant quantities. Chien et al. [6] implemented a system that involves statistical classifiers as well as logic inferences to solve MWPs. However this approach is limited only to Chinese language. The main drawback of these approaches is the limitations of the available logic inferencing solutions.

In recent past, statistical methods with templates have provided the most accurate results. Here, a set of manually defined templates are being used to train a statistical classifier. Kushman et al. [4] provided the first system of this kind. Their system was able to handle questions containing two or three unknowns. However, their approach was computationally expensive. Zhou et al.'s [16] approach addresses this limitation through a quadratic programming approach. A further improved solution was presented by Hevapathige et al. [18]. However, all these techniques require handmade templates, and consume a large feature space to define their features.

Hosseini et al. [11] proposed a container-entity based graph approach that solves the mathematical question sentence with a state transition sequence. But they faced irrelevant information mapping issues, and parsing issues.

Wang et al. [10] designed a hybrid model that combines a Recurrent Neural Network (RNN) model and a similarity-based retrieval model to achieve additional performance improvements. However, this hybrid model also uses some templates. Also the other issue is this deep learning approach requires a large dataset for the training purpose.

4   Data Preparation

Compare type, change type and whole-part type of elementary MWPs (containing 2–3 sentences) and their formulae are the data sources of this research. They have two known variables (expressed numerically or textually) and one unknown variable (expressed textually).

A dataset of 1713 questions was created. Out of these, 782 questions were newly created by O/L Mathematics teachers in Sri Lanka. Others were extracted from available question bases. Table 1 shows the statistics of the dataset.

Labeling the samples is manually done by two domain experts, who are GCE O/L Mathematics teachers in Sri Lanka. The Kappa statistic measurement of this collected dataset is 0.8598.

Table 1. Data sources and number of samples

Data source	#of samples
O/L teachers in Sri Lanka	782
Add-Sub dataset [14]	389
ARIS dataset [9]	112
Roy et al.'s dataset [3]	230
SingleOP [12] (Used only in the testing set)	200
Total	**1713**

There are 8 types of labels/classes that are identified and associated with our data set. Those labels/classes are X-Y, X+Y, Y-X, Y+Z, Y-Z, Z-Y, Z+W and Z-W. The X-Y and Y-X labels are for compare type of questions, and Y-Z, Y+Z, Z-Y are for change type of questions. Z+W and Z-W classes are associated with the whole-part type of questions. For instance, Q1 presented in Sect. 2 can be associated with X+Y, Q2 can be associated with Y-X, and Q3 can be associated with Z+W. This particular labeling approach is introduced to easily distinguish question types.

In the above labeling approach, letter 'X' represents the 1st variable and letter 'Y' represents the 2nd variable in a compare type of question. For change type of questions, the 1st variable is represented by letter "Y" and the 2nd variable is represented by letter "Z" to differentiate the variety of question types. For whole-part type of questions, the 1st variable is represented by letter "Z" and the 2nd variable is represented by letter "W". The relevant plus or minus operation is represented between these 1st and 2nd variables. Table 2 shows the number of questions we have per each type.

5   Our Approach

Figure 1 shows the system architecture. System takes the annotated training samples as input in the training phase and do pre-process on it. The feature

Table 2. Amount of questions collected

Label	Question type	No. of questions taken	No. of Qs used	
			Train set	Test set
X-Y	compare	200	160	40
X+Y	compare	115	85	30
Y-X	compare	250	230	20
Y-Z	change	320	295	25
Y+Z	change	250	227	23
Z-Y	change	200	180	20
Z+W	whole-part	200	182	18
Z-W	whole-part	178	154	24

extractor module works towards extracting the features. A classifier is then trained using these feature vectors. Multiple classification algorithms were experimented with. The trained model is used to determine the formula type of unseen questions. Finally, the answer solver generates the answer.

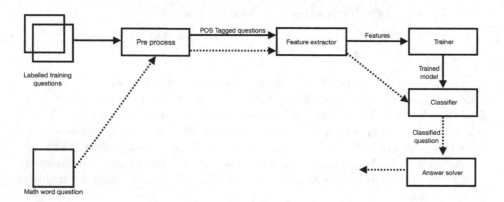

Fig. 1. System architecture

5.1 Pre-processing

In some of the questions, the numeric value can appear in word-based format. In those cases, word-to-number paraphrase converter takes the tokenized plain input sentence and converts the word-based numeric values into a numerical value. The formats of the word-based numeric values are in standard format [17]. Next, the parser outputs a POS(part—of—speech) tagged sentence.

5.2 Feature Extraction

The feature extractor component takes the POS tagged representation of a question as input. The output from this module is a feature vector that can be passed to the trainer module. This POS tag based feature extraction is based on the work by Roy et al. [3]. But there it has been used for extraction of very simple features. In contrast our system used this technique to extract some new features (both complex and simple).

We have introduced 20 new features and 10 features are derived from Amnueypornsakul and Bhat's [5] approach. The remaining 11 features are taken from Roy et al. 's [3] approach. 5 out of Roy et al.'s [3] 11 features are overlapping with Amnueypornsakul and Bhat 's [5] 10 features. Therefore, 36 features are considered as unique features in this research. Below list shows the newly introduced 20 features (complex features are in boldface).

1. Index position, distance between comparative adjective word and noun in a threshold value.
2. 1st and 2nd sentences having the same proper nouns.
3. 1st and 3rd sentences having the same proper nouns.
4. Question in the 2nd sentence is about the actions that have been mentioned in the 1st sentence.
5. 1st numerical value is greater than the 2nd numerical value.
6. Question contains exactly two numeric values in 1st or 2nd sentences.
7. Question contains the same proper nouns in all the sentences.
8. **The action mentioned in the last sentence is actually performed by the proper nouns in the 2nd sentence.**
9. Question having exactly two proper nouns.
10. Matching proper nouns that exist in the same order in the sentences.
11. **The final value of the question is related to the main proper noun in the question.**
12. **Possessive pronoun of the main proper noun exists near the noun/quantity changer in a question.**
13. The change action has made a negative impact on a quantity.
14. The change action made a positive impact on quantity in the question.
15. Action maker/proper noun exists with the nearest verb.
16. Action/quantity change happened for the same entity in a question.
17. **Co-reference resolution to identify the existence of the proper noun in the following sentences.** By checking the personal pronouns in the second or third sentences in a question, the proper noun in the first sentence is mapped for this feature.
18. Collective noun exists in sentences.
19. **Question explicitly mentions some keywords related to an addition operation (The keywords are derived from training set as well as web).**
20. **Proper nouns in the questions separately perform the same action for quantity changes.**

Also a list of positive and negative words is defined in the system to identify the positive and negative impact of a sentence in the question. (ie positive words:-add, join, sum, together, negative words:- damaged, expired)

5.3 Training and Classification

Different types of machine learning techniques were used to create classifiers. Initially, we have considered Gaussian Naïve Bayes, Random Forest, Decision tree, Perceptron and SVM (Support vector machines). Later, these classifiers were combined to create classifier ensembles. There are various ways to create ensemble classifiers. Combining the same type of classifiers with different parameters, or combining the different types of classifiers with different parameters are some of the known ways of creating ensemble classifiers. However, it is noted that we considered avoiding multistage classifiers or dependency classifiers in our approach.

5.4 Generation of the Answer

The final answer to the question is generated by the answer solver module. The inputs for this component are the predicted formula/class (from the classifier) and the two numerical values derived from the question itself. The formula is predicted by the classifier, and it posts that formula to the answer solver component. After that, the first and second numerical values are derived from the question context and those are passed to this component for further calculation. At the end, the numeric values are plotted in the predicted formula and the final answer is generated. Since this module is working totally independent from the other modules, SymPy kind of answer solver packages can be easily plugged-in with this module.

6 Evaluation

Evaluation was separately done for individual and ensemble classifiers. 10 fold cross-validation and hold-out based evaluation methods were used to evaluate the system. Data samples were shuffled two times before being passed to the evaluator module to randomize the samples and ensure that the samples are well spread in the training feature vector. We separately evaluated the accuracy of the system for Roy et al. 's [3] feature set (11 features), Roy et al. 's [3] feature set plus our newly introduced 20 features (total is 31 features), Amnueypornsakul and Bhat's [5] feature set (10 features), Amnueypornsakul Bhat's [5] feature set plus our newly introduced feature set (total is 30 features), and the combination of all 36 features. All these feature combinations were tested with all the selected classifiers.

Apart from that, SingleOP dataset [14] was used to test our system for another hold-out evaluation. Table 3 represents some of the notable accuracy values for different feature sets with different type of classifier algorithms considered in this research. Here the 10-fold and hold-out values represent the results for the testing set. For 10 fold and hold-out evaluation, 10% of the dataset is kept as test dataset and the remainder was considered as the training set. When compared with other experiment results reported in Table 3 (only the feature combinations having the best results are shown, due to space limitations, d- means the number of dimensions, or the features), all the classifiers show a drop in accuracy for all feature combinations with respect to the SingleOP dataset. This is expected, since this data set is completely independent from the training data set. However, combination that includes our features is still the best. This shows that our newly introduced features add more robustness to the classification process.

Table 3. Classification accuracy with different feature sets

Classifier	Feature set	10-fold (%)	Hold-out (%)	SingleOP data (%)
SVM	30 d	**94.2**	**92.88**	87.5
	36 d	**94.2**	92.32	**88.5**
Decision tree	30 d	94.4	**96.24**	87
	36 d	94.7	95.76	**88.7**
Gaussian NB	30 d	76.8	**78.0**	39.5
	36 d	77.1	**88.08**	38.5
RandomForest	30 d	**84.5**	83.28	63.5
	31 d	84.14	**91.67**	71
	36 d	83.5	**86.48**	**72**

As can be seen, all the classifiers perform the best when our new features are combined with Roy et al.'s [3] and/or Amnueypornsakul and Bhat's [5] features. Decision tree reported the best accuracy. The F-measure of decision tree classifier with different feature sets is noted Table 4. As a result of maximum number of questions for label Y-Z in the dataset, it obtained the highest F-measure.

6.1   Ensemble Classifier Evaluation

Two different evaluation criteria were used in ensemble based evaluations. Since decision tree shows the highest accuracy, first, the decision tree classifiers trained with different parameters were combined. After that, in another experiment, GaussianNB, SVM, RandomForest, decision tree classifiers were combined with different parameters. Due to space limitations, results only the result related to the SingleOP dataset are shown.

Ensemble-Based Evaluation Results for the Same Classifier with SingleOP Dataset: As mentioned earlier the decision tree based ensemble classifier is used for this evaluation with SingleOP dataset. The accuracy is identified

Table 4. F-measure for decision tree for 10-fold cross validation

Class	10 dim	11 dim	20 dim	30 dim	31 dim	36 dim
X-Y	0.8186	0.8878	0.8800	**0.9433**	0.9253	0.9430
X+Y	0.6726	0.8196	**0.9376**	0.9056	0.8900	0.9163
Y-X	0.5490	0.7582	0.9318	0.9580	0.9486	**0.9602**
Y+Z	0.8662	0.8820	0.9003	0.9341	**0.9463**	0.9432
Y-Z	0.9270	0.9104	0.9116	0.9631	0.9623	**0.9657**
Z-Y	0.1128	0.0959	0.7679	**0.8753**	0.7409	0.8220
Z+W	0.6604	0.7175	0.8803	0.9410	0.8687	**0.9523**
Z-W	0.1142	0.03	0.8564	**0.9505**	0.9457	0.9380
Average	0.5901	0.6376	0.8832	**0.9338**	0.9034	0.9300

Table 5. Ensemble based same type of classifier evaluation results for the SingleOP dataset

Class	Precision		Recall		F-Measure	
	ensemble1	ensemble2	ensemble1	ensemble2	ensemble1	ensemble2
X-Y	0.9375	0.8461	0.9375	0.6875	0.9375	0.7586
X+Y	**1.0**	0.4444	0.5	**1.0**	0.6666	0.6153
Y-X	**1.0**	0.0	**1.0**	0.0	**1.0**	0.0
Y+Z	0.6477	0.9272	0.9344	0.8360	0.7651	0.8793
Y-Z	0.8484	0.875	0.4375	0.9843	0.5773	**0.9264**
Z-Y	0.9166	**1.0**	0.6875	0.6875	0.7857	0.8148
Z+W	0.9259	0.8205	0.7142	0.9142	0.8064	0.8648
Z-W	0.1111	0.2	**1.0**	**1.0**	0.2	0.12
Average	0.1296	0.6391	**0.7763**	0.7636	0.7173	0.6224

as 88.9% for this evaluation. Precision, recall and F-measure for this ensemble classifier are as in the "ensemble1" column in Table 5. According to this result, Table 5, X+Y, Y-X formulae obtained the highest precision. Also Y-X and Z-W formulae show the highest recall. Y-X formula shows the highest F-measure value.

Ensemble-Based Evaluation Results for Different Classifiers with the SingleOP Dataset: For this experiment, an accuracy of 89.2% was reported with the SingleOP dataset. Precision, recall and F-measure for this ensemble classifier are as in the "ensemble2" column in Table 5. Based on the result from "ensemble2" column in Table 5, the Z-Y formula obtained the highest precision which is more precise than the result obtained from "ensemble1" classifier. Also, X+Y and Z-W formulae show the highest recall which are better than the values obtained from "ensemble1" classifier evaluation. On the other hand, the Y-Z formula shows the highest F-measure value than the value obtained from the "ensemble1" classifier evaluation.

When comparing these results with those in Table 3, it is clear that both ensembles perform better than all of the individual classifiers.

7   Conclusion and Future Work

Previous research has reported different approaches to solve different types of MWPs. Out of these, template-based approaches and deep learning approaches have shown much promise. However, for simple MWPs in elementary Mathematics, such techniques are an not sufficient for solve all types of math problems.

We show that answers to such simple MWPs can be generated using simple statistical classification techniques. This does not require any manually written templates (unlike in template-based approaches), nor a very large data set (unlike in deep learning approaches). We presented and evaluated a comprehensive feature set that can be used to train these statistical classifiers and their ensembles. These new features were able to achieve a greater result than the features used in previous research. Also this feature set is also more robust as it performs well on unseen questions as well.

In future, multiplication and division type of questions can be integrated with this system. Also we can reduce the classes/labels by considering only the operation type that we perform in a question. For example, the formula "X+Y" and the formula "Z+W" perform the same addition operation. In future, these labels can be merged as one label/class.

The assumption we considered is that the quantity changes of a particular entity/variety in a question are always associated with the same noun. However, in some cases, this assumption will not work when the second noun entity is referred by some other synonymous word. For example, consider the question "Dan has 9 pills and gave Sara 4 of the pills. How many pills does Dan have now?". In this case, we assume that the "pill" is a noun entity and the quantities 9 and 4 are only related to that "pill" entity. But in some cases, this assumption will not work when the second noun entity is referred by some other synonymous word(ie tablet). In the future, this issue can be resolved by a word similarity measurement technique. Also in future the classifiers can be used to predict positive and negative impact of a sentence.

Acknowledgement. This research was funded by a Senate Research Committee (SRC) Grant of University of Moratuwa.

References

1. Erabadda, B., Ranathunga, S., Dias, G.: Computer aided evaluation of multi-step answers to algebra questions. In: 2016 IEEE 16th International Conference on Advanced Learning Technologies (ICALT) 1993. Austin, TX, pp. 199–201 (2016)

2. Kadupitiya, J.C.S., Ranathunga, S., Dias, G.: Automated assessment of multi-step answers for mathematical word problems 2016 Sixteenth International Conference on Advances in ICT for Emerging Regions (ICTer), Negombo, pp. 66–71 (2016)
3. Roy, S.I., Vieira, T.J.H., Roth, D.I.: Reasoning about quantities in natural language. Trans. Assoc. Comput. Linguist. **3**, 1–13 (2015)
4. Kushman, N., Artzi, Y., Zettlemoyer, L., Barzilay, R.: Learning to automatically solve algebra word problems. In: 52nd Annual Meeting of the Association for Computational Linguistics, pp. 271–281, December 2014
5. Amnueypornsakul, B., Bhat, S.: Machine-Guided Solution to Mathematical Word Problems, ACL, pp. 111–119 (2014)
6. Huang, C.T., Lin, Y.C., Su, K.Y.: Explanation generation for a math word problem solver. Int. J. Comput. Linguist. Chin. Lang. Process. (2015). The 2015 Conference on Computational Linguistics and Speech Processing ROCLING 2015, pp. 64–70
7. Liang, C.C., Hsu, K.Y., Huang, C.T., Li, C.M., Miao, S.Y., Su, K.Y.: A Tag-based english math word problem solver with understanding, reasoning and explanation. In: HLT-NAACL Demos, pp. 67–71 (2016)
8. Matsuzaki, T.: The most uncreative examinee: a first step toward wide coverage natural language math problem solving. In: AAAI, pp. 1098–1104, July 2014
9. Dellarosa, D.: A computer simulation of children's arithmetic word-problem solving. Behav. Res. Meth. Instrum. Comput. **18**(2), 147–154 (1989)
10. Wang, Y., Liu, X., Shi, S.: Deep neural solver for math word problems. In: Conference on Empirical Methods in Natural Language Processing, pp. 856–865, September 2017
11. Hosseini, M.J., Hajishirzi, H., Etzioni, O., Kushman, N.: Learning to solve arithmetic word problems with verb categorization. In: Conference on Empirical Methods on Natural Language Processing, pp 523–533, October 2014
12. Robert Sweetland: Types of Addition and Subtraction Problems Examples with whole numbers (1992). http://www.homeofbob.com/math/numVluOp/wholeNum/addSub/adSubTypsChrt.html
13. Koncel-Kedziorski, R.: MaWPS: A Math Word Problem Repository, HLT-NAACL, pp 1152–1157, June 2016
14. MaWPS: A Math Word Problem Repository (2016). http://lang.ee.washington.edu/MAWPS/datasets/SingleOp.json
15. Morton, K., Yanzhen, Q.: A novel framework for math word problem solving. Int. J. Inf. Educ. Technol. **3**(1), 88–93 (2013)
16. Zhou, L., Dai, S., Chen, L.: Learn to solve algebra word problems using quadratic programming. In: EMNLP The Association for Computational Linguistics, pp. 817–822, September 2015
17. Furey and Edward. Numbers to Words Converter. https://www.calculatorsoup.com
18. Hevapathige A., Wellappili D., Kankanamge G.U., Dewappriya N., Ranathunga S.: A two-phase classifier for automatic answer generation for math word problems. In: 18th International Conference on Advances in ICT for Emerging Regions (ICTer), pp. 1–6 (2018)

Learning Mobile App Embeddings Using Multi-task Neural Network

Ahsaas Bajaj$^{(\boxtimes)}$, Shubham Krishna, Hemant Tiwari, and Vanraj Vala

Samsung R&D Institute, Bengaluru, India
{ahsaas.bajaj,shubham.k1,h.tiwari,vanraj.vala}@samsung.com

Abstract. Last few years have seen a consistent increase in the availability and usage of mobile application (apps). Mobile operating systems have dedicated stores to host these apps and make them easily discoverable. Also, app developers depict their core features in textual descriptions while consumers share their opinions in form of user reviews. Apart from these inputs, applications hosted on app stores also contain indicators such as category, app ratings, and age ratings which affect the retrieval mechanisms and discoverability of these applications. An attempt is made in this paper to jointly model app descriptions and reviews to evaluate their use in predicting other indicators like app category and ratings. A multi-task neural architecture is proposed to learn and analyze the influence of application's textual data to predict other categorical parameters. During the training process, the neural architecture also learns generic app-embeddings, which aid in other unsupervised tasks like nearest neighbor analysis and app clustering. Various qualitative and quantitative experiments are performed on these learned embeddings to achieve promising results.

1 Introduction

We live in an era where mobile applications have become really popular platforms to provide various utilities and information. With an increase in demand of smart phones, the number of available apps have also constantly grown [1]. Google play store and Apple app store are the most common providers of these applications. Majority of applications are developed by third-party developers and private organizations to fulfill the needs of their customers. Browsing through the Google Play Store [2], one can see the abundance of applications available. Apart from the description and user reviews, they also have many associated tags (or indicators) like category, content ratings, review ratings, price, number of downloads, etc. Generally, the popular apps have well-defined descriptions and large number of reviews which make them rich in information. However, it is difficult to tag this information for the newly added apps or the unpopular ones. Once these indicators are available, they can optimize many different tasks related to mobile apps. For example, some apps have content (age) rating of

---

A. Bajaj and S. Krishna—Equal contribution.

© Springer Nature Switzerland AG 2019
E. Métais et al. (Eds.): NLDB 2019, LNCS 11608, pp. 29–40, 2019.
https://doi.org/10.1007/978-3-030-23281-8_3

18+ and it is important not to show these apps when queried by a user below this age bracket. Also, there are apps with category indicator as education or games which can help boost their ranking when queried by a teenager browsing the play store (or app store). Similarly, apps with poor ratings can be prevented from populating the top few results. Moreover, the app category might help the user to properly arrange them in his/her device after installation. Some users might want to have a separate folder for each category of the downloaded apps. Therefore, the additional indicators (tags) associated with each application serve very important purpose to optimize its search and also the arrangement (ordering) in users' mobile devices.

Most of the information regarding a particular app is mentioned in its description and reviews. These textual sources, in combination, can serve a great deal to derive useful insight about a mobile application. Therefore, it is crucial to model this data in order to provide useful predictions for rest of the correlated indicators. Generally, app developers talk about their most important features in their description which are often reinforced by users sharing their experience with these features in the review section. Users not only share their sentiment about the app but also talk about app's features from their perspective. This adds up to the information given by the developers in the app description. Therefore, this paper attempts to jointly model applications' description and reviews in a multi-task learning fashion in order to predict various indicators related to the applications. As most of output indicators (or their prediction tasks) are correlated with each other, multi-task learning is the most effective approach, further confirmed by the experiments performed in this paper. As a result of this learning, generic application embeddings are developed to solve various tasks related to mobile applications.

An overview of related work in the domain of mobile apps and multi-task learning is presented in Sect. 2. The methodology of predicting the app indicators and learning app embeddings are discussed in Sect. 3. Section 4 elaborates experiments and results using the proposed architecture for various tasks. Finally Sect. 5 recapitulates the proposed approach with current applications and scope of future extensions.

2   Related Work

In recent years, multiple mobile applications have become available to feed our information needs. This tremendous amount of data has called for its efficient utilization to provide a pleasant user experience by optimizing retrieval or ranking. Since names of mobile apps are ambiguous, it is difficult to figure out the exact functionality of the apps from their names. An approach to enrich this information by exploiting the additional knowledge from the web and real-world logs was proposed in [3]. Using this extra information, they studied the problem of automatic app classification. In 2015, researchers studied the domain of mobile app retrieval using topic modeling [4]. They proposed a probabilistic model, named *AppLDA*, which combines app description and reviews to extract relevant topics from mobile apps. They also released a test collection of app data which can be used for its quantitative evaluation. Further, work done in [5] explored the role of social media texts in order to capture implicit user intent and its usage for mobile app retrieval. Researchers have also used Learning-to-Rank algorithms to optimize retrieval of mobile notifications triggered by installed apps [6].

Recently, there have been studies to facilitate classification of apps into pre-defined interest taxonomies. Using language modeling on smart phone logs, [7] proposed a neural approach to learn app embeddings in a low-dimensional space. Inspired by the deep learning approaches like Word2vec [8], vectorized representations of mobile applications were generated to capture semantic relationship between apps [9]. They proposed an application recommender tool by building a similarity function based on metrics like popularity, security, usability, etc. They also studied the importance of these additional parameters to determine usefulness of an app for a particular user. Researchers have recently tried to develop relevance-based application embeddings to facilitate various information retrieval tasks related to mobile applications [10].

The research done so far has focused on learning mobile application embeddings suitable for a specific task, like classification, recommendation, and others. Most of these tasks are loosely related to each other and have some commonality. If a model is trained for a specific task, it often fails to generalize well and captures data-dependent noise [11]. In academia, multi-task learning [12] has proved to be useful for enhancing the performance of learning algorithms by modeling multiple tasks jointly. It has found to be of significant use for multiple domains like natural language processing [13], computer vision [14], speech recognition [15], and many more. Recently, [16] proposed an approach called MRNet-Product2Vec for learning generalized embeddings of Amazon products. They employed the use of multi-task recurrent neural network to model diverse set of product characteristics like weight, size, color, price, etc. The learned low-dimensional embeddings were demonstrated to be as good as sparse and high-dimensional representations.

The authors observe that the technique of multi-task learning has not been employed to learn embeddings for mobile applications. As in the case of e-commerce products [16], mobile applications also have multiple indicators like popularity, content ratings, category which can be jointly modeled to create the app embeddings in a more generic sense. This paper proposes a novel method to learn dense app embeddings using app descriptions and reviews while optimizing for different tasks of predicting app indicators. The learned app embeddings are then evaluated with different tasks like app recommendation and clustering.

3   Proposed Method

The purpose of learning application embedding is to develop a generic representation of the available apps. This facilitates various tasks related to mobile applications such as app recommendation, retrieval and categorization. In order to learn these embeddings, it is crucial to capture diverse characteristics related to applications. Each of these characteristics can be modeled as a regression or classification task of its own, where the text data acts as input and different app indicators (category, ratings, etc.) as the output. As discussed in the previous section, multi-task learning has the capability to learn a generalized feature representation by exploiting commonalities and differences among various tasks. While optimizing losses for different tasks in a single network, multi-task learning introduces inductive bias which helps the architecture to learn a trade-off between the multiple losses which eventually helps in achieving generalized solutions. This representation aids to improve the accuracy of multiple tasks while also being time and resource efficient as there is no need to train and store different models for each task to be performed.

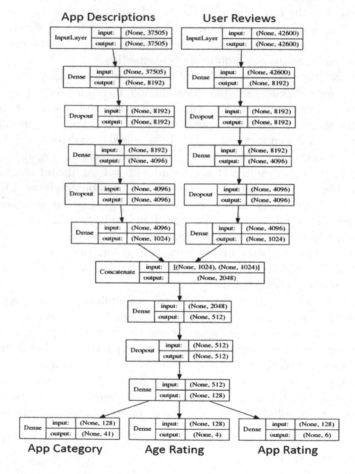

Fig. 1. Multi-task neural architecture

There are two known ways to perform multi-task learning, namely, soft parameter sharing and hard parameter sharing [11]. The proposed architecture is based on the concept of hard parameter sharing with each task having its specific output layer and some shared hidden layers. This architecture is based on the intuition that if multiple tasks are to be learned simultaneously, the model eventually finds a representation that captures important information pertaining to all the tasks. This reduces noise and the chance of over-fitting for a specific task.

3.1 Architecture

As depicted in Fig. 1, the network contains two inputs, one for app description and the other for user reviews. The input data is vectorized using the TF-IDF representation [17,18] to generate weighted term vectors [19] which are then passed to the neural network. As tf-idf methodology is based on bag-of-words representation, the proposed method is language agnostic and works for description and reviews in any language. The authors experimented with various number of layers and different encoding sizes in each layer. After hyper-parameter tuning, the best possible configuration is proposed as the final architecture in this paper. Overall, the network contains nine hidden layers to encode meaningful word representations. Initially, both the inputs are dealt with separate hidden layers and later concatenated to form an unified representation at the sixth hidden layer. After concatenation, this representation is further condensed and fine tuned to generate a dense embedding, before feeding forward the data to the task-specific output layers. The ReLU non-linearity [20] is applied as an activation to each layer's output. Dropout technique [21] is also employed by setting to zero the output of each hidden neuron with probability of 0.35. These choices are motivated by their popularity in academia.

3.2 Training Process

The purpose of this architecture is to predict different app indicators such as App Category, Age Rating, App Rating and learn generic app embeddings by doing so. These prediction tasks are multi-class classification problems and are jointly solved using the multi-task learning paradigm. The output comprises of task specific layers, one layer for each of the three tasks. The softmax function [22] is applied to predict the probability distribution over different classes. The categorical cross entropy loss is calculated separately for each output layer.

$$L_1(\hat{\boldsymbol{y_1}}, \boldsymbol{y_1}) = -\sum_{i=1}^{N}\sum_{j=1}^{C_1} \boldsymbol{y_1}_i^j \log(\hat{\boldsymbol{y_1}}_i^j),$$

$$L_2(\hat{\boldsymbol{y_2}}, \boldsymbol{y_2}) = -\sum_{i=1}^{N}\sum_{j=1}^{C_2} \boldsymbol{y_2}_i^j \log(\hat{\boldsymbol{y_2}}_i^j),$$

$$L_3(\hat{\boldsymbol{y_3}}, \boldsymbol{y_3}) = -\sum_{i=1}^{N}\sum_{j=1}^{C_3} \boldsymbol{y_3}_i^j \log(\hat{\boldsymbol{y_3}}_i^j)$$

where y_k is the true label and $\hat{y}_k$ is the predicted probability for the kth task. For the architecture given in this paper, $k = 3$. The total number of training examples are given by N. C_1, C_2, C_3 are the total number of classes for each of the tasks respectively. The final loss is the sum of above three losses which is back-propagated for learning the parameters of the network. The back-propagation algorithm with Adam optimizer is used for updating the parameters. The final loss is given by:

$$L(\hat{y}, y) = L_1(\hat{y_1}, y_1) + L_2(\hat{y_2}, y_2) + L_3(\hat{y_3}, y_3) \tag{1}$$

Before training, the entire data is split in the ratio of 75-25 for training and testing. Python-based deep learning library: Keras [23] is used to carry out the training process. After the training is complete, the dense representation given at the last hidden layer (say, layer t) gives the application embeddings. This is given by:

$$AppEmb = ReLU(W_t \times Z_{t-1} + b_t) \tag{2}$$

where, W_t and b_t are the respective weight matrix and bias vector for the layer t. Both of these are learned during the training process with Z_{t-1} being the output from the previous layer $t - 1$. For the architecture given in Fig. 1, $t = 9$.

4   Experimental Details

To evaluate the performance of the proposed method, various experiments are performed to evaluate the multi-task architecture and the benefits of the learned application embeddings. A publicly available apps dataset is used to test the performance of methodologies discussed in the previous section. This dataset includes textual data in form of app descriptions and user reviews, which are useful to learn various parameters related to mobile applications. Analysis is performed using these textual inputs to model categorical app indicators using the multi-task architecture. As per the knowledge of the authors, there are no existing methods which perform multi-task learning on this dataset. Therefore, comparisons are performed with single-task and single-input versions of the same approach. Moreover, the proposed model generates the application embeddings after the training is complete. These can be useful for various use-cases and experiments like nearest neighbor analysis and app clustering show the capability of these embeddings vectors for unsupervised tasks. Results indicate superior performance as compared to existing state-of-the-art method like Doc2Vec.

4.1   Dataset

Data Set for Mobile App Retrieval [4] is used for evaluation of the proposed methods. This data includes information about 43,041 mobile apps including the title, description, category, package name, user-reviews, and other app indicators. For sanity, apps without description or user reviews are not considered during evaluation. With the above mentioned preprocessing, the number of unique apps comes down to 39,700 with a vocabulary size of 37,505 and 42,600 unique words for description and reviews respectively. To perform the tasks discussed in Sect. 3, the following indicators are considered to form the output layer of the multi-task neural architecture.

- App Category: Total 41 categories including Music, Family, Racing, Trivia, Weather, Productivity, etc.
- Age (Content) Rating: Low Maturity, Medium Maturity, High Maturity, and Everyone.
- App Rating: Star ratings between 1–5 are mapped to five brackets, namely, Very Low, Low, Medium, High, Very High.

The choice of these three indicators is based on the intuition to find all the app characteristics that could be derived from the textual data in form of app description and reviews. For example, there is no direct link between an app's textual data and its price, number of downloads, number of reviewers, developer, date of publishing, etc. Many of these parameters may be based on developer quality, app's software, design or services. However, indicators like category and ratings could be learned from the choice of words and their sentiment used in the associated text.

4.2 Analysis with Multi-task Learning

This section details different experiments which are performed to quantify the benefits of the proposed architecture using the dataset discussed in the previous section. Categorical Accuracy [23] is the metric which is used for evaluating the performance of the model. It calculates the mean accuracy rate across all predictions for multi-class classification problems.

Comparison with Single Task Predictions. In this section, we present our results for predicting different app indicators using the proposed multi-task network discussed in Sect. 3. Table 1 shows the results for predicting the app indicators like app category, age rating and app rating by using app's description and reviews as the input sources. To evaluate the working of the proposed approach, single-task models are also built to tackle each of the tasks separately. These models are built using the structure similar to the Fig. 1 but with a single output layer for the specific task to be performed. The results show that our multi-task model out-perform the single-task approaches which were supposed to be tuned for their specific task. This shows the capability of the proposed architecture to learn a generic representation while also being time and resource efficient.

Comparison with Single Input Predictions. For further analysis, we modified the input sources in Fig. 1 by toggling between the description and reviews to understand their respective usefulness for predicting different app indicators.

Table 1. Evaluation results for single and multi-task learning

Model	Accuracy for different tasks		
	App category	Age rating	App rating
Single task	0.630	0.685	0.402
Proposed multi task	**0.632**	**0.694**	**0.725**

Multi-task learning is employed to perform this analysis and the results are shown in Table 2. It can be seen that the app description significantly impacts the prediction of app category and age rating. Whereas, user reviews play a major role to estimate app ratings. This also makes sense because an app developer talks about the features of his/her app in the description and consumers discuss these features in their reviews. Therefore, both of these can help to model app category and age rating which are mostly feature driven. On the other hand, app ratings are mostly based on the user sentiment depicted in their reviews. Therefore, reviews directly impact the prediction of app rating. As both the description and reviews are jointly modeled in the proposed technique, it performs significantly well on all the given tasks.

Table 2. Evaluation results for different multi-task architectures

Text sources	Accuracy for different tasks		
	App category	Age rating	App rating
App description	0.615	0.688	0.638
App reviews	0.544	0.615	0.716
Description and reviews	**0.632**	**0.694**	**0.725**

As shown in the results, the proposed network learns the useful correlations in a noise-free manner. This helps to out-perform single-task as well as single-input models. Results were also tested for statistical significance and achieved $p - value = 0.0$ (for majority results) in hypothesis testing. This proves that the results are not obtained due to randomness. The architecture discussed in this paper is robust and also scalable. By changing or augmenting the input and output data, it can adapt accordingly for performing different tasks in a suitable manner. This architecture also gives the ability to generate dense application embeddings as discussed further.

4.3 Analysis with Learned App Embeddings

The application embedding in Eq. 2 gives the low-dimensional representation for the apps which are built during the process of multi-task learning. Since the training was performed to optimize several classification tasks together, it achieves generic embeddings at the last hidden layer. This section shows the novelty of these embeddings with their usefulness for tasks related to app clustering and recommendation.

Table 3. Evaluation results on Clustering Techniques

Embeddings	DBSCAN		k-Means (k = 41)	
	Silhouette score	Davies bouldin score	Silhouette score	Davies bouldin score
Doc2Vec	−0.126	1.671	−0.014	3.997
AppEmb	**0.324**	**0.99**	**0.411**	**1.647**

App Clustering. For validating the usefulness of these generic app embeddings, app clustering has been carried out using k-Means and DBSCAN algorithms. These algorithms are commonly used clustering methods, where k-Means is a common baseline and DBSCAN is a well-known density based clustering technique. Proposed embeddings (*AppEmb*) has been compared with the Doc2Vec embeddings [24], trained on the same corpus of app descriptions and reviews. For evaluating the performance of clustering, Silhouette [25] and Davies Bouldin [26] scores have been calculated and are shown in Table 3. It is well known that silhouette score is a measure of consistency within clusters of data. A high positive value indicates that the current assigned cluster is the best match for that data point and vice versa. Silhouette score is positive for the proposed embeddings and is negative for the Doc2vec embeddings. This indicates a better cluster assignment using *AppEmb*. For further validation, Davies Bouldin score has also been calculated. It is defined as the ratio of within-cluster to between-cluster distances. So, a value closer to zero indicates a better clustering and our embeddings have relatively better Davies Bouldin score than Doc2Vec embeddings. It is clear from these two metrics that our proposed app embeddings outperform Doc2Vec embeddings.

Nearest Neighbor Analysis. Qualitative analysis has also been carried out for proving the worth of the proposed app embeddings. Multiple apps for serving a simple purpose are being developed and in that case our generic app embeddings (*AppEmb*) can be used for app recommendation task. Nearest neighbor analysis finds the closest neighbors of a specific application, and *AppEmb* (Eq. 2) can be used to represent the apps. Results in Table 4 show that for application of a particular category (query), the closest matched apps mostly belong to the same category. The results for the nearest neighbor analysis have also been represented as a 2-D plot in Fig. 2 using t-SNE visualization [27]. The visualization also shows accurate grouping for different categories of applications.

Table 4. Qualitative results with Nearest Neighbors Analysis using *AppEmb*

Application	Predicted nearest applications - name (Category)				
Job Search **(Business)**	Freelancer - Hire & Work (Business)	Tech Job Search by Dice (Business)	JobStreet (Business)	Job Interview (Business)	Confident Interview (Business)
9GAG - Funny pics and videos **(Entertain)**	Spider in phone funny joke (Entertain)	Best Vines (Entertain)	Discovery Channel (Entertain)	DIRECTV for Tablets (Entertain)	Helium Voice Change (Entertain)
U.S. Bank **(Finance)**	Bank of the West Mobile (Finance)	Western Union (Finance)	Esurance Mobile (Finance)	Bitcoin Wallet (Finance)	Wallet (Finance)
Mobile Tracker **(Productivity)**	Mobile Number Call Tracker (Productivity)	Cell Tracker (Productivity)	Family Mobile Tracker (Productivity)	GPS Location Tracker (Social)	Tornado - American Red Cross (Weather)
Kids Learn Write Letters **(Education)**	Kids Multiplication Tables (Education)	SchoolWay -formerly schConnect (Education)	ABC for Kids All Alphabet (Education)	The Wheels On The Bus (Education)	ABCs Kids Tracing Cursive (Education)

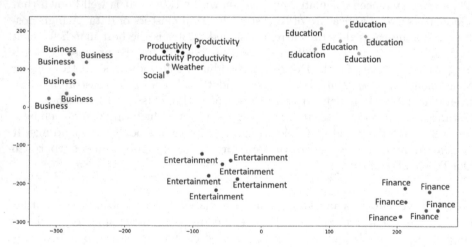

Fig. 2. t-SNE projection for applications and their categories given in Table 4

5　Conclusions and Future Work

In this paper, generic application embeddings are learned with our multi-task neural network architecture using the description and user reviews of mobile applications. These embeddings are developed by keeping in mind the increasing usage of mobile applications and the difficulties faced to find out relevant ones from a large collection. The learning of these embeddings is carried out based on correlation of app indicators with the textual data available for the apps. These embeddings have outperformed Doc2vec on tasks like k-Means and DBSCAN clustering. The results show that the predictions on this multi-task network outperform their single-task and single-input counterparts. These predictions which are based on language-agnostic modeling of apps' text data can be useful for the users, app developers and also the companies which host applications on their websites. In future, app data from stores of different regions can be evaluated to solidify this claim. Another extension of this work can be to modify the neural network as per the insights developed from the results of this paper. Using different forward and back-propagation paths, a selective multi-task learning can be performed where each input text affects different output indicators in an independent way.

References

1. Statista: Number of available applications in the Google Play Store from December 2009 to December 2018. https://www.statista.com/statistics/266210/number-of-available-applications-in-the-google-play-store/
2. Google: Play Store. https://play.google.com/store
3. Zhu, H., Chen, E., Xiong, H., Cao, H., Tian, J.: Mobile app classification with enriched contextual information. IEEE Trans. Mob. Comput. **13**, 1550–1563 (2014)
4. Park, D.H., Liu, M., Zhai, C., Wang, H.: Leveraging user reviews to improve accuracy for mobile app retrieval. In: Proceedings of the 38th International ACM SIGIR Conference on Research and Development in Information Retrieval, pp. 533–542. ACM (2015)
5. Park, D.H., Fang, Y., Liu, M., Zhai, C.: Mobile app retrieval for social media users via inference of implicit intent in social media text. In: Proceedings of the 25th ACM International on Conference on Information and Knowledge Management, pp. 959–968. ACM (2016)
6. Bajaj, A., Tiwari, H., Vala, V.: Enhanced learning to rank using cluster-loss adjustment. In: Proceedings of the ACM India Joint International Conference on Data Science and Management of Data, pp. 70–77. ACM (2019)
7. Radosavljevic, V., et al.: Smartphone app categorization for interest targeting in advertising marketplace. In: Proceedings of the 25th International Conference Companion on World Wide Web, International World Wide Web Conferences Steering Committee, pp. 93–94 (2016)
8. Mikolov, T., Sutskever, I., Chen, K., Corrado, G.S., Dean, J.: Distributed representations of words and phrases and their compositionality. In: Advances in Neural Information Processing Systems, pp. 3111–3119 (2013)
9. Rustgi, P., Fung, C., Rashidi, B., McInnes, B.: Droidvisor: an android secure application recommendation system. In: 2017 IFIP/IEEE Symposium on Integrated Network and Service Management (IM), pp. 1071–1076. IEEE (2017)
10. Bajaj, A., Krishna, S., Rungta, M., Tiwari, H., Vala, V.: Relemb: A relevance-based application embedding for mobile app retrieval and categorization arXiv:1904.06672 [cs.IR] (2019). http://arxiv.org/abs/1904.06672
11. Ruder, S.: Multi-Task Learning Objectives for Natural Language Processing. http://ruder.io/multi-task-learning-nlp

12. Caruana, R.: Multitask learning. Mach. Learn. **28**, 41–75 (1997)
13. Collobert, R., Weston, J.: A unified architecture for natural language processing: deep neural networks with multitask learning. In: Proceedings of the 25th International Conference on Machine Learning, pp. 160–167. ACM (2008)
14. Girshick, R.: Fast R-CNN. In: Proceedings of the IEEE International Conference on Computer Vision, pp. 1440–1448 (2015)
15. Deng, L., Hinton, G., Kingsbury, B.: New types of deep neural network learning for speech recognition and related applications: an overview. In: 2013 IEEE International Conference on Acoustics, Speech and Signal Processing (ICASSP), pp. 8599–8603. IEEE (2013)
16. Biswas, A., Bhutani, M., Sanyal, S.: MRNet-Product2Vec: a multi-task recurrent neural network for product embeddings. In: Altun, Y., et al. (eds.) ECML PKDD 2017. LNCS (LNAI), vol. 10536, pp. 153–165. Springer, Cham (2017). https://doi.org/10.1007/978-3-319-71273-4_13
17. Luhn, H.P.: A statistical approach to mechanized encoding and searching of literary information. IBM J. Res. Dev. **1**, 309–317 (1957)
18. Sparck Jones, K.: A statistical interpretation of term specificity and its application in retrieval. J. Documentation **28**, 11–21 (1972)
19. Salton, G., Wong, A., Yang, C.S.: A vector space model for automatic indexing. Commun. ACM **18**, 613–620 (1975)
20. Nair, V., Hinton, G.E.: Rectified linear units improve restricted boltzmann machines. In: Proceedings of the 27th International Conference on Machine Learning (ICML 2010), pp. 807–814 (2010)
21. Srivastava, N., Hinton, G., Krizhevsky, A., Sutskever, I., Salakhutdinov, R.: Dropout: a simple way to prevent neural networks from overfitting. J. Mach. Learn. Res. **15**, 1929–1958 (2014)
22. Bridle, J.S.: Probabilistic interpretation of feedforward classification network outputs, with relationships to statistical pattern recognition. In: Soulie, F.F., Herault, J. (eds.) Neurocomputing, vol. 68, pp. 227–236. Springer, Heidelberg (1990). https://doi.org/10.1007/978-3-642-76153-9_28
23. Chollet, F., et al.: Keras (2015). https://keras.io
24. Le, Q., Mikolov, T.: Distributed representations of sentences and documents. In: International Conference on Machine Learning, pp. 1188–1196 (2014)
25. Rousseeuw, P.J.: Silhouettes: a graphical aid to the interpretation and validation of cluster analysis. J. Comput. Appl. Math. **20**, 53–65 (1987)
26. Davies, D.L., Bouldin, D.W.: A cluster separation measure. IEEE Trans. Pattern Anal. Mach. Intell. (2), 224–7 (1979)
27. Maaten, L.V.D., Hinton, G.: Visualizing data using t-SNE. J. Mach. Learn. Res. **9**, 2579–2605 (2008)

Understanding User Query Intent and Target Terms in Legal Domain

Sachin Kumar(✉) and Regina Politi(✉)

LexisNexis, Raleigh, USA
{sachin.kumar.1,regina.politi}@lexisnexis.com

Abstract. Lexis Advance is a legal research service provided by LexisNexis that can respond to natural language queries. It includes a module called Lexis Answers which implements advanced Natural Language Processing (NLP) capabilities to improve understanding of the intent of the user's queries. Lexis Answers can respond to natural language questions concerning legal question types such as statute of limitations, elements of a claim, definition of legal terms, and others. Herein, we report on the successful use of advanced NLP approaches for detecting not only named entities, but entire legal phrases, a skill previously requiring domain knowledge and human expertise. We have utilized the Conditional Random Fields (CRFs) approach that employs hand-engineered features combined with word2vec embeddings trained on legal corpus. Furthermore, to reduce our dependency on hand-engineered features, we have also implemented deep learning architecture comprising of bidirectional Long Short-Term Memory (BiLSTM) and linear chain CRF. Both approaches were benchmarked against a rule-based approach for different types of legal questions. We find that both CRF and BiLSTM-CRF can identify query intents and legal concepts with comparable precision but much higher recall and F-score than the baseline. The resulting models have been employed in Lexis Answers as critical improvement in our natural language query understanding.

Keywords: Named Entity Recognition · Query intent · Target terms · Conditional random fields · Bidirectional LSTM · Natural language processing · Deep learning · Legal domain

1 Introduction

Electronic legal research using LexisNexis products involves searching for the most relevant information within large databases of legal content, including cases, treatises, statutes and regulations, etc. Researchers are generally required to convert the legal question they need to answer into a search query that the legal databases can effectively use. Lexis Advance, from LexisNexis, is typical of modern legal research platforms where users enter search queries in various

Supported by LexisNexis, USA.

© Springer Nature Switzerland AG 2019
E. Métais et al. (Eds.): NLDB 2019, LNCS 11608, pp. 41–53, 2019.
https://doi.org/10.1007/978-3-030-23281-8_4

forms in order to find what they need as efficiently as possible. The volume of search queries handled by platforms such as Lexis Advance is usually very large. For instance, Lexis Advance handles up to 2 million queries a day. While most of those queries are still either looking for a very specific known document, or are in a pattern-matching form (e.g., keywords with Boolean logic), newer generations of researchers are asking for better natural language support. It is becoming clear that a modern, robust legal research system should be able to go beyond simple document retrieval or explicit pattern matching toward truly recognizing a user's intent in a natural language query. There is a significant commercial pressure to correctly understand natural language query intent and deliver accurate and comprehensive answers so that attorneys can spend less time looking for relevant information and more time wielding it in support of their clients.

The underlying problem of finding the direct answer to legal queries is very complex and challenging, primarily because of the complex nature of legal queries. Queries submitted to the search engine can be very short and unclear in terms of user real intent, i.e., *"rape limitations period"*. Or, they can be very complex and linked with Boolean connectors as in *"statute /s limitation or repose /s Alabama /s negligent or defect! /s construction"*. To be able to understand user's research interest we divide the task into two main categories: identification of *query intent* and identification of *target terms*. Ambiguity is involved in these type of tasks since not always the phrase has a legitimate target term. There is even disagreement between Subject Matter Experts (SMEs) concerning the identification of the correct target terms or query intents. Consider the following query: *"What are the elements of supplemental jurisdiction?"*, this query from the linguistic prospective may look legitimate as it asks about certain elements. However, there is no such a thing as elements of supplemental jurisdiction and expert knowledge is required to make this observation. Finding effective ways of using advanced technologies to solve these challenges is a key to future success of legal market research.

To address these challenging requirements, Lexis Answers service was established within the Lexis Advance system to interpret searches and mine the answers in order to provide concise responses to legal research questions. Indeed, the system goes beyond simply providing documents with potentially relevant sections highlighted: it actually extracts and delivers direct answers to legal questions in combination with a results list that enables deeper research.

The technology behind Lexis Answers is powered by one of the most widely used tasks of Information Extraction (IE). Its called Named Entity Recognition (NER), and is a fundamental building block of complex NLP tasks. NER is defined as a process of classifying entities in unstructured text into predefined categories. It is widely used to process news, corporate files, medical records, government documents, court hearings and social media. A recent survey [15] summarized impressive results obtained with NER to identify persons, organizations, locations and many other miscellaneous named entity types. Many other domains such as biomedical, chemical [4] as well as financial [2] have used

NER to locate information concerning personal patient data, or identify such terms as proteins, DNA's, RNA, drugs, stock market. etc. The success of these applications of NER was enabled by the history of creating extensive dictionaries and lookup tables complemented with large datasets essential for training of statistical models capable of recognizing respective named entities. Despite such versatile and growing usage of NER in multiple domains, its application in legal domain remains challenging mostly because of ambiguity and complexity of legal terms that had to be recognized, which requires the involvement of domain experts.

The original implementation of NER in Lexis Answers was highly innovative in its ability to go beyond simple retrieval of relevant documents and provide more specific answers to user queries as described above. However, this implementation was not without certain deficiencies associated with the use of rule based approaches for term retrieval. Such approaches are computationally effective but require a lot of manually annotated development data, together with significant input from experienced rule writers, but still culminating in high precision but low recall. In addition, maintenance of such rule sets can be a challenge, since often the rules have intricate inter-dependencies that are easy to forget, thus making modification risky.

Herein, we describe the development of novel NER models that can correctly identify user intent and target terms. Thus delivering the most relevant answer card for the user's specific search request, without requiring domain knowledge expertise. In this paper, we explore Conditional Random Fields (CRFs), a machine learning method that employs hand-engineered features such as: part of speech tagging, contextual features, gazetteers, etc. These hand-engineered features were combined with word representations obtained from the word2vec models trained on a large legal corpus. Furthermore, to reduce our dependency on hand-engineered features we have also explored Recurrent Neural Networks that have become highly popular in the machine learning community recently. We have implemented a framework that consisted of bidirectional Long Short-Term Memory (BiLSTM) and linear chain CRF. Both approaches were benchmarked against the baseline, rule-based, approach. Our results show that both CRF and BiLSTM-CRF outperformed the rule-based approach in identification of both query intent as well as target terms. We have also shown that no significant improvement was achieved in the target term recognition using BiLSTM-CRF as opposed to CRF. This observation most probably results from the fact that deep learning models normally require large amount of training data [8]. However, manual tagging of such large amounts of data is impractical in the company setting. The resulting models have been implemented within Lexis Answers as critical improvement in the ability of the system to understand and respond to natural language queries.

2   Related Work

Many papers have been published in recent years that employed NER approaches in legal domain. In Dozier et al. [2], NER was applied to legal documents with a machine learning based model to obtain named entities like person, place or organization etc. With the advent of deep learning, many papers reported the application of this approach for NER tasks. Most notably, Shankar and Buddarapu [10] experimented with the application of multiple deep learning models like LSTM, GRU, CNN to identify judges names in user queries. Previous studies also used convolutional neural networks to detect the intent of queries [3]. Inspired by Sutton et al. [12] we have implemented a CRF based machine learning approach. Additionally, studies described in [5,11] inspired us to implement the BiLSTM-CRF architecture based deep learning model. In this paper, we use CRF and BiLSTM-CRF models not only to detect search queries intent but also the target terms within the query.

3   Methods

3.1   Data

Table 1 shows counts of different types of unique user queries collected in this study and used for both training and external evaluation. The collected queries consisted of different types of legal questions like statutes of limitations (SOL), elements (ELO), doctrine (DOC), definitions (DEF) and jurisdiction related (JURI) queries. In addition to those, other types of user queries which were not associated with any of the types listed, called negatives, were collected as well. Negative samples allow representation of discriminating queries, and ensure the learning algorithm is not just focused on the desired queries. Dataset collected for JURI did not require negative queries since no intent had to be recognized in these queries but only target terms. All legal question type queries can have jurisdiction mentioned and were part of our training dataset.

Table 1. Number of user queries used for training and test data sets for each legal question type

Query type	Training	Test
SOL	5000	1327
Negative to SOL	3469	791
ELO	4698	1311
Negative to ELO	2348	587
DOC	3639	695
Negative to DOC	1766	612
DEF	2358	590
Negative to DEF	1753	439
JURI	9457	2339

For each specific legal question type dataset, for instance ELO, user queries identified as ELO were treated as positives and all other user queries were treated as negatives. The negative queries for each query type were randomly picked from the pool of all negatives queries for that particular query type. Every query was annotated by SMEs to identify phrases associated with query intent and target term. An examples of annotated queries of each type are given in Fig. 1a. A number of preprocessing steps were performed in order to standardize queries. For instance: conversation to lowercase, removal of punctuation and other special characters, removal of Boolean notation etc.

3.2   Tagging Scheme

The main goal of named entity recognition is to tag each token in a sentence with an appropriate entity label. An entity can include more than one token, for instance, when this entity is a target term. For example, *reasonable attorneys fees* within *"what is the definition of reasonable attorneys fees"*. In this study we use IOB tagging (Inside, Outside, Beginning) to label every token of the sentence as "B-tag" if the token represents beginning of an entity, "I-tag" for token inside the entity and "O" for tokens outside the entity. Annotations created by SMEs allowed accurate labeling of tokens in the query with IOB tags as shown in Fig. 1b.

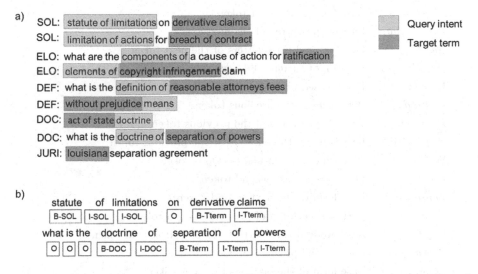

Fig. 1. Examples of annotated user queries. Figure 1a shows the annotations performed by SMEs when phrases associated with query intent are colored blue, target terms and jurisdictions are colored pink. Figure 1b shows an example of IOB tagging created for SOL and DOC query types, following SMEs annotations.

Two types of entities, i.e., intent (SOL, ELO, DOC, DEF) and target term, have to be identified in the query in order to consider a given query actionable for providing a relevant answer by Lexis Answers service. For the jurisdiction recognizer, the query had only one entity, jurisdiction. Recognition of jurisdiction in the queries allows the search to be more specific to that location.

3.3 Named Entity Recognition with CRFs

Our first approach to identify query intent and target terms employed a linear-chain factor graph known as Conditional Random Fields (CRFs) [12]. The CRF algorithm implementation used in this paper was provided by sklearn-crfsuite library.

Table 2. List of hand-engineered features.

Name	Description
Word	Lower cased current token
Lemma	Base form of current token
Suffix	Last two and three letters of the current token
Is digit	Checking if current token is a digit
POS-tag	Part of speech tag of the current token
$Word_1$	Lower cased next token
$Lemma_1$	Base form of the next token
$POS-tag_1$	Part of speech tag of the next token
$Suffix_1$	Last two and three letters of the next token
WNW	Bi gram of current and next token
$Word_{-1}$	Lower cased previous token
$Lemma_{-1}$	Base form of the previous token
$POS-tag_{-1}$	Part of speech tag of the previous token
$Suffix_{-1}$	Last two and three letters of the previous token
WPW	Bi gram of current and previous token
$Word_2$	Lower cased next second token
$Lemma_2$	Base form of the next second token
$POS-tag_2$	Part of speech tag of the next second token
$Suffix_2$	Last two and three letters of the next second token
$Word_{-2}$	Lower cased previous previous token
$Lemma_{-2}$	Base form of the previous previous token
$POS-tag_{-2}$	Part of speech tag of the previous previous token
$Suffix_{-2}$	Last two and three letters of the previous previous token
Gazetteers	Geographical dictionary associated with jurisdictions in US

In CRFs the input is a sequence of words(tokens) in a query $X = (x_1, ...x_n)$ represented using a vector of features and corresponding tags $Y = (y1, ...y_n)$.

The conditional probability represents the probability of obtaining the output
Y given the input X and is given by

$$P(Y/X) = \frac{1}{Z(X)} \prod_{i=1}^{X} exp \sum_{n=1} \lambda_k f_k(Y_{i-1}, Y_i, X, i)$$

where $f_k(Y_{i-1}, Y_i, X, i)$ is a feature function, λ_k is the weight learned about the
feature and Z(X) is a normalization function.

Features Used: In this study, we used a combination of both hand-engineered
and word embedding features. Table 2 lists hand-engineered features used for
recognition of query intent and target terms within SOL, ELO, DEF, and DOC
queries as well as features used for recognition of jurisdictions. Dictionary of geographical areas associated with jurisdictions in the US, so called gazetteers, were
used as one of the features in jurisdiction recognizer only. These features were
combined with vector representation of the word, also known as word embedding [7,13]. The word vectors were created by training a word2vec continuous
bag of words (CBOW) model. To train the model, we randomly extracted 1M
legal corpus headnotes, which are the brief summaries of a particular point of
law that appears at the beginning of every case law document.

3.4   Bidirectional LSTM CRF Model

In traditional neural networks inputs and outputs are not independent of each
other and in order to predict the next word in a sentence, there is a need to know
which words came before it. So, our choice of neural networks had to capture the
information that had been calculated so far. Recurrent Neural Networks(RNN)
fits best with a property of having a "memory" to capture information calculated.
Moreover, in sequence tagging task, since both past features (via forward states)
and future input features (via backward states) for a given time are needed, we
used a bidirectional lstm [5]. Furthermore, we combined our bidirectional LSTM
with linear chain CRF to form a bidirectional LSTM-CRF network (Fig. 2).
Bidirectional LSTM enables to get past input and future input features, whereas
CRF layer provides sentence level tag information. At the architecture level, our
model can be broken into three components [5,6,11]:

- Dense word Representation:
  For a dense representation of each word, we had the same previously used
  Word2vec word embeddings of 100 dimensions which was trained on a corpus
  of a 1M headnotes.
  In detail: For each word, we captured meaning and relevant features. This was
  achieved by building a vector formed by concatenation of word embeddings
  and vector containing character level features to capture the word morphology. Word embeddings $w_{word2vec}$ were extracted from word2vec and character level features by using bidirectional LSTM over the sequence of character
  embeddings and subsequent concatenation of final states to obtain a fixed size
  vector w_{char}. As a result, each word was represented as $w = w_{char} + w_{word2vec}$

- Contextual Word Representation:
  We ran our bidirectional LSTM over the sequence of word vectors obtained in the previous part, to obtain another sequence of vectors that represented the concatenation of two hidden states.

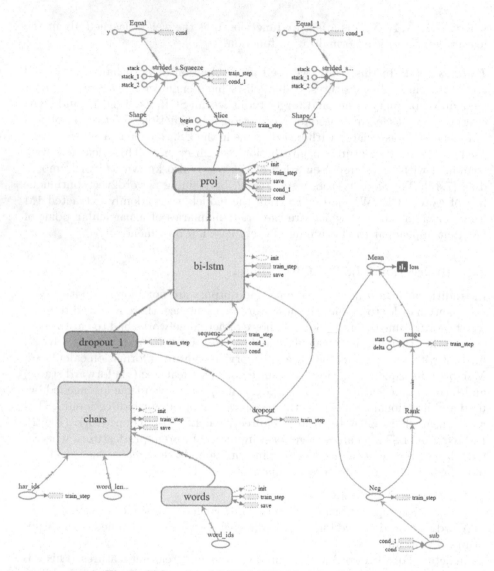

Fig. 2. Bidirectional LSTM-CRF Model architecture

Decoding:
In this phase for the contextual information obtained in previous state, linear chain CRF was used to make final prediction [12,14,16]. So for any given

sequence of words $w_1, ...w_m$, sequence of score vectors $s_1, ...s_m$ and sequence of tags $y_1, ...y_m$, linear chain CRF defined a global score C, such that

$$C(y_1, ...y_m) = b[y_1] + \sum_{t=1}^{m} s_t[y_t] + \sum_{t=1}^{m-1} T[y_t, y_{t+1}] + e[y_m]$$
$$= \text{begin} + \text{scores} + \text{transitions} + \text{end}$$

where T is a transition matrix and b is a vector of scores that captures the cost of beginning or ending with a given tag.

3.5   Rule-Based Model

We compared our approaches to rule-based model, which was part of the original Lexis Answers service. The rule-based model used a set of hand crafted regular expressions to label queries with their intent and target terms. This model can successfully identify the query intent and target terms for well structured queries in a specific format. For example, in the following query *"what is statute of limitations for mail fraud"*, the intent will be recognized as SOL and the target term will be *mail fraud* because of the manually collected keywords used by experts. However, when the pattern changes slightly, this model is incapable of recognizing the intent, or the target term, or both.

3.6   Evaluation Metrics

We utilize standard measures to evaluate the performance of our recognizers, i.e. precision, recall and F1-measure for the intent and for each one of the target term' tags. F1 is the harmonic average of precision and recall which is defined as $F1 = 2PR/(P+R)$. For rule based model, variation of rules was used to identify query intent and target terms. To calculate the measures, we referred to the gold dataset (SMEs annotated dataset). For rule-based approach only full matching target term phrases were considered as true positives, partially matching target terms were considered as false positives, not recognized target terms were count as false negatives and queries that did not have target terms at all were identified as true negatives.

4   Results

Table 3 summarizes the results of applying different types of models. They were obtained by training with over 20% of queries set aside for evaluation from each specific legal question type dataset. The table also shows the results of the baseline, rule-based, approach. CRF and BiLSTM-CRF models clearly outperformed baseline model especially in measures of recall as well as F-scores in both intent and target terms. Rule-based models are very specific and are based on strict rules resulting in high precision scores, and low recall as a result of high number of false negatives. This is applicable to both query intent and target terms.

Table 3. Results of testing different types of models over different types of user queries.

Type of model	Type of query	Type of entity/tag	Precision	Recall	F1-score
Rule based	ELO	Intent	99.6	23.1	37.5
		Target terms	63.3	29.3	40.0
	SOL	Intent	95.6	32.7	48.7
		Target terms	57.7	29.3	44.4
	DEF	Intent	100.0	48.1	65.0
		Target terms	96.5	60.0	74.0
	DOC	Intent	95.4	45.0	61.2
		Target terms	95.8	63.8	76.6
	JURI	Target terms	100.0	9.9	18.1
CRF	ELO	Intent	98.0	99.5	98.8
		B-Tterm	94.9	91.6	93.2
		I-Tterm	96.2	94.7	95.4
	SOL	Intent	99.0	99.2	99.1
		B-Tterm	95.9	96.3	96.1
		I-Tterm	95.0	96.5	95.7
	DEF	Intent	96.8	94.9	95.8
		B-Tterm	93.5	84.5	88.8
		I-Tterm	91.0	80.2	85.2
	DOC	Intent	94.9	95.45	95.1
		B-Tterm	91.6	91.3	91.4
		I-Tterm	92.7	92.1	92.4
	JURI	B-Tterm	99.0	99.0	99.0
		I-Tterm	99.0	100.0	99.0
BiLSTM-CRF	ELO	Intent	99.8	99.8	99.8
		B-Tterm	97.3	97.3	97.3
		I-Tterm	96.2	96.2	96.2
	SOL	Intent	95.4	95.4	95.4
		B-Tterm	99.2	99.2	99.2
		I-Tterm	98.3	98.3	98.3
	DEF	Intent	91.4	91.4	91.4
		B-Tterm	94.1	94.1	94.1
		I-Tterm	82.8	82.8	82.8
	DOC	Intent	94.2	94.2	94.2
		B-Tterm	97.4	97.4	97.4
		I-Tterm	88.7	88.7	88.7
	JURI	B-Tterm	99.4	99.4	99.4
		I-Tterm	99.8	99.8	99.8

So overall, the coverage of machine learning and deep learning approaches is better compared to rule based approach, which results in overall higher F-scores.

Correct identification of query intent using CRF and BiLSTM-CRF does not require recognition of both B- and I-tags. An existence of only B-tag allowed the identification of query intent. As a result, the table does not list B-intent and I-intent measures but just an average of those, defined as *Intent*.

Our results show that the recognition of intents for every factoid are highly accurate, reaching F-scores between 95.1 to 99 for most of the legal question types. However, recognition of target terms for B- and I-target term tags were comparatively lower. This happens because of the natural difficulties related to these type of entities. Sometimes a phrase from a linguistic point of view based on patterns learned by the model should be considered as the target term, however SMEs will not identify those as such or only part of the phrase will be considered as the target term. Target terms are vague and there is a need in expert knowledge in order to say if the phrase is a target term or not. These conflicts were resolved in a post processing phase where a crosscheck was performed against a dictionary of legitimate legal concepts.

The Jurisdiction model mentioned in this paper is very different from other legal question type queries. This model was solely trained to recognize particular geographic areas as a target term. Unlike target terms, geographic areas are very well defined thus making the recognition process easier overall. As described in Methods in Sect. 3.3 jurisdiction gazetteers were used as part of features for jurisdiction recognizer contributing significantly to the accuracy of the CRF model. However, we also observed that BiLSTM-CRF model have also reached high precision and recall (99.5%) without using gazetteers.

Finally, deep learning approaches have been as effective as CRF in NER tasks for legal domain text. In this work, using BiLSTM-CRF, we observed state-of-the-art metrics [10,11] for some but not all of the legal question types, most probably due to limited amount of high quality SME-tagged data available for some legal questions types used in training the models. These results show that the decision on what approach to use is data driven and not always metrics improvement will be reached when moving from machine learning to one of recurrent neural networks such as Bi-LSTM.

5 Conclusion and Future Work

In this paper, for the first time, we have successfully utilized CRF and BiLSTM-CRF to recognize user's query intent and legal phrases as target terms. We started from CRF model using hand-engineered features and moved to contextual vector representation learned and leveraged by BiLSTM-CRF architecture. We show that both methods significantly outperformed previously used rule-based approach, specifically on recognition of target terms.

In future studies, we plan to expand datasets used for training and to explore NER using several context sensitive embeddings such as ELMO [9] and BERT [1]. In addition, to make the recognition of terms more domain sensitive we want to improve post processing steps to extract the correct legal concepts out of the identified target terms in user search queries.

References

1. Devlin, J., Chang, M., Lee, K., Toutanova, K.: BERT: pre-training of deep bidirectional transformers for language understanding. CoRR abs/1810.04805 (2018)
2. Dozier, C., Kondadadi, R., Light, M., Vachher, A., Veeramachaneni, S., Wudali, R.: Named entity recognition and resolution in legal text. In: Francesconi, E., Montemagni, S., Peters, W., Tiscornia, D. (eds.) Semantic Processing of Legal Texts. LNCS (LNAI), vol. 6036, pp. 27–43. Springer, Heidelberg (2010). https://doi.org/10.1007/978-3-642-12837-0_2
3. Hashemi, H.B.: Query intent detection using convolutional neural networks. In: International Conference on Web Search and Data Mining, Workshop on Query Understanding (2016)
4. Hemati, W., Mehler, A.: Lstmvoter: Chemical named entity recognition using a conglomerate of sequence labeling tools. J. Cheminformatics **11** (2019). https://doi.org/10.1186/s13321-018-0327-2
5. Huang, Z., Xu, W., Yu, K.: Bidirectional LSTM-CRF models for sequence tagging. CoRR abs/1508.01991 (2015)
6. Lample, G., Ballesteros, M., Subramanian, S., Kawakami, K., Dyer, C.: Neural architectures for named entity recognition. CoRR abs/1603.01360 (2016)
7. Mikolov, T., Sutskever, I., Chen, K., Corrado, G., Dean, J.: Distributed representations of words and phrases and their compositionality. In: Proceedings of the 26th International Conference on Neural Information Processing Systems, vol. 2, pp. 3111–3119. NIPS 2013, Curran Associates Inc., USA (2013)
8. Najafabadi, M.M., Villanustre, F., Khoshgoftaar, T.M., Seliya, N., Wald, R., Muharemagic, E.: Deep learning applications and challenges in big data analytics. J. Big Data **2**(1) (2015) https://doi.org/10.1186/s40537-014-0007-7
9. Peters, M.E., et al.: Deep contextualized word representations. CoRR abs/1802.05365 (2018)
10. Shankar, A., Buddarapu, V.N.: Deep ensemble learning for legal query understanding. In: CIKM (2019)
11. Sreelakshmi, K., Rafeeque, P.C., Sreetha, S., Gayathri, E.S.: Deep bi-directional lstm network for query intent detection. Procedia Comput. Sci. **143**, 939–946 (2018). https://doi.org/10.1016/j.procs.2018.10.341. 8th International Conference on Advances in Computing & Communications (ICACC-2018)
12. Sutton, C., McCallum, A.: An introduction to conditional random fields. Found. Trends Mach. Learn. **4**(4), 267–373 (2012). https://doi.org/10.1561/2200000013
13. Turian, J., Ratinov, L., Bengio, Y.: Word representations: a simple and general method for semi-supervised learning. In: Proceedings of the 48th Annual Meeting of the Association for Computational Linguistics, pp. 384–394. ACL 2010, Association for Computational Linguistics, Stroudsburg, PA, USA (2010)
14. Xu, P., Sarikaya, R.: Convolutional neural network based triangular CRF for joint intent detection and slot filling. In: 2013 IEEE Workshop on Automatic Speech Recognition and Understanding, pp. 78–83 (2013). https://doi.org/10.1109/ASRU.2013.6707709

15. Yadav, V., Bethard, S.: A survey on recent advances in named entity recognition from deep learning models. In: Proceedings of the 27th International Conference on Computational Linguistics, pp. 2145–2158. Association for Computational Linguistics (2018)
16. Yao, K., Peng, B., Zweig, G., Yu, D., Li, X., Gao, F.: Recurrent conditional random field for language understanding. In: 2014 IEEE International Conference on Acoustics, Speech and Signal Processing (ICASSP), pp. 4077–4081 (2014). https://doi.org/10.1109/ICASSP.2014.6854368

Bidirectional Transformer Based Multi-Task Learning for Natural Language Understanding

Suraj Tripathi[✉], Chirag Singh[✉], Abhay Kumar[✉],
Chandan Pandey[✉], and Nishant Jain[✉]

Samsung R&D Institute, Bengaluru, India
{suraj.tri,c.singh,abhayl.kumar,chandan.p,
nishant.jain}@samsung.com

Abstract. We propose a multi-task learning based framework for natural language understanding tasks like sentiment and topic classification. We make use of bidirectional transformer based architecture to generate encoded representations from given input followed by task-specific layers for classification. Multi-Task learning (MTL) based framework make use of a different set of tasks in parallel, as a kind of additional regularization, to improve the generalizability of the trained model over individual tasks. We introduced a task-specific auxiliary problem using the k-means clustering algorithm to be trained in parallel with main tasks to reduce the model's generalization error on the main task. POS-tagging was also used as one of the auxiliary tasks. We also trained multiple benchmark classification datasets in parallel to improve the effectiveness of our bidirectional transformer based network across all the datasets. Our proposed MTL based transformer network improved state-of-the-art overall accuracy of Movie Review (MR), AG News, and Stanford Sentiment Treebank (SST-2) corpus by 6%, 1.4%, and 3.3% respectively.

Keywords: Bidirectional transformer · Sentiment classification · Multi-task learning · Unsupervised learning

1 Introduction

The learning of representation is the cornerstone of every task of machine learning. Deep learning became widely popular due to the very effective learning of representation through error backpropagation. The main issue with deep learning based methods is that they require a large amount of labeled data to generalize well on unseen data but in many Natural Language Processing (NLP) tasks, labeled data is scarce, so usually, pre-training for a language model on unsupervised data is used for learning universal language representations and transfer learning [1, 2].

Another widely used approach for feature learning is Multi-Task learning [3]. The learning behavior of humans also inspires MTL since we are capable of capturing general idea across tasks and easily transfer knowledge acquired from one task to

---

A. Kumar, C. Pandey and N. Jain—Equal Contribution.

E. Métais et al. (Eds.): NLDB 2019, LNCS 11608, pp. 54–65, 2019.
https://doi.org/10.1007/978-3-030-23281-8_5

another task. This is because human learning generalizes well and is not focused on learning specific patterns of a task too well. It is useful for multiple tasks to be jointly trained so that the features learned in one task can benefit other related tasks. Recently, there is a lot of interest in applying MTL for representation learning especially using deep neural networks [4, 5]. MTL helps by augmenting the dataset of all the tasks involved since we are training multiple tasks in parallel. It also helps in reducing the generalization error by preventing overfitting to a specific task. In an MTL setting, it is crucial to select the relevant and related task, but to encourage cross-task learning, diversity is also essential. We contend that if we combine the language pre-training with MTL, both can help to learn even better representation for general Natural Language Understanding (NLU) tasks. Following that, we decided to demonstrate the effectiveness of our proposed networks on the tasks of Sentiment analysis and Topic modeling, which are among the widely used tasks of NLP.

2 Related Work

Existing systems on Sentiment analysis are mostly deep learning based or some other supervised learning approaches like Support Vector Machines (SVMs) over a carefully constructed feature set. In spite of being one of the most explored tasks, it is still very challenging due to the inherent ambiguity of natural language and the complexity of human emotions. Work on Sentiment analysis can be broadly classified into Traditional approaches and Deep learning based methods.

Traditional approaches [6, 7] are based on engineering features like Bag of words model or a combination of words and their sentiment strength scores. These sentiment scores are assigned to words by an algorithm and some manual engineering as well. As expected these methods are cumbersome and error-prone. Manually covering all relevant features is very difficult in practice. Also, a slight modification in dataset or problem definition requires repetition of the whole process all over again.

Feature learning through back-propagating errors is one of the key strengths of deep neural networks. Emergence of Deep networks like Convolutional Neural Networks (CNNs) and Recurrent Neural Networks (RNNs) and word embedding methods like Word2Vec, GloVe, etc. marked the new era of machine learning techniques where there is no need for manual feature engineering. Word embeddings along with deep neural network methods have outperformed traditional methods as shown in [8]. CNN based methods capture the N-Gram features via convolutions whereas RNNs capture sequential information and dependencies. Architectures based on BiLSTM-CRF also have been employed for NLU task of sentiment analysis [9]. McCann et al. [10] makes use of contextualized word vectors to capture sentiment present in text utterances whereas Radford et al. [11] utilized byte-level recurrent language models for sentiment analysis.

Deep learning methods are widely employed, but most of them contain either just sequential information or structural information. Even after that, they are not able to capture the complete meaning and context of the sentence. Also, deep networks lack generalization over unseen input instances because of model complexity. Another problem with word embeddings is that they capture only distributional information but not the polarity. Sentiment analysis is heavily dependent on polarity information. The lexicon-based polarity information has been integrated by Shin et al. [12] with word embedding which is an improvement that shows in their result but again their network architecture is not general and suffers from specific task limitations. Also, most of these models are shallow and unidirectional. Bi-directionality, if any is at a very shallow level.

Recent work on topic classification [13–15] is based on a variant of Long-short term memory (LSTM) or CNN network. Howard et al. [13] have used fine-tuning over a pre-trained language model whereas Johnson et al. [15] employs a deep pyramid convolutional neural network for text categorization. Our proposed MTL based Transformer encoder outperforms the state-of-the-art architectures of sentiment and topic classification by a significant margin for three benchmarks NLU datasets. Our model and motivation behind it are described below.

3 Bidirectional Transformer Network

3.1 Motivation

Every computational step in a deep neural network involves an approximation, and that is the primary source of error propagation. Longer the computational path more the error introduced. That is where self-attention [16] is helpful. Transformer model is first introduced by Google [16] which makes use of self-attention layers that helps in reducing computational path length in a deep network. The Transformer network is essentially an Encoder-Decoder architecture for seq2seq learning tasks like Machine Translation, Named-entity recognition (NER) tagging, etc. We are using just the truncated (12 layers) encoder part of the Transformer since ours are classification tasks only and hence do not require a decoder. Encoder stack in our model consists of 12 layers of encoders arranged in a stacked manner. Each Encoder layer, refer Fig. 1, in turn, consists of two sublayers:

- Bidirectional Self-Attention layer
- Fully Connected layer

Self-attention layer enables context-aware learning for the encoder. Encoder layer makes use of a self-attention mechanism which looks at the whole sequence of words in a sentence while learning the embedding for a single word. This way encoder can capture the sentence level context in embedding for each word. In contrast to RNNs, encoder employs direct short circuit peep into the whole sentence context which relieves the network of the burden of approximating long-term path dependency. It also makes the attention deeply learn bidirectional context since the encoder is looking in both directions. The bi-directionality in other models based on CNNs or LSTMs [17] is shallow because they are just adding/concatenating forward and backward unidirectional context representations rather than utilizing a single bidirectional attention process.

The Transformer encoder is pre-trained for general language modeling task on a huge corpus such as Wikipedia and publicly available book-corpus. It is an

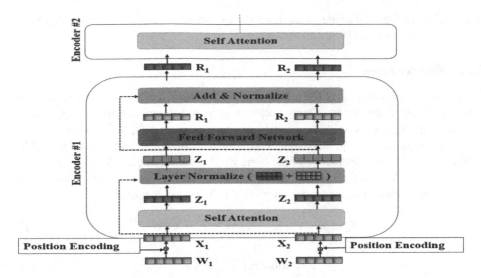

Fig. 1. Encoder layer of transformer network

unsupervised training for a masked language model task. This pre-training is important because language model is the most general form of language understanding and therefore the knowledge acquired is easily transferred across multiple tasks, which is very important for multi-task learning. We now describe the Transformer based model in detail.

3.2 Word Embedding Layer

Embedding layer is before the first encoder layer. Each word token in the sentence is a vector W, which is obtained by indexing a word-embedding matrix. After that W is transformed into vector X, refer Eq. 1, by adding position encoding to it. The dimensionality of X, W, P is 1024, which is the same as the hidden size H of an encoder. We experimented with different values of hidden size H and finalized it to be 1024 based on the performance on the validation set. We utilize position vectors to encode the relative sequential ordering of words in a sentence. For word embedding W, we used sub-word tokens as done by [18], with the vocabulary of 30,000 tokens. It is representative of that word for the current sentence, which captures bi-directional context information due to self-attention. So, the model learns different word embedding W for a word depending on the different context it appears in.

$$X = W + P \tag{1}$$

Where W is a sub-word vector, P is position vector, and X is the final representation of a single token and is input to the first encoder layer. The first token in any training example is always a CLS token of the size of H; it is essentially a special classification token containing the pooled (sum along the sequence) representation of the sequence.

CLS token in the final hidden state (i.e., the output of Transformer) is the sequence representation used in classification tasks.

3.3 Encoder Stack

The sequence of tensors X is fed into the first encoder layer, where self-attention layer projects it into three vectors, Query vector, a Key vector, and a Value vector. These vectors are obtained using three separate attention matrices W^Q, W^K, and W^V which are learned during training, and together they form an attention head. The Q, K, and V vectors are a different abstraction of the same word, each playing a different role during self-attention. Attention score of the current word with respect to each word in the sentence is calculated using these vectors in the following manner:

$$Q = X \cdot W^Q; K = X \cdot W^K; V = X \cdot W^V \qquad (2)$$

$$Z = softmax\left(\frac{Q \cdot K^T}{\sqrt{d_k}}\right)V \qquad (3)$$

Query representation of current word W_1 (i.e., Q_1) dot product with Key vector of some other word say W_2 (i.e., K_2), gives an importance score of W_2 with respect to current word W_1. We then make use of softmax normalization to get the attention scores of all words in the sentence with respect to the current word W_1.

These attention scores are further multiplied with the value vectors to give weighted importance to each word and then summed along the sentence length to get Z vector, which is the output of self-attention layer for the current word/token. This score determines how much attention/focus to give other parts of the sentence while encoding the current word. Residual connection and layer normalization are done after each self-attention and feedforward layer.

$$Z = LayerNorm(X + Z) \qquad (4)$$

Z is fed to feed-forward layer, which further processes it and forwards it to next encoder layer. The feed-forward layer is of size 4H (H is hidden layer size of Encoder as described in Sect. 3.2). Dropout is used for regularization after each sublayer before adding it to the original input to the sublayer by a residual connection. The dropout rate is a hyper-parameter and trained as part of the training process.

3.4 Combining the Multi-headed Attention

Each self-attention layer has multiple attention heads. The number of attention heads is a hyperparameter. We used 12 heads in each self-attention layer for our tasks. Each attention head of a self-attention layer calculates its representation Z, as explained above, which are concatenated and multiplied with a weight matrix W^O, to get the final output for the fully connected output layer. Each head focusses on different parts of the sentence to learn different dependencies and context rather than learning a particular peculiarity, which is then combined to form the complete embedding of the word/token for better generalization.

3.5   Feed-Forward Layer

The output of self-attention layer, Z is input to this layer which is nothing but one layer fully connected network with hidden size 4H and output size H.

$$R = feedforward(Z) \tag{5}$$

$$R = LayerNormalize(R+Z) \tag{6}$$

R is the final output of one encoder layer which is fed to the encoder stacked above.

4   Multi-Task Learning Approach

A simple technique for performing MTL is to train the target and auxiliary tasks simultaneously. The model parameters are shared between tasks in this setting, pushing the model to learn the representation of features that generalize better across tasks.

4.1   Auxiliary Task Definition

We defined a task-specific auxiliary problem to be used in parallel during training time with our main natural language processing task of sentiment, and topic classification. We employ unsupervised k-means algorithm in conjunction with input n-grams for the auxiliary sub-task definition. Following subsections will briefly discuss the auxiliary sub-task definition for different main NLU tasks.

Movie Review Auxiliary Task-1 Definition: Movie review (MR) corpus is composed of text sequences with either positive or negative polarity. Each text sequence represents a single sentence consisting of words from the English dictionary. The polarity of any text sequence is highly dependent on the consisting words and context of those words, where the context of a word is defined by the content before and after the input word. To explicitly model individual words meaning and its context, we make use of word-based n-grams in conjunction with k-means algorithm, where n is a hyperparameter which is tuned as part of the training process.

For an input data instance $X = \{x_1, x_2, \ldots, x_l\}$, where x_i represents word at the ith position and l denotes the length of the input text, we generate a set of n-grams. For example, if $n = 2$, the generated set will consist of all unigrams $\{x_i, i = 1, 2, \ldots, l\}$ and bigrams $\{[x_i, x_{i+1}], i = 1, 2, \ldots, l-1\}$. Similarly, this process is repeated for all data instances present in the MR corpus which results in an exhaustive list of n-grams. The list of n-grams generated from the training part of MR dataset is being used as an input to the k-means clustering algorithm.

For MR corpus, we assumed that polarity of each n-gram could be categorized into 3 clusters named as positive, negative and neutral as utterances in MR corpus belongs to either positive or negative sentiment. We make use of the K-means clustering algorithm to generate an assignment of each n-gram to a particular cluster. Embeddings extracted from Google's pre-trained word2vec model is used as input to the clustering algorithm. A 300 dimension embedding replaces each n-gram, and n-grams with more than one words are replaced by the average of the individual word embeddings. To initialize the centroids of 3 clusters (positive, negative and neutral), we selected m words relevant to each cluster, where m is a tunable parameter, and used the average of the word embeddings obtained using word2vec as centroid values.

We will briefly describe the k-means algorithm used for generating the assignment of individual n-grams. As mentioned above, we make use of a vector of 300 dimensions to represent data and centroid points. After initialization of all the points, the algorithm works in two phases: The first phase includes using a distance metric like Euclidean distance $(d(x,y))$ to assign each data point to a particular cluster. When all the data points are assigned to some cluster, the second phase starts which recalculate the centroid of all the clusters by taking the average of the data points assigned to it. This two-process keeps on repeating until there is no changing in the cluster assignment in some iteration.

$$d(x,y) = \left[\sum_{i=1}^{k}(x_i - y_i)^2\right]^{\frac{1}{2}} \tag{7}$$

Euclidean Distance $d(x,y)$ is used for calculating the distance between a data point and centroids, where k denotes the length (300) of vector assigned to datapoints and centroids. After the convergence of the k-means algorithm, a list of n-grams and its corresponding cluster (positive, negative and neutral) assignment is generated. We make use of this n-gram and cluster assignment to define our auxiliary task as follows:

- For each text sequence $\{x_1, x_2, \ldots, x_l\}$, where l denotes the length of the sequence in the corpus. We assign a cluster class to each word x_i based on the majority class of all of its n-grams. In case of a tie, we break it by choosing the cluster class of bigger n-gram. For example, if we are using $n = 2$ and unigram of a word outputs cluster label as neutral and its bigram outputs positive then will take positive as our final cluster class for that particular word x_i assuming that bigger n-gram can capture the context in a better way.
- This process generates a set of data points $<x,y>$, where $x = \{x_1, x_2, \ldots, x_l\}$ represents input text sequence and $y = \{y_1, y_2, \ldots, y_l\}$ represents its corresponding cluster label sequence. This novel sequence to sequence mapping problem is used as an auxiliary task in our proposed network.

Movie Review Auxiliary Task-2 Definition: Part-of-speech (POS) tagging of a word make use of both its definition and context-i.e., the relationship of the word with content before and after the word. Therefore, a single word can have different POS tag based on the context in which it is used, for example in the sentence "Tell me your answer," answer is a noun whereas in "Answer the question," it is being used as a verb. This indicates that POS tagging can help in understanding and extracting relationship present in the given input text instance. Following the same intuition, we decided to make use of POS tagging as an auxiliary task in parallel with our main tasks. We make use of NLTK POS tagger to generate POS tags for each data instance present in the MR corpus. For input instance $x = \{x_1, x_2, \ldots, x_l\}$, where l denotes the length of the sequence, we will have a corresponding POS tag set represented by $p = \{p_1, p_2, \ldots, p_l\}$ of the same length. We make use of a set of 15 POS tags as our output class corresponding to each word.

Auxiliary Tasks Definition for AG News and SST-2 Corpus: AG News consists of text utterances that belong to one of the four news categories, making it a multi-class classification problem whereas SST-2 is a binary sentiment classification task like MR classification problem. For AG News corpus, auxiliary task-1 assigns each word to one of the five clusters representing world news, business news, sci-tech news, sports news and neutral nature using a similar strategy as defined in auxiliary task 1 definition for MR corpus. Following the same definition of auxiliary task-1 and task-2 of MR task, we define sub-tasks for AG News and SST-2 corpus.

MTL based learning framework, refer Fig. 2(a), is trained by optimizing joint loss L_{MTL}, refer Eq. 8, which consists of $L_{main-task}$, loss function of main task and $L_{auxiliary-task}$, loss function of auxiliary task. The hyperparameter λ is used to control the effect of auxiliary task loss function on the MTL loss function, and it is optimized as part of the training process.

$$L_{MTL} = L_{main-task} + \lambda * L_{auxiliary-task} \tag{8}$$

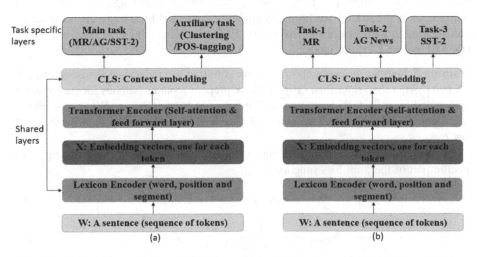

Fig. 2. (a) Auxiliary task-based MTL framework. (b) Learning multiple tasks in parallel

4.2 Single-Task Model

We also analyzed the performance of the standalone Transformer encoder stack for each task separately. The final output of the last layer of Transformer encoder is pooled (summed), which is effectively CLS token from the final layer (as described in Sect. 3.2), and is further input to a fully connected and then to a softmax layer. Softmax output is a conditional class distribution over the vocabulary of output labels. We make use of Cross-Entropy loss function, which is nothing but Negative log likelihood of model output distribution (P) under the empirical distribution of the data (X).

$$Loss = -E_{p_{data(y|x)}} log p(y|x) \qquad (9)$$

Where $p_{data(y|x)}$ is empirical distribution of the data(X).

4.3   Multi-task Model

Training related tasks in parallel leads to learning robust shared features [19]. It is a form of inductive bias like other regularization techniques, which causes the model not to learn peculiarities of a single task but rather prefer models, which can explain multiple tasks. This leads the learning to converge to models with better generalization errors. We used hard parameter sharing method for multi-task learning approach, which is generally applied by sharing the hidden layers among all tasks while keeping the output and softmax layers different and task specific.

We are inspired by [19], which showed that chances of overfitting are inversely proportional to the number of tasks jointly being learned by hard parameter sharing. Multi-task learning leads to implicit Data Augmentation and Transfer Learning. Different task dataset has different kind of inherent noise which if learned in a stand-alone manner leads to overfitting to corresponding data. Whereas, training two or more tasks jointly reduces the risk of overfitting by enabling the model to average the noise rather than sticking to particular data distribution. Following this, refer Fig. 2(b), we trained AG, MR and SST-2 joint model:

Training Schedule: In every epoch, we are shuffling the individual datasets for each sub-tasks. After that, refer Eqs. (10)–(12), we prepare separate batches for each dataset represented by B_k, which leads to a total number of batches N_{Batch}. We merge all the batches in a single file which is then randomly shuffled and used to train on the joint loss L_{total}.

Each mini-batch is task-specific and updates model weights by minimizing task-specific part of the joint loss function L_{total}.

$$D_i = Shuffle(D_i), for\ i = 1, 2, 3 \qquad (10)$$

$$N_{Batch} = \frac{1}{Batch\_Size} \sum\nolimits_{i=1}^{3} size(D_i) \qquad (11)$$

$$\cup_{k=1}^{N_{Batch}} B_k \qquad (12)$$

Joint loss function, which is a summation of individual loss, functions for each of the subtasks is defined as

$$L_{total} = L_{AG} + L_{MR} + L_{SST} \qquad (13)$$

Where L_{AG}, L_{MR}, L_{SST} are cross-entropy loss functions for individual subtasks and L_{total} is the effective joint loss used for optimizing model parameters of our proposed network.

5 Datasets

Sentiment Analysis: Our model is trained separately on two datasets for sentiment analysis: Stanford Sentiment Treebank (SST) [20] and Movie Review (MR) [7]. Both dataset consists of movie reviews and their sentiment. We use each dataset's binary version. SST-2 contains 76961 phrases, 6919 sentences for training and 1820 sentences for testing. MR corpus consists of 10,662 reviews belonging to either positive or negative sentiment, with 5331 reviews of each class. We make use of 5-fold cross-validation to demonstrate the effectiveness of our proposed approaches.

Topic Classification: :The dataset is AG News corpus [21] in which the articles are divided into four categories. Four categories represent news of different domains, where domains are World, Business, Sports, and SciTech. This dataset comprises of 120,000 training and 7,600 test data instances.

6 Discussion

The presented accuracy Table 1 compares the performance of our proposed models with the current state-of-the-art architectures and indicates a significant reduction in the generalization error. We introduced four networks based on bidirectional transformer based encoder, MTL technique and joint training of various NLU tasks and all of them achieved start-of-the-art accuracies on the tasks of MR, AG News, SST-2 classification. The inclusion of shared learning through MTL and joint training of multiple tasks showed consistent improvement over the performance of standalone architecture.

Table 1. Datasets classification accuracies

Model architecture	MR	AG news	SST-2
BiLSTM-CRF [9]	82.3	-	-
WCCNN [14]	83.8	-	-
KPCN [14]	-	88.4	-
DPCNN [15]	-	93.1	-
BCN + Char + CoVe [10]	-	-	90.3
bmLSTM [11]	-	-	91.8
Transformer Network (TN)	87.7	93.9	92.3
TN + POS auxiliary task	88.8	94.1	93.7
TN + Clustering auxiliary task	89.1	**94.5**	94.6
TN + All tasks trained in parallel (MR + AG News + SST-2)	**89.8**	94.2	**95.1**

For MR corpus, our best model achieved more than 6% improvement in overall accuracy over the state-of-the-art accuracies mentioned in [9, 14] whereas for AG News and SST-2 corpus our best model achieved 1.4% and 3.3% overall accuracy improvement respectively compared to the current state-of-the-art results.

7 Conclusion

In this paper, we analyzed the effectiveness of MTL based bidirectional transformer architecture for sentiment classification (MR and SST-2) and topic classification (AG News) and showcased consistent improvement over current state-of-the-art architectures. We introduced MTL based learning by proposing clustering based sequence-to-sequence auxiliary task as well as by using POS-tagging as one of the auxiliary task to further enhance the performance of our transformer network. We observed significant improvement with the addition of MTL framework when compared with the standalone network. We also analyzed the performance of our transformer based network by training it with a combined corpus of MR, AG News, and SST-2, which led to the improvement in generalization ability of our network across all the tasks. Our best model improved state-of-the-art overall accuracy of MR, AG News and SST-2 corpus by 6%, 1.4%, and 3.3% respectively. For future work, we will work on proposing other related auxiliary tasks that can be jointly trained with the main tasks to improve model's generalization ability.

References

1. Gao, J., Galley, M., Li, L.: Neural approaches to conversational AI. In: The 41st International ACM SIGIR Conference on Research & Development in Information Retrieval, pp. 1371–1374. ACM (2018)
2. Devlin, J., Chang, M.W., Lee, K., Toutanova, K.: BERT: pre-training of deep bidirectional transformers for language understanding, arXiv preprint arXiv:1810.04805 (2018)
3. Zhang, Y., Yang, Q.: A survey on multitask learning, arXiv:1707.08114 [cs], July 2017. http://arxiv.org/abs/1707.08114
4. Liu, X., Gao, J., He, X., Deng, L., Duh, K., Wang, Y.: Representation learning using multi-task deep neural networks for semantic classification and information retrieval. In: Proceedings of NAACL (2015)
5. Luong, M., Le, Q., Sutskever, I., Vinyals, O., Kaiser, L.: Multitask sequence to sequence learning. In: Proceedings of ICLR, pp. 1–10 (2016)
6. Mullen, T., Collier, N.: Sentiment analysis using support vector machines with diverse information sources. In: Proceedings of the 2004 Conference on Empirical Methods in Natural Language Processing (2004)
7. Pang, B., Lee, L.: Seeing stars: exploiting class relationships for sentiment categorization with respect to rating scales. In: Proceedings of the 43rd Annual Meeting on Association for Computational Linguistics, pp. 115–124. Association for Computational Linguistics, June 2005
8. Kim, Y.: Convolutional neural networks for sentence classification. In: Proceedings of the 2014 Conference on Empirical Methods in Natural Language Processing (EMNLP), Doha, Qatar, October 2014, pp. 1746–1751. Association for Computational Linguistics (2014)

9. Chen, T., Xu, R., He, Y., Wang, X.: Improving sentiment analysis via sentence type classification using BiLSTM-CRF and CNN. Expert. Syst. Appl. **72**, 221–230 (2017)
10. McCann, B., Bradbury, J., Xiong, C., Socher, R.: Learned in translation: contextualized word vectors. In: NIPS (2017)
11. Radford, A., Jozefowicz, R., Sutskever, I.: Learning to Generate Reviews and Discovering Sentiment, arXiv:1704.01444 [cs], April 2017. http://arxiv.org/abs/1704.01444
12. Shin, B., Lee, T., Choi, J.D.: Lexicon integrated CNN models with attention for sentiment analysis. In: Proceedings of the 8th Workshop on Computational Approaches to Subjectivity, Sentiment and Social Media Analysis, pp. 149–158 (2017)
13. Howard, J., Ruder, S.: Universal language model fine-tuning for text classification. In: Proceedings of ACL, pp. 328–339 (2018)
14. Wang, J., Wang, Z., Zhang, D., Yan, J.: Combining knowledge with deep convolutional neural networks for short text classification. In: Proceedings of IJCAI (2017)
15. Johnson, R., Zhang, T.: Deep pyramid convolutional neural networks for text categorization. In: Proceedings of the 55th Annual Meeting of the Association for Computational Linguistics (Volume 1: Long Papers), vol. 1, pp. 562–570 (2017)
16. Vaswani, A., et al.: Attention is all you need. In: Advances in Neural Information Processing Systems, pp. 5998–6008 (2017)
17. Peters, M.E., et al.: Deep contextualized word representations. In: Proceedings of the 2018 Conference of the North American Chapter of the Association for Computational Linguistics: Human Language Technologies (NAACL: HLT), New Orleans, Louisiana (2018)
18. Wu, Y., et al.: Google's neural machine translation system: bridging the gap between human and machine translation. arXiv preprint arXiv:1609.08144 (2016)
19. Baxter, J.: A Bayesian/information theoretic model of learning to learn via multiple task sampling. Mach. Learn. **28**(1), 7–39 (1997)
20. Socher, R., et al.: Recursive deep models for semantic compositionality over a sentiment treebank. In: Proceedings of EMNLP, pp. 1631–1642 (2013)
21. Zhang, X., Zhao, J., LeCun, Y.: Character-level convolutional networks for text classification. In: Advances in Neural Information Processing Systems, pp. 649–657 (2015)

LSVS: Link Specification Verbalization and Summarization

Abdullah Fathi Ahmed[1]([✉]), Mohamed Ahmed Sherif[1,2],
and Axel-Cyrille Ngonga Ngomo[1,2]

[1] Data Science Group, Paderborn University,
Pohlweg 51, 33098 Paderborn, Germany
afaahmed@mail.upb.de
[2] Department of Computer Science, University of Leipzig, 04109 Leipzig, Germany
{mohamed.sherif,axel.ngonga}@upb.de

Abstract. An increasing number and size of datasets abiding by the Linked Data paradigm are published everyday. Discovering links between these datasets is thus central to achieve the vision behind the Data Web. Declarative Link Discovery (LD) frameworks rely on complex Link Specification (LS) to express the conditions under which two resources should be linked. Understanding such LS is not a trivial task for non-expert users, particularly when such users are interested in generating LS to match their needs. Even if the user applies a machine learning algorithm for the automatic generation of the required LS, the challenge of explaining the resultant LS persists. Hence, providing explainable LS is the key challenge to enable users who are unfamiliar with underlying LS technologies to use them effectively and efficiently. In this paper, we address this problem by proposing a generic approach that allows a LS to be verbalized, i.e., converted into understandable natural language. We propose a summarization approach to the verbalized LS based on the selectivity of the underlying LS. Our adequacy and fluency evaluations show that our approach can generate complete and easily understandable natural language descriptions even by lay users.

Keywords: Open linked data · Verbalization · Link discovery ·
Link specification · NLP · Text summarization

1 Introduction

With the rapid increase in the number and size of RDF datasets comes the need to link such datasets. Declarative Link Discovery frameworks rely on complex Link Specification to express the conditions necessary for linking resources within these datasets. For instance, state-of-the-art LD frameworks such as LIMES [13] and SILK [9] adopt a property-based computation of links between entities. For configuring LD frameworks, the user can either (1) manually enter a LS or (2) use machine learning for automatic generation of LS.

© Springer Nature Switzerland AG 2019
E. Métais et al. (Eds.): NLDB 2019, LNCS 11608, pp. 66–78, 2019.
https://doi.org/10.1007/978-3-030-23281-8_6

There are a number of machine learning algorithms that can find LS automatically, by using either supervised, unsupervised or active learning. For example, the EAGLE algorithm [15] is a supervised machine-learning algorithm able to learn LS using genetic programming. In newer work, the WOMBAT algorithm [19] implements a positive-only learning algorithm for automatic LS finding based on generalization via an upward refinement operator. While LD experts can easily understand the generated LS from such algorithms, and even modify if necessary, most lay users lack the expertise to proficiently interpret those LSs. In addition, these algorithms have been so far unable to explain the LS they generate to lay users. Consequently, these users will face difficulty to (i) assess the correctness of the generated LS, (ii) adapt their LS, or (iii) choose in an informed manner between possible interpretations of their input.

In this paper, we address the readability of LS in terms of natural language. To the best of our knowledge, this is the first work that shows how to verbalize LS. As a result, it will help people who are unfamiliar with the underlying technology of LS to understand and interact with it efficiently. The contribution of this paper is twofold. First, we address the readability of LS and propose a generic rule-based approach to produce natural text from LS. Second, we present a selectivity-based approach to generate a summarized verbalization of LS. Our approach is motivated by the pipeline architecture for natural language generation (NLG) systems performed by systems such as those introduced by Reiter and Dale [18].

The rest of this paper is structured as follows: First, we introduce our basic notation in Sect. 2. Then, we give an overview of our approach underlying LS verbalization in Sect. 3. We then evaluate our approach with respect to the *adequacy* and *fluency* [5] of the natural language representations it generates in Sect. 4. After a brief review of related work in Sect. 5, we conclude our work with some final remarks in Sect. 6.

Throughout the rest of the paper, we use the following LS shown in Listing 1 as our running example, which is generated by the EAGLE algorithm to link the ABT-BUY benchmark dataset from [10], where the source resource x will be linked to the target resource y if our running example's LS holds.

```
1    OR(jaccard(x.name,y.name)|0.42,trigrams(x.name,y.description)|0.61)
```

Listing 1. Running example.

2   Preliminary

In the following, we present the core of the formalization and notation necessary to implement our LS verbalization. We first give an overview of the grammar that underlies LS. Then, we describe the notation of LS verbalization.

2.1   Link Specification

The link discovery problem is formally defined as follows: Given an input relation ρ (e.g., owl:sameAs), a set of source resources S and a set of target resources

T, the goal of link discovery is to discover the set $\{(s,t) \in S \times T : \rho(s,t)\}$. Declarative link discovery frameworks define the conditions necessary to generate such links using LS. Several grammars have been used for describing LS in previous work [9,15,19]. In general, these grammars assume that a LS consists of two types of atomic components: *similarity measures m*, which allow the comparison of property values of input resources and *operators op*, which can be used to combine these similarities to more complex specifications. Without loss of generality, we define a similarity measure m as a function $m : S \times T \to [0,1]$. We use *mappings* $M \subseteq S \times T$ to store the results of the application of a similarity function to $S \times T$ or subsets thereof. We define a *filter* as a function $f(m, \theta)$. We call a specification *atomic LS* when it consists of exactly one filtering function. A complex specification (*complex LS*) can be obtained by combining two specifications L_1 and L_2 through an *operator op* that allows the results of L_1 and L_2 to be merged. Here, we use the operators $\sqcap$, $\sqcup$ and $\backslash$ as they are complete and frequently used to define LS [19]. A graphical representation of our running example's complex LS from Listing 1 is given in Fig. 1.

We define the semantics $[[L]]_M$ of a LS L w.r.t. a mapping M as given in Table 1. Those semantics are similar to those used in languages like SPARQL, i.e., they are defined extensionally through the mappings they generate. The mapping $[[L]]$ of a LS L with respect to $S \times T$ contains the links that will be generated by L. We define the *selectivity score* of a sub-LS $L_s \in L$ as a function $\sigma(L)$ that returns the F-Measure achieved by the mapping $[[L_s]]$ of L_s by considering the mapping $[[L]]$ generated by the original LS L as its reference mapping.

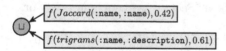

Fig. 1. Our running example complex LS. The filter nodes are rectangles while the operator nodes are circles.

Table 1. Link specification syntax and semantics.

LS	$[[LS]]_M$
$f(m, \theta)$	$\{(s,t)\|(s,t) \in M \wedge m(s,t) \geq \theta\}$
$L_1 \sqcap L_2$	$\{(s,t)\|(s,t) \in [[L_1]]_M \wedge (s,t) \in [[L_2]]_M\}$
$L_1 \sqcup L_2$	$\{(s,t)\|(s,t) \in [[L_1]]_M \vee (s,t) \in [[L_2]]_M\}$
$L_1 \backslash L_2$	$\{(s,t)\|(s,t) \in [[L_1]]_M \wedge (s,t) \notin [[L_2]]_M\}$

2.2 Link Specification Verbalization

Our definition of realization function ζ relies on the formalization of the LS declared in the previous Section. Let $\mathcal{A}$ be the set of all *atomic LS* that can be combined in a *complex LS L*. Let C^S resp. C^T be two sets of constraints that specify the sets S resp. T. Let $\mathcal{M}$ be a set of similarity measures and $\mathcal{T}$ a set of thresholds. In General, a constraint C is a logical predicate. Constraints in LS could state, for example, the rdf:type of the elements of the set they describe, i.e., $C(x) \leftrightarrow x$ rdf:type someClass, or the features that each element in the set must have, e.g., $C(x) \leftrightarrow (\exists y : x$ someProperty $y)$. Each $s \in S$ must abide by each of the constraints $C_1^S \ldots C_m^S$, while each $t \in T$ must abide by each of the constraints $C_1^T \ldots C_k^T$. We call $z \in \mathcal{A} \cup C^S \cup C^T \cup \mathcal{M} \cup \mathcal{T}$ an *atom*. We define the realization function $\zeta : \mathcal{A} \cup C^S \cup C^T \cup \mathcal{M} \cup \mathcal{T} \to Language$, where *Language* is our target language. In turn, this realization function ζ maps each atom to a word or sequence of words in our target language. Formally, the goal of this paper: first is to construct the extension of ζ to the entire LS so that all *atoms* z can be mapped to their realization $\zeta(x)$. Second : how these atomic realizations can be combined. For the sake of simplicity, we denote the extension of ζ by the same label ζ. We adopt a rule-based approach to achieve this goal, where the rule extending ζ to the entire LS is expressed in a conjunctive manner. This means that for premises $P_1, \ldots, P_n$ and consequences $K_1, \ldots, K_m$ we write $P_1 \wedge \ldots \wedge P_n \Rightarrow K_1 \wedge \ldots \wedge K_m$. The premises and consequences are clarified by using an extension of the Stanford dependencies[1]. Notably, we build on the constructs explained in Table 2. For example, dependency between a *verb* and its *object* is represented as dobj(*verb, object*).

Table 2. Dependencies used by LS verbalization.

Dependency	Explanation
amod	Represents the *adjectival modifier* dependency
	For example, amod(ROSE,WHITE) stands for white rose
dobj	Dependency between a verb and its *direct object*
	For example, dobj(EAT,APPLE) expresses ''to eat an/the apple''
nn	The *noun compound modifier* is used to modify a head noun by the means of another noun
	For instance, nn(FARMER,JOHN) stands for farmer John
poss	Expresses a possessive dependency between two lexical items
	For example, poss(JOHN,DOG) express John's dog
prep_X	Stands for the preposition X, where X can be any preposition, such as via, of, in and between
subj	Relation between *subject* and verb
	For example, subj(PLAY,JOHN) expresses John plays

3   Approach

We have now introduced all ingredients necessary for defining our approaches for LS verbalization and summarization. Our goal is to generate a complete and correct natural language representation of an arbitrary LS. Our approach is motivated by the pipeline architecture for natural language generation (NLG)

---

[1] For a complete description of the vocabulary, see http://nlp.stanford.edu/software/dependencies_manual.pdf.

systems as introduced by Reiter and Dale [18]. The NLG architecture consists of three main stages: *document-planner*, *micro-planner* and *surface realizer*. Since this work is the first step towards the verbalization of LS, our efforts will be focused on *document-planner* (as explained in Sect. 3.1) with an overview of the tasks carried out in the *micro-planner* (Sect. 3.2). The *surface realizer* is used to create the output text.

3.1 Document-Planner

The *document-planner* consists of the content determination process to create messages and the document structuring process that combines those messages. We focus on document structuring to create independently verbalizable messages from the input LS and to decide on their order and structure. These messages are used for representing information. This part is carried out in the preprocessing and processing steps.

Preprocessing: The goal of the preprocessing step is to extract the central information of LS. This step mainly relies on the *atomic LS* where the necessary information can be extracted. The input for this step is the *atomic LS* while the output is the realization of each individual part of the *atomic LS*. To this end, we break down the *atomic LS* into its individual parts, consisting of properties p (for each *atomic LS* there are two properties - 1. p_s for the resource $s \in S$ and 2. p_t for the resource $t \in T$), threshold θ and similarity measure m. After that, on each part of the *atomic LS* we apply the dependency rule introduced in Table 1. We start with the realization of similarity measure m (e.g. *jaccard* as stated in our running example in Listing 1) as follows:

1. $\zeta(\texttt{m}) \Rightarrow \texttt{nn(m,similarity)}$

Now, we can combine $\zeta(\mathrm{m})$ and $\zeta(\theta)$.

2. $\zeta(m,\theta) = \zeta(\mathrm{m}) \wedge \zeta(\theta) \Rightarrow \texttt{prep\_of}(\zeta(\theta),\zeta(\mathrm{m}))$

Furthermore, if θ equals 1, we replace its value by *"exact match"* and in cases where θ is equal to 0, we replace it by *"complete mismatch"*. Otherwise, we keep the θ value (e.g., in the case of our running example). Regarding the properties p_s and p_t, we move the explanation into the processing step since they play an important role in the construction of a *subject* to be used later in sentence building.

Processing: In this step, we aim to map all *atoms* z into their realization function $\zeta(z)$ and to define how these atomic realizations are to be combined. The input for this step is the LS and the output is the verbalization of the LS at hand. Given our formalization of LS in Sect. 2.1, any LS is a binary tree, where the root of the tree is an operator *op* and each of its two branches are LSs. Therefore, we recursively in-order apply our processing step at the LS tree at hand. As the complete verbalization of an atomic LS mainly depends on the properties p_s and p_t, we here distinguish two cases: a *first case* where p_s and p_t are equal, so we only need to verbalize p_s. In this case the realization function of an atomic LS $a \in \mathcal{A}$ is constructed as follows:

3. ζ(a)$\Rightarrow$subj(have,nn(prep_of((ζ(p_s), ζ(source and target)),
   ζ(resources))) $\wedge$ dobj (have,ζ(m,θ))

The *second case* is where the p_s and p_t are not equal. Here, both properties need to be verbalized as follows:

4. ζ(p_s,p_t)$\Rightarrow$ ζ(p_s) $\wedge$ ζ(p_t)

3.2   Micro-planer

The micro-planner is divided into three processes: *lexicalization, referring expression generation* and *aggregation*. We explain each process in the following.

Lexicalization: Within the lexicalization process we decide what specific words should be used to express the content. In particular, we choose the actual nouns, verbs, adjectives and adverbs to appear in the text from a lexicon. Also, we decide which particular syntactic structures to use, for example, whether to use the phrase `the name of the resource` or `resource's name`.

5. ζ(p_s)$\Rightarrow$ prep_of(poss(ζ(resource), p_s),ζ(source))
6. ζ(p_t)$\Rightarrow$ prep_of(poss(ζ(resource), p_t),ζ(target))
7. ζ(a)$\Rightarrow$ subj(have,ζ(p_s,p_t)) $\wedge$ dobj(have,ζ(m,θ))

Applying *preprocessing* and *processing* steps followed by *Lexicalization* step on our running example from Listing 1 generates the following verbalization: `The name of source and target resources has a 42% of Jaccard similarity or the resource's name of the source and the resource's description of the target have a 61% of Trigrams similarity.` Note that our running example contains both cases.

Referring Expression Generation: Here we carry out the task of deciding which expressions should be used to refer to entities. Considering the example, `the source and the target have a resource's name and they have a 45% of Jaccard similarity,` they is referring to the expression `the source and the target`. However, we avoid such a construction in our verbalization because we aim to generate a simple yet readable text that contains the central information of the LS at hand.

Aggregation: The goal of aggregation in NLG is to avoid duplicating information that has already been presented. In our LS verbalization, we mainly focus on the *subject collapsing*, defined in [4] as the process of *"collecting clauses* *with common elements and then collapsing the common elements"*. Formally, we define *subject* subj(v_i, s_i) as s_i, *object* dobj(v_i, o_i) as o_i

8. $\zeta(s_1) = \zeta(s_2) = \ldots = \zeta(s_n) \Rightarrow$ subj$(v_1, s_1) \wedge$ dobj$(v_1,$ coord$(o_1, o_2, \ldots, o_n))$

In the Listing 2, we present a second example LS where grouping is applicable.

```
1   OR(jaccard(x.name,y.name)|0.42,qgrams(x.name,y.name)|0.61)
```

Listing 2. Grouping example.

The original verbalization of LS from Listing 2 is: The name of source and target resources has a 42% of Jaccard similarity or the name of source and target resources has a 61% of Qgrams similarity. And after applying grouping, our verbalization will become more compact as follows: The name of source and target resources has a 42% of Jaccard similarity or a 61% of Qgrams similarity.

3.3   Summarization

We propose a sentence-scoring-based LS summarization approach. The basic idea behind our summarization approach is to simplify the original LS tree by, in order, pruning LS sub-trees that achieve the minimum *selectivity score*. i.e., keep the information loss minimum. Given an input LS L_i, our summarization approach first generates an ordered list **L** of simplified LSs of L_i, where **L** is ordered by the selective score of each of its elements in descending order. This step is carried out by iteratively pruning the sub-tree of L_i with the minimum selectivity score.

In cases where a summarization threshold $\tau \in [0, 1]$ is given, the output of our summarization algorithm will be generated by applying our LS verbalization approach to the LS $L \in \mathbf{L}$ with the highest selectivity score $\sigma(L) \leq \tau$. Otherwise, the output of our summarization approach will be a list of the verbalization of the whole list **L**.

4   Evaluation

We evaluated our approaches for LS verbalization and summarization in order to elucidate the following questions:

Q_1: Does the LS verbalization help the user to better understand the conditions sufficient to link the resources in comparison to the original LS?

Q_2: How fluent is the generated LS verbalization? i.e., how good is the natural language description of the LS verbalization in terms of comprehensibility and readability?

Q_3: How adequate is the generated LS verbalization? i.e., How well does the verbalization capture the meaning of the underlying LS?

Q_4: How much information do we lose by applying our summarization approach?

4.1 Experimental Setup

To answer the first three questions, we conducted a *user study* in order to evaluate our LS verbalization. Therefore, we used our approach to verbalize a set of five LSs automatically generated by the EAGLE algorithm [15] for the benchmark datasets of Amazon-GP, ABT-BUY, DBLP-ACM, and DBLP-Scholar from [10]. Our user study consists of four tasks, where each task consists of five multiple choice questions[2]. Altogether, we have a group of 18 participants in our user study from the DICE[3] and AKSW[4] research groups. In the following, we explain each task:

- *Task 1:* This task consists of five identical sub-tasks. For each we present the survey participant a LS and three pairs of source and target resources represented by their respective concise bounded descriptions (CBD)[5] graph. These pairs are matched together based on the provided LS with different degrees of confidence. To this end, the participant is asked to find the best matched pair, and we measure the response time for each participant.
- *Task 2:* This task also consists of five identical sub-tasks. We again follow the same process in *Task 1* of presenting the participant with the CBDs of matched resources, but this time we give the survey participant the verbalization of the LSs. Again, we record the response time of each participant.
- *Task 3:* Within this task, a survey participant is asked to judge the fluency of the provided verbalization. We follow here the machine translation standard introduced in [5]. Fluency captures how good the natural language description is in terms of comprehensibility and readability according to the following six ratings: (6) Perfectly clear and natural, (5) Sounds a bit artificial, but is clearly comprehensible. (May contain minor grammatical flaws.), (4) Sounds very artificial, but is understandable (although may contain significant grammatical flaws), (3) Barely comprehensible, but can be understood with some effort, (2) Only a loose and incomplete understanding of the meaning can be obtained, and (1) Completely not understandable at all.
- *Task 4:* In this task, we provide a survey participant with a LS and its verbalization. They are then asked to judge the adequacy of the verbalization. Here we follow the machine translation standard from [5]. Adequacy addresses how well the verbalization captures the meaning of the LS, according to the following six ratings: (6) Perfect, (5) Mostly correct, although maybe some expressions don't match the concepts very well, (4) Close, but some information is missing or incorrect, (3) There is significant information missing or incorrect, (2) Natural Language (NL) description and LS are only loosely connected, and (1) NL description and LS are in no conceivable way related.

For answering the last question, we conducted an experiment on the benchmark datasets from [10]. We ran the supervised version of the WOMBAT algorithm to generate an automatic LS for each dataset. We again used [19] to

---

[2] The survey interface can be accessed at https://umfragen.uni-paderborn.de/index.php/186916?lang=en.

[3] https://dice.cs.uni-paderborn.de/about/.

[4] http://aksw.org/About.html.

[5] https://www.w3.org/Submission/CBD/.

configure WOMBAT. Afterwards, we applied our summarization algorithm to each of the generated LSs. Because of the space limitation, we present only the verbalization of the original LS (the ones generated by WOMBAT) as well as the first summarization of it for the `Amazon-GP` and `DBLP-Scholar` datasets in Table 3. The complete results are available on the project website[6].

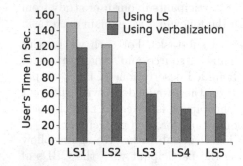 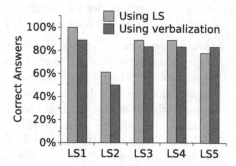

Fig. 2. Average response time of our user study.   **Fig. 3.** Correct answers of our user study.

4.2   Results and Discussion

After collecting all the responses of our user study, we filtered out those survey participants who were unlikely to have thoroughly executed the survey (i.e., the ones who took notably less time than the average response time of all other participants) or who were likely distracted while executing it (i.e., the ones who took notably more time than the average time of all other participants). This process reduces the number of valid participants to 16. Our final accepted time window was $3.5 - 38$ min for *Task 1 & 2*. Accordingly, we start our evaluation by comparing the user time required to find the best matched source-target pair using LS (*Task 1*) against using the verbalization of the provided LS (*Task 2*).

As shown in Fig. 2, the average user response time with LS verbalization is less than the ones for LS in all the 5 LSs in our users study. On average, using verbalization is 36% faster than using LS. Additionally, we also compared the error rates of participants in *Task 1 & 2*, i.e. the number of incorrect answers per question. As shown in Fig. 3, using verbalization we have a higher error rate (5% mean squared error) than when using LS. These results show that using LS verbalization decreases the average response time, which is an indicator that our participants were able to better understand underlying LS using verbalization. Still, using the LS verbalization does not always lead our participants to select the correct answer. This is due to the complexity involved in the underlying LSs, which leads to verbalization that is too long. This answers Q_1. Using our simplification approach on the same LS verbalization leads our participants to achieve better results.

---

[6] https://bit.ly/2XKDpKZ.

The results of *Task 3* (see Fig. 4) show that the majority of the generated verbalizations (i.e., the natural language descriptions) were fluent. In particular, 87% of the cases achieved a rating of 3 or higher. On average, the fluency of the natural language descriptions is 5.2 ± 1.8. This answers Q_2.

For *Task 4*, the average adequacy rating of our verbalization was 5 ± 2.55 (see Fig. 5), which we consider to be a positive result. In particular, 40% of

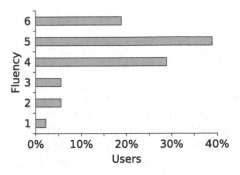

Fig. 4. Fluency results.   **Fig. 5.** Adequacy results.

Table 3. Verbalization of different summarization of a LS for the DBLP-SCHOLAR and Amazon-GP dataset together with respective **F**-measure.

Dataset	F	Verbalization
DBLP-SCHOLAR	1	The link will be generated if the title of the source and the target resources has a 66% of Cosine similarity or the resource's title of the source and the resource's author of the target has a 43% of Jaccard similarity or the resource's author of the source and the resource's title of the target has a 43% of Trigram similarity
DBLP-SCHOLAR	0.88	The link will be generated if the title of the source and the target resources has a 66% of Cosine similarity
Amazon-GP	1	The link will be generated if the resource's title of the source and the resource's name of the target has a 48% of Cosine similarity or the description of the source and the target resources has a 43% of Cosine similarity or the resource's title of the source and the resource's description of the target has a 43% of Jaccard similarity
Amazon-GP	0.97	The link will be generated if the resource's title of the source and the resource's name of the target has a 48% of Cosine similarity

all verbalizations were judged to be perfectly adequate and 83% of the cases achieved a rating of 3 or higher. This answers Q_3.

As we can see in Table 3, applying our summurization approach reduces the verbalization of the original LS to more that half of its original size. At most our summarization approach loses an F-Measure of 12% of the original description, which we conceder a fair price given the high summarization rate. This clearly answer our last question.

5 Related Work

While we believe that this is the first work that shows how to verbalize LS, related work comes from three research areas: declarative link discovery approaches, verbalization of Semantic data and text summarization.

Declarative Link Discovery frameworks rely on complex LS to express the conditions necessary for linking resources within RDF datasets. For instance, state-of-the-art LD frameworks such as LIMES [13] and SILK [9] adopt a property-based computation of links between entities. All such frameworks enable their users to manually write LS and excute it against source-target resources. In recent years, the problem of using machine learning for the automatic generation of accurate LS has been addressed by most of the link discovery frameworks. For example, the SILK framework [9] implements a batch learning approach for the discovery LS, based on genetic programming, which is similar to the approach presented in [3]. For the LIMES framework, the RAVEN algorithm [14] is an active learning approach that treats the discovery of specifications as a classification problem. In RAVEN, the discovery of LS is done by first finding class and property mappings between knowledge bases automatically. It then uses these mappings to compute linear and boolean classifiers that can be used as LS. EAGLE [15] has addressed the readability of LS alongside accuracy and efficiency. However,the generated LS is still expressed in a declarative manner. Recently, the WOMBAT algorithm [19] has implemented a machine leaning algorithm for automatic LS finding by using generalization via an upward refinement operator.

With the recent demand on new explainable machine learning approaches, comes the need for *the verbalization of semantic data* involved within such approaches. For example, [17] expands on an approach for converting RDF triples into Polish. The authors of [12] espouse a reliance on the Linked Data Web being created by reversing engineered structured data into natural language. In their work [20], the same authors show how this approach can be used to produce text out of RDF triples. Yet another work, [11], generated natural language out of RDF by depending on the BOA framework [7,8] to compute the trustworthiness of RDF triples using the Web as background knowledge. Other approaches and concepts for verbalizing RDF include [16] and [22]. Moreover, approaches to verbalizing first-order logics [6] are currently being devised. In very recent work [21], the authors have addressed the limitations of adapting rule-based approaches to generate text from semantic data by proposing a statistical model for NLG using neural networks.

The second fold of our approach is the summarization of LS, which is related to work in the area of *text summarization* with a focus on sentence scoring techniques. The work [1] surveys many sentence scoring techniques. Furthermore, the survey [2] addresses many text summarization methods. However, in our summarization technique the summarization score is user-defined.

6   Conclusions and Future Work

In this paper, we presented LSVS, an approach for verbalizing LS. LSVS produces both a direct literal verbalization of the content of the LS and a more natural aggregated version of the same content. We presented the key steps of our approach and evaluated it with a user study. Our evaluation shows that the verbalization generated by our approach is both complete and easily understandable. Our approach not only accelerates the understanding of LS by expert users, but also enables non-expert users to understand the content of LS. Still, our evaluation shows that the fluency of our approach is worse when the LS gets more complex and contains different operators. In future work, we will thus improve upon our aggregation to further increase this fluency. Moreover, we will devise a consistency checking algorithm to improve the correctness of the natural language generated by our approach.

Acknowledgement. This work has been supported by the BMVI projects LIMBO (GA no. 19F2029C) and OPAL(no. 19F2028A), Eurostars Project SAGE (GA no. E!10882) as well as the H2020 projects SLIPO (GA no.731581).

References

1. Assessing sentence scoring techniques for extractive text summarization. Expert Systems with Applications (2013)
2. Allahyari, M., et al.: Text summarization techniques: a brief survey. CoRR (2017)
3. Carvalho, M.G., Laender, A.H.F., Gonçalves, M.A., da Silva, A.S.: Replica Identification Using Genetic Programming. ACM, New York (2008)
4. Dalianis, H., Hovy, E.: Aggregation in natural language generation. In: Adorni, G., Zock, M. (eds.) EWNLG 1993. LNCS, vol. 1036, pp. 88–105. Springer, Heidelberg (1996). https://doi.org/10.1007/3-540-60800-1_25
5. Doddington, G.: Automatic evaluation of machine translation quality using n-gram co-occurrence statistics. In: Proceedings of HLT, pp. 138–145 (2002)
6. Fuchs, N.E.: First-order reasoning for attempto controlled english (2010)
7. Gerber, D., Ngomo, A.-C.N.: Extracting multilingual natural-language patterns for rdf predicates. In: EKAW, pp. 87–96 (2012)
8. Gerber, D., Ngonga Ngomo, A.-C.: Bootstrapping the linked data web. In: 1st Workshop on Web Scale Knowledge Extraction @ ISWC 2011 (2011)
9. Isele, R., Jentzsch, A., Bizer, C.: Efficient multidimensional blocking for link discovery without losing Recall. In: WebDB (2011)
10. Köpcke, H., Thor, A., Rahm, E.: Comparative evaluation of entity resolution approaches with fever. Proc. VLDB Endow. **2**(2), 1574–1577 (2009)

11. Lehmann, J., Gerber, D., Morsey, M., Ngonga Ngomo, A.-C.: Defacto - deep fact validation. In: ISWC (2012)
12. Mellish, C., Sun, X.: The semantic web as a linguistic resource: opportunities for natural language generation (2006)
13. Ngomo, A.-C.N., Auer, S.: Limes - a time-efficient approach for large-scale link discovery on the web of data. In: IJCAI (2011)
14. Ngonga Ngomo, A.-C., Lehmann, J., Auer, S., Höffner, K.: Raven - active learning of link specifications. In: Proceedings of OM@ISWC (2011)
15. Ngonga Ngomo, A.-C., Lyko, K.: EAGLE: efficient active learning of link specifications using genetic programming. In: Simperl, E., Cimiano, P., Polleres, A., Corcho, O., Presutti, V. (eds.) ESWC 2012. LNCS, vol. 7295, pp. 149–163. Springer, Heidelberg (2012). https://doi.org/10.1007/978-3-642-30284-8_17
16. Piccinini, H., Casanova, M.A., Furtado, A.L., Nunes, B.P.: Verbalization of rdf triples with applications. In: ISWC - Outrageous Ideas Track (2011)
17. Pohl, A.: The polish interface for linked open data. In: Proceedings of the ISWC 2010 Posters & Demonstrations Track, pp. 165–168 (2011)
18. Reiter, E., Dale, R.: Building Natural Language Generation Systems. Cambridge University Press, New York (2000)
19. Sherif, M.A., Ngonga Ngomo, A.-C., Lehmann, J.: WOMBAT – a generalization approach for automatic link discovery. In: Blomqvist, E., Maynard, D., Gangemi, A., Hoekstra, R., Hitzler, P., Hartig, O. (eds.) ESWC 2017. LNCS, vol. 10249, pp. 103–119. Springer, Cham (2017). https://doi.org/10.1007/978-3-319-58068-5_7
20. Sun, X., Mellish, C.: An experiment on "free generation" from single rdf triples. In: Association for Computational Linguistics (2007)
21. Vougiouklis, P., et al.: Neural wikipedian: generating textual summaries from knowledge base triples. J. Web Semant. 52–53, 1–15 (2018)
22. Wilcock, G., Jokinen, K.: Generating Responses and Explanations from RDF/XML and DAML+OIL (2003)

Deceptive Reviews Detection Using Deep Learning Techniques

Nishant Jain$^{(\boxtimes)}$, Abhay Kumar$^{(\boxtimes)}$, Shekhar Singh$^{(\boxtimes)}$,
Chirag Singh$^{(\boxtimes)}$, and Suraj Tripathi$^{(\boxtimes)}$

Samsung R&D Institute India, Bangalore, India
{nishant.jain,abhay1.kumar,s.singh02,c.singh,
suraj.tri}@samsung.com

Abstract. With the increasing influence of online reviews in shaping customer decision-making and purchasing behavior, many unscrupulous businesses have a vested interest in generating and posting deceptive reviews. Deceptive reviews are fictitious reviews written deliberately to sound authentic and deceive the consumers. Traditional deceptive reviews detection methods are based on various handcrafted features, including linguistic and psychological, which characterize the deceptive reviews. However, the proposed deep learning methods have better self-adaptability to extract the desired features implicitly and outperform all traditional methods. We have proposed multiple Deep Neural Network (DNN) based approaches for deceptive reviews detection and have compared the performances of these models on multiple benchmark datasets. Additionally, we have identified a common problem of handling the variable lengths of these reviews. We have proposed two different methods – Multi-Instance Learning and Hierarchical architecture to handle the variable length review texts. Experimental results on multiple benchmark datasets of deceptive reviews have outperformed existing state-of-the-art. We evaluated the performance of the proposed method on other review-related task-like review sentiment detection as well and achieved state-of-the-art accuracies on two benchmark datasets for the same.

Keywords: Deceptive reviews · Fake reviews · Deep learning ·
Convolutional neural network · Recurrent neural network · Word embedding

1 Introduction

In recent years, there has been a dramatic increase in the number of online user-generated reviews for a plethora of products and services across multiple websites. These reviews contain the subjective opinion of the users along with various detailed information. We rely a lot on these user-reviews before making up our mind, like which restaurant to go, what to buy, which hotel to stay in, and so on. Given the increased influence of these reviews in shaping customer's decision making, there is an incentive and opportunities for unscrupulous business to generate and post fake

N. Jain and A. Kumar—equal contribution. S. Singh, C. Singh and S. Tripathi—equal contribution.

E. Métais et al. (Eds.): NLDB 2019, LNCS 11608, pp. 79–91, 2019.
https://doi.org/10.1007/978-3-030-23281-8_7

reviews, either in favor of themselves or in disapproval of competition rivals. Deceptive reviews are deliberately written to sound authentic and help businesses to gain financial advantage and enhance their reputation. In addition, online reviews are of varied writing styles, linguistic types, content and review lengths, making it difficult for human readers to identify themselves as well.

With ever-increasing instances of deceptive reviews, there has been a series of research works to identify deceptive/fake reviews using different linguistic and psychological cues. In a marketing research study by Spiegel[1], it has been shown that nearly 95% of shoppers make a purchase after reading online reviews and that the product with at least five reviews has 270% greater likelihood to be purchased than the products with no reviews. This shows the necessity of robust deception detection methods to maintain the reliability and facticity of online reviews. Recently, this has captured the attention of both businesses and research community; giving rise to state-of-the-art results.

2 Related Work

Spam detection has been historically researched extensively in the contexts of e-mails [1] and web-texts [2]. In recent years, researchers have proposed various approaches for deceptive or manipulative reviews detection. Jindal et al. [3] proposed a supervised classifier (Logistic Regression) using features based on review content, reviewer profile, and the product descriptions. Yoo et al. [4] presented the comparative study of language structure of truthful and deceptive reviews using deception theory and demonstrated the difficulty of detecting deceptive reviews based on the structural properties, i.e. lexical complexity. Ott et al. (2011) [5] employed Turkers to write deceptive reviews and created a benchmark dataset of 800 reviews (400 gold-standard deceptive reviews and 400 truthful reviews) to be used in subsequent works. They modeled it as n-gram based text categorization task and proposed a Support Vector Machine (SVM) classifier exploiting the computational linguistics and psychological approaches for detecting deceptive reviews. They have also framed it as a genre classification task, exploiting the writing style difference between informative and imaginative reviews for truthful and deceptive reviews respectively. Additionally, they have assessed the human performance for the task; the average accuracy of three human judges were meager 57.3% as compared to 89.9% accuracy of their proposed classifier. Feng et al. [6] investigated the syntactic stylometry approach and achieved better performance by using syntactic features from context-free-grammar parse trees. Most of these works were focused on extracting the richer textual features to improve deception detection performance. However, the difficulty of creating human-labeled data and the inability of hand-crafted features to capture non-local semantic information over a discourse solicited various alternative approaches, like semi-supervised learning approach, approaches exploiting the user behavioral aspects, etc.

---

[1] https://spiegel.medill.northwestern.edu/online-reviews/.

Feng et al. [7] used aspect based profile compatibility measure to compare the test review with the product profile, built from a separate collection of reviews for the same product. Mukherjee et al. [8] modeled the spamicity of an author using various observed reviewer's behavior to identify spammer. Apart from the textual features, many other works also have been focused on the behavioral aspects (like extreme ratings, too many reviews in short time, duplicate content or ratings, etc.) of the spammers. Ren et al. [9] proposed and Fusilier et al. [20] improved the semi-supervised learning method to detect deceptive reviews.

In recent years, Deep Neural Network (DNN) models have been used to learn better semantic representations for improved performance in various NLP tasks. Kim [10] introduced Convolutional Neural Network (CNN) model for text classification to capture the frame-based semantic features. Ren et al. [11] explore a neural network model to learn document-level representation for detecting deceptive reviews. They make use of gated recurrent neural network model with attention mechanism for detecting deceptive opinion; by capturing non-local discourse information over sentence vectors. Zhao et al. [12] use a convolution neural network model by embedding the word order characteristics in its convolution and pooling layer; this makes the model more efficient in detecting deceptive opinions.

3 Datasets

We evaluate our architectures quantitatively on three different benchmark datasets for deceptive reviews detection (Sects. 3.1 to 3.3). Additionally, we evaluated our proposed architecture for addressing variable length text sequences on another related task, i.e. Review Sentiment Detection. We evaluate our proposed model on two additional datasets (Sects. 3.4 and 3.5) for review sentiment detection to show the scalability of the proposed network for various text classification task. The statistics of these datasets is summarized in Table 1.

3.1 Deceptive Opinion Spam Corpus v1.4 (DOSC)

Deceptive opinion dataset [5] consists of real and fake reviews about 20 separate hotels in Chicago. It contains 400 real and 400 fake reviews of both positive and negative sentiments respectively. The truthful reviews have been collected from online websites like TripAdvisor[2], Expedia[3] etc., while the deceptive opinions have been collected using Amazon's Mechanical Turk. It also provides a predetermined five folds for 5-fold cross-validation.

---

[2] https://www.tripadvisor.com.
[3] https://www.expedia.com/.

3.2   Four-City Dataset

Four-city Dataset [13] consists of 40 real and 40 fake reviews for each of the eight hotels in four different cities. The real reviews were chosen using random sampling on positive 5-star reviews. Amazon's Mechanical Turk was used to write fake reviews to get gold-standard deception dataset.

3.3   YelpZip Dataset

YelpZip dataset [14] consists of real world reviews of restaurants and hotels sampled from yelp along with near ground truth as provided by the Yelp review filter. YelpZip consists of reviews from 5044 hotels by 260,277 reviewers from various New York State zip codes. There are 608598 total reviews with 528141 true and 80457 deceptive review dataset. Due to high data-imbalance, Fontanarava et al. [15] created a balanced dataset by under-sampling the truthful reviews. We follow the same setting as well.

3.4   Large Movie Review Dataset

Large Movie Review Dataset (LMRD) [16] is a sentiment classification benchmark dataset that contains 50,000 reviews from IMDB, with no more than 30 reviews per film. The dataset provides a train-test split with both training and testing split containing 25,000 reviews respectively. Both splits are further evenly divided into positive and negative reviews with 12,500 reviews in each category.

3.5   Drug Review Dataset

Drug Review Dataset (DRD) [17] is a review sentiment dataset with 215063 reviews from drugs.com website. The reviews in this dataset have three different polarities – positive, negative and neutral. The dataset provides a 75%–25% train-test split using stratified random sampling.

Table 1. Dataset statistics

Dataset name	Total number of reviews	Word-length of review text			
		Minimum	Maximum	Mean	Standard deviation
YelpZip	608598	1	5213	115	106
DOSC	1600	26	784	149	87
Four-city	640	4	413	138	47
DRD	215063	1	1857	86	46
LMRD	50000	4	2470	234	173

3.6   Dataset Statistics and Visualization

We have shown (in Fig. 1) the word-cloud diagrams to visualize the word frequencies in truthful and deceptive reviews. Both the word-clouds have occurrences of very similar words. We found that the frequencies of the top hundred words per review in both truthful and deceptive reviews are similar. This makes it very difficult for traditional methods like *tf-idf* to identify the deceptive reviews correctly. This calls for deep learning based approaches to learn the semantic and syntactic differences between the truthful and deceptive reviews. We have also shown (in Fig. 2) the frequency distribution of data with respect to the length for two different datasets.

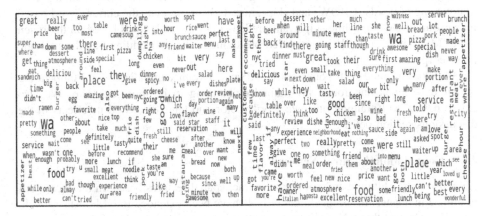

Fig. 1. Word-cloud of frequently used words in truthful and deceptive reviews

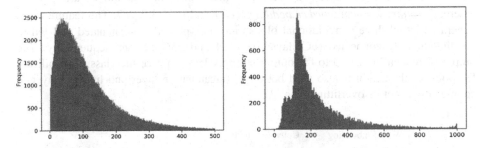

Fig. 2. Histogram plot of sequence length of reviews (for Yelp & LMRD datasets respectively)

4 Neural Network Models

For both Convolutional Neural Network (CNN) and Recurrent Neural Network (RNN), due to computational limitations, we decided to trim the input sequence to 150 words (average word-length for all the datasets derived from the statistic shown in Table 1) since this led to reasonable convergence time and memory footprint.

Also, inputs (reviews) are represented using word embedding as discussed below.

4.1   Word Embedding

Word embedding gained popularity through its use in various NLP tasks say, language modelling, text classification and sentiment analysis in recent past. Word embedding is a distributed representations of a word in an n-dimensional space using high-dimensional vectors (say, 200 to 500 dimensions).

Mikolov et al. [18] proposed an unsupervised architecture to learn the distributed representation of words in a corpus. He gave two different architecture namely continuous bag of words (CBOW) and skip-gram. CBOW predicts the current word based on surrounding words while skip-gram predicts the surround words based on the given word. Mikolov et al. [18] also provided pre-trained word embedding based on google news corpus.

We experimented with both pre-trained and randomly initialized word embedding in our experiment and found pre-trained embedding to be more accurate. We also experimented with training our own word vector and have discussed its effects in the discussions section below.

4.2   Convolutional Neural Network

CNN uses convolutional and pooling to extract spatial features based on the locality of reference in images. In recent studies, CNNs have been extended to NLP tasks as well. Kim [10] showed that CNN can be used effectively for the text classification task and gives promising results. In our model, we use three parallel convolutional layers with 100 filters each. The kernel shape for the three convolutional blocks are-$f_i^h \times f_i^w$, where, $f_i^w = dimension\ of\ word\ embedding\ vector$ and $f_i^h = 3, 5, \& 7$. The feature maps generated by all three convolutional blocks are max-pooled, concatenated and fed to the hidden fully connected (FC) layer of 1024 and 256 neurons sequentially. The output of hidden FC is fed to the output softmax layer to give the class probabilities. Dropout regularization is also used between concatenation layer and first FC layer to counter the effect of overfitting (Fig. 3).

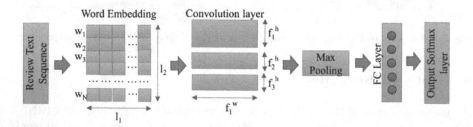

Fig. 3. Architecture diagram of convolutional neural network

4.3   Recurrent Neural Network

CNN is a great tool to extract features from a local region but less effective while learning long term dependencies. To overcome this obstacle, RNNs are used which are capable of learning features from long term dependencies through their recurrent structure. But the vanilla RNN in practice suffers from problems of short-term memory. They are not able to retain information over longer sentences due to vanishing gradient problem. To handle this issue, we use Gated Recurrent Unit (GRU) [19] in our experiments, which can overcome this problem by regulating the flow of information through them.

The architecture consists of a single GRU of 1024 units along with attention module. The attention module assigns a value between zero and one to each word; depicting its importance or relevance to the context vector. The output of GRU weighted by attention values is fed to the FC layer, which outputs the final class probability distribution. The GRU unit can be defined as below:-

$$z_t = \sigma_g \left(W_z x_t + U_z h_{t-1} + b_z \right) \tag{1}$$

$$r_t = \sigma_h \left(W_r x_t + U_r h_{t-1} + b_r \right) \tag{2}$$

$$h_t = (1 - z_t) \odot h_{t-1} + z_t \odot \sigma_h (W_h x_t + U_h (r_t \odot h_{t-1}) + b_h) \tag{3}$$

where x_t is input vector and h_t is output vector at time t. z_t is the update gate vector and r_t is the reset gate vector. W, U and b are learnable matrices and vectors. σ_g and σ_h represents activation functions and $\odot$ is Hadamard operator and $h_0 = 0$.

5 Proposed Methods for Handling Variable Length Reviews

We observe a large variation in the sequence length of review texts, and therefore we need to decide the maximum sequence length of the input review texts. Usually, we set a maximum sequence length to get optimal computation cost and accuracies. If the max-sequence length is small, then some part of the text is trimmed and not exposed to the network. This affects the accuracies worst in those cases, where the user summarize their opinion in the last few sentences. Selecting larger sequence length leads to high computation cost and slow convergence. Having observed this common problem across various text classification task especially review-text related tasks, we proposed two different architectures to address the problem. The proposed models are described in the following sub-sections.

5.1 Hierarchical Model Architecture

We propose a hierarchical model architecture with CNN followed by GRU to take variable length text sequences as input. The input review text is divided into multiple instances, each having twenty words. The total number of instances is a variable depending on the actual word length of the given review. Each instance is fed to a CNN network to extract localized regional features from the words in the instances. These features from CNN are able to capture lexical n gram characteristics by using convolutional kernels of different shapes. We use three different convolution blocks with dimension $f_i^h \times f_i^w$, where $f_i^w = dimension\ of\ word\ embedding\ vector$, and $f_i^h = 3, 5, \&7$. Different convolutional kernels are capable of capturing n-gram semantics of varying granularities, i.e., tri-gram, five-gram, and seven-gram. The convolutional layer features are fed to max-pooling layer and the output is flattened. The flattened output from the CNN network acts as instance representation. The CNN block is the same as the one described in Sect. 4.2.

At the next stage, different instance representations are passed to a GRU network. The GRU model is good at capturing long-range discourse structures among the instance representations. The output representation of the GRU is fed to the FC layer and subsequently to the softmax layer to predict the class label probabilities. The detailed architecture of the proposed hierarchical CNN-GRU network is shown in Fig. 4.

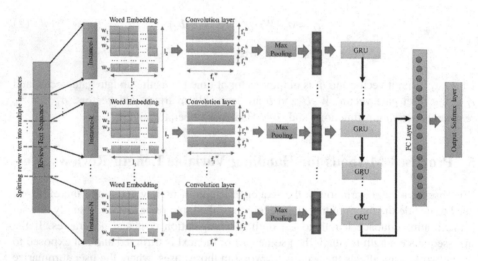

Fig. 4. Schematic diagram of hierarchical CNN-GRU model

5.2 Multi-Instance Learning (MIL)

We also propose a simplistic way to handle variable length review text using any deep learning architectures. In the multi-instance paradigm, we split the input text sequence into multiple instances of a fixed length (shown in Fig. 5). These different splits are fed as different instances to the model and act as different training examples with the same labels. We discard the last instance if its word-length is less than fifteen. During test time, we evaluate the class probabilities for all instances and assign the label by taking max-vote of predictions of all instances.

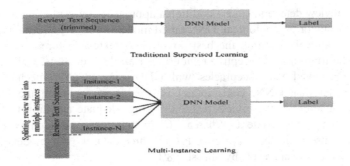

Fig. 5. Schematic diagram of traditional and multi-instance learning.

6 Evaluations and Discussions

We have performed a series of experiments for the deceptive review detection and the experimental results are presented and discussed in the following sub-sections. In case of no predefined train-test splits available for the dataset, we have split the data using stratified k-fold method and presented 5-fold cross validation scores.

6.1 Evaluation of Various DNN Based Models for Deceptive Review Detection

We have experimented with different model architectures for deceptive review detection. We have presented the 5-fold cross validation in Table 2. Baseline model accuracy for each dataset is also shown in the table for comparative study. Most literature

Table 2. Experimental results for various deceptive review datasets.

Dataset	Models	Accuracy (%)	Precision	Recall
Deceptive opinion spam corpus v1.4	Baseline [5]	86.5	0.86	0.87
	CNN	89.6	0.89	0.89
	GRU	90.3	0.91	0.90
	MIL	**90.1**	**0.90**	**0.90**
	CNN-GRU	**91.9**	**0.92**	**0.91**
Four-city dataset	Baseline [13]	80.1	0.79	0.82
	CNN	82.4	0.82	0.81
	GRU	82.9	0.83	0.83
	MIL	**82.8**	**0.83**	**0.83**
	CNN-GRU	**84.7**	**0.85**	**0.85**
YelpZip dataset	Baseline [15]	54.2	0.63	0.48
	CNN	63.8	0.59	0.61
	GRU	64.2	0.63	0.62
	MIL	**64.6**	**0.60**	**0.62**
	CNN-GRU	**66.4**	**0.67**	**0.65**

on deceptive review detection covers traditional approaches by exploiting hand-crafted features- linguistic or psychological. DNN based models have outperformed traditional methods by adaptively learning the best possible contextual features, responsible to distinguish the truthful and deceptive reviews. For all three datasets, CNN and GRU models have achieved better accuracies, with GRU using attention mechanism being marginally better than CNN. GRU model is capable of capturing the long range dependencies inherent in the review text and giving adequate attention to the words aligned with the review context. Whereas, CNN fails to capture context-dependent semantic relationships in the long texts. For all three datasets, our models have outperformed state-of-the-art result by atleast 2.8%.

6.2   Evaluation of Proposed Models for Handling Variable Length Reviews

As evident from Table 1, the review texts are varied in its length and follow a long-tail distribution as shown in Fig. 2. Although majority of the reviews are less than roughly hundred words, there are many reviews with much larger word-length. Usually, the decision to trim the long review text assumes that the underlying semantic and syntactic features, responsible for distinguishing truthful and deceptive reviews are present throughout the review text. But, this assumption ignores the human tendency or standard writing structure to conclude important factors at the end of review texts. Owing to this fact, we don't want to discard any valuable part of the review text and expose the complete review during the model training process. Our claims are verified by the increased performances of both our proposed models as compared to the vanilla CNN and GRU models. In Table 2, we have made a comparative study of two proposed approaches for handling variable length deceptive reviews. Hierarchical CNN-GRU model outperforms CNN or GRU models by at least 1.6% for all three datasets.

6.3   Evaluation on Proposed Models for Handling Variable Length Reviews on Another Task (Review Sentiment Detection)

To show the effectiveness and scalability of our proposed models in handling variable length text sequences, we evaluated them on two benchmark review sentiment detection task as well. In Table 3, we have made a comparative study of proposed

Table 3. Accuracy for the proposed models for review sentiment detection datasets

Dataset	Models	Accuracy (%)	Precision	Recall
LMRD	CNN	86.5	0.87	0.86
	GRU	86.8	0.87	0.87
	MIL	**87.1**	**0.87**	**0.87**
	CNN-GRU	**88.9**	**0.88**	**0.89**
DRD	CNN	76.8	0.77	0.77
	GRU	76.3	0.76	0.76
	MIL	**78.2**	**0.78**	**0.78**
	CNN-GRU	**83.8**	**0.84**	**0.83**

approaches and standard DNN models. Hierarchical CNN-GRU model outperforms standard CNN/GRU models by 2.1% and 7% on LMRD and DRD datasets respectively. Improved performance of the proposed models on two different tasks and 5 different datasets illustrates the importance of considering the entire review to get the complete context and not missing out any important and distinguishing aspects.

6.4   Discussions

Effect of Different Lengths of Review Texts: Review texts vary a lot in its word length, ranging from 1 to 5213 in our datasets. In the vanilla CNN and GRU architectures, we need to decide the maximum sequence length of the input review text, and the text is either trimmed or zero-padded accordingly. However, by restricting the review text to a smaller fixed length, the models are not exposed to the complete review and hence perform poorly in learning the overall context of the reviews. In addition, adopting a larger maximum sequence length increases both the number of learnable parameters and computational cost. Our proposed models take a different number of instances depending on the length of an input review text and learns the complete context of the review. The better performance of our proposed models; confirms the hypothesis that discriminative semantic and syntactic features are not evenly distributed throughout the review texts and could be even present in the concluding sentences.

Effect of Pre-trained Word Embeddings: We used word2vec models to get word-embedding vector. We experimented with both pre-trained word2vec on Google News corpus and pre-trained on review dataset (combined corpus of all dataset mentioned in Sect. 3). There is a marginal improvement in accuracies by using pre-trained word2vec embedding on review datasets. Empirically, we find that 300-dimensional embedding performs better than 150-dimensional one.

7   Conclusions

In the paper, we have experimented with various deep learning based models for identifying deceptive reviews. We presented a comparative study of the experimental results of the various models on four different benchmark datasets. Additionally, we have identified a common problem across all datasets, i.e., variable length of the reviews. In other text classification task, we usually trim the long text sequences and zero-pad the short ones. However, we lose a significant part of the review's semantic information. We have proposed two different approaches to handle the high variance of review textual lengths. Multi-Instance Learning approach is based on feeding different instances of the same training example to the same model. Hierarchical CNN-GRU model is based on extracting n-gram like semantic features using Convolutional Neural Network (CNN) and learning semantic dependencies among the extracted features from CNN modules. Both these models are capable of handling very long reviews texts and are better at deception detection. We have demonstrated that the proposed MIL and Hierarchical CNN-GRU models outperform the classical CNN and RNN models on all four benchmark datasets. For the future work, we will consider adding metadata of the reviews in our proposed models to make it more robust and accurate. We will also study and analyse the effect of our models on different NLU tasks; involving long and variable length features.

References

1. Gyöngyi, Z., Garcia-Molina, H., Pedersen, J.: Combating web spam with trustrank. In: Proceedings of the 13th International Conference on Very Large Data Bases, vol. 30, pp. 576–587 (2004)
2. Ntoulas, A., Najork, M., Manasse, M., Fetterly, D.: Detecting spam web pages through content analysis. In: Proceedings of the 15th International Conference on World Wide Web, pp. 83–92. May 2006 (2016)
3. Jindal, N., Liu, B.: Opinion spam and analysis. In: Proceedings of the 2008 International Conference on Web Search and Data Mining, pp. 219–230, February 2008
4. Yoo, K.H., Gretzel, U.: Comparison of deceptive and truthful travel reviews. Information and Communication Technologies in Tourism. Springer, Vienna (2009)
5. Ott, M., Choi, Y., Cardie, C., Hancock, J.T.: Finding deceptive opinion spam by any stretch of the imagination. In: Proceedings of the 49th Annual Meeting of the Association for Computational Linguistics: Human Language Technologies, vol. 1, pp. 309–319 (2011)
6. Feng, S., Banerjee, R., Choi, Y.: Syntactic stylometry for deception detection. In: Proceedings of the 50th Annual Meeting of the Association for Computational Linguistics: Short Paper, vol. 2, pp. 171–175, July 2012
7. Feng, V.W., Hirst, G.: Detecting deceptive opinions with profile compatibility. In: Proceedings of the Sixth IJCNLP, pp. 338–346 (2013)
8. Mukherjee, A., et al.: Spotting opinion spammers using behavioral footprints. In: Proceedings of ACM SIGKDD International Conference on Knowledge Discovery and Data Mining, pp. 632–640 (2013)
9. Ren, Y., Ji, D., Zhang, H.: Positive unlabeled learning for deceptive reviews detection. In EMNLP, pp. 488–498 (2014)
10. Kim, Y.: Convolutional neural networks for sentence classification. arXiv preprint arXiv: 1408.5882(2014)
11. Ren, Y., Ji, D.: Neural networks for deceptive opinion spam detection: an empirical study. Inf. Sci. **385**, 213–224 (2017)
12. Zhao, S., Xu, Z., Liu, L., Guo, M., Yun, J.: Towards accurate deceptive opinions detection based on word order-preserving CNN. Math. Probl. Eng., **2018** (2018)
13. Li, J., Ott, M., Cardie, C.: Identifying manipulated offerings on review portals. In: Proceedings of the 2013 Conference on Empirical Methods in Natural Language Processing, pp. 1933–1942 (2013)
14. Rayana, S., Akoglu, L.: Collective opinion spam detection: bridging review networks and metadata. In: Proceedings of the 21 thacmsigkdd International Conference on Knowledge Discovery and Data Mining, pp. 985–994, August 2015
15. Fontanarava, J., Pasi, G., Viviani, M.: Feature analysis for fake review detection through supervised classification. In: IEEE International Conference on Data Science and Advanced Analytics, pp. 658–666, October 2017
16. Maas, A.L., Daly, R.E., Pham, P.T., Huang, D., Ng, A. Y., Potts, C.: Learning word vectors for sentiment analysis. In: Proceedings of the 49th Annual Meeting of the Association for Computational Linguistics: Human Language Technologies, pp. 142–150, June 2011

17. Gräßer, F., Kallumadi, S., Malberg, H., Zaunseder, S.: Aspect-based sentiment analysis of drug reviews applying cross-domain and cross-data learning. In: Proceedings of the 2018 International Conference on Digital Health, pp. 121–125, April 2018
18. Mikolov, T., Chen, K., Corrado, G., Dean, J.: Efficient estimation of word representations in vector space. arXiv preprint arXiv:1301.3781(2013)
19. Chung, J., Gulcehre, C., Cho, K., Bengio, Y.: Empirical evaluation of gated recurrent neural networks on sequence modeling. arXiv preprint arXiv:1412.3555(2014)
20. Fusilier, D.H., Montes-y-Gómez, M., Rosso, P., Cabrera, R.G.: Detecting positive and negative deceptive opinions using PU-learning. Inf. Process. Manag. 51, 433–443 (2015)

JASs: Joint Attention Strategies for Paraphrase Generation

Isaac K. E. Ampomah$^{(\boxtimes)}$, Sally McClean, Zhiwei Lin, and Glenn Hawe

Faculty of Computing, Engineering and Built Environment,
Ulster University, Belfast, UK
{ampomah-i,si.mcclean,z.lin,gi.hawe}@ulster.ac.uk

Abstract. Neural attention based sequence to sequence (seq2seq) network models have achieved remarkable performance on NLP tasks such as image caption generation, paraphrase generation, and machine translation. The underlying framework for these models is usually a deep neural architecture comprising of multi-layer encoder-decoder sub-networks. The performance of the decoding sub-network is greatly affected by how well it extracts the relevant source-side contextual information. Conventional approaches only consider the outputs of the last encoding layer when computing the source contexts via a neural attention mechanism. Due to the nature of information flow across the time-steps within each encoder layer as well flow from layer to layer, there is no guarantee that the necessary information required to build the source context is stored in the final encoding layer. These approaches also do not fully capture the structural composition of natural language. To address these limitations, this paper presents several new strategies to generating the contextual feature vector jointly across all the encoding layers. The proposed strategies consistently outperform the conventional approaches to performing the neural attention computation on the task of paraphrase generation.

Keywords: Neural attention · Source context ·
Multi-layer encoder-decoder

1 Introduction

The goal of a paraphrase generator is to generate an output sentence that conveys the same/similar meaning as the input sentence but with different words and expressions. Paraphrase generation plays a fundamental role in many NLP tasks such as summarization, question answering, query re-writing in web searches and machine translation. For example under machine translation task, paraphrase generator can be used to rewrite or simplify complex input sentences to further enhance their translations [21]. However, the task of generating diverse and accurate paraphrases poses a big challenge. This can be attributed to the diversity and complexity of natural language.

Several neural based sequence to sequence (seq2seq) architectures have been proposed for the task of paraphrase generation. At the core of these models is

© Springer Nature Switzerland AG 2019
E. Métais et al. (Eds.): NLDB 2019, LNCS 11608, pp. 92–104, 2019.
https://doi.org/10.1007/978-3-030-23281-8_8

the encoder-decoder framework. The basic seq2seq model first encodes an input sentence of arbitrary length into a fixed length hidden representation which is then processed by the decoder in order to generate a satisfactory target sentence. As observed by [1], learning the fixed length hidden representation creates a bottleneck during the decoding phase. The local contextual information gets diluted as information flows through the encoder network which results in the loss of information and consequently poor target sentence generation. To improve the performance, attention neural mechanism [1,12] was introduced to compute the alignment between the encoder network's output and the current decoding step. That is, instead of learning a fixed length vector, the contextual information for every word is kept and later referenced by the decoder. During decoding, this mechanism makes it possible for the decoder to peek through the encoder to utilize the local contextual information for a better mapping of the input sequence to the output sequence.

The encoder and decoder sub-networks employed by many state of the art models are composed of multiple RNN layers. Generally, deeper networks are expected to outperform shallow ones [18,20]. But simply stacking multiple layers do not always lead to better performance. Consider a multi-layer encoding sub-network, the existing attention mechanism is performed across only the outputs of the top-most encoding layer. As we increase the number of encoding layers, the conventional/vanilla attention failed to train efficiently. A common problem with very deeper networks is vanishing/exploding gradients which affects the convergence of the model. Also, the accuracy gets saturated and degrades as the depth increases [9]. Furthermore, due to the nature of information flow across the time steps and the multiple layers, there is no guarantee that all the necessary features needed to capture the local context information is stored within the last layer (even with the addition of residual/highway connections between the encoding layers). This is because the memory of each encoding layer is shared among the multiple time steps and as such is prone to the recency bias problem [4]. Each layer within the encoding sub-network learns/extracts a particular set of features which are then passed to the upper layers for further processing and feature extraction. Therefore it seems reasonable to perform the attention mechanism collaboratively across all the layers within the encoding sub-network. For instance, computing the attention/alignment over the embedding layer's output vector allows the decoder a direct access to the source tokens and this as shown by [5] can further improve the performance.

In this work, we investigate different strategies to performing neural attention mechanism jointly across the multiple encoding layers as presented in Fig. 1. During each decoding step, the decoder is exposed to the entire encoding network. This approach called *JASs* enhances the flow of error signal along the depth of the encoder as well as the time-step within each encoding layer. Evaluations on two paraphrase datasets show that unlike the Vanilla approach (which performs the alignment computation across only the final encoding layer's output), the joint attention mechanism across the entire encoding sub-network produces a better performance in terms of the evaluation metrics (Translation Edit Rates

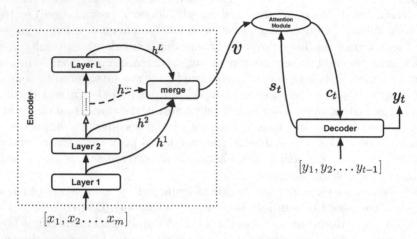

(a) Naive Joint Attention: performs the attention mechanism across a joint encoder output vector **v** generated via the *merge* module.

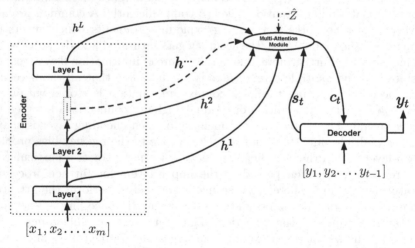

(b) Multi-Attention Mechanism: performs the neural alignment computation directly across each encoding layers via the *Multi-Attention Module* based on the binary vector $\hat{\mathbf{Z}} = [\hat{Z}_1, \hat{Z}_2]$, where $\hat{Z}_i \in \{0, 1\}$.

Fig. 1. Joint attention strategies for generating the overall source context vector $\mathbf{c}_t$ across a multi-layer encoding sub-network, where **x** and **y** are the source and target sequence respectively. y_t is the target token at decoding time-step t.

(TER) [16], BLEU [14] and METEOR [11]). In some cases, the gain in performance was achieved with no increase in the number of the model's parameters and only a small increase in the computational overhead. The contributions of this work are:

1. Proposing new approaches to perform attention computations efficiently across all the layers of a multi-layer encoding sub-network for the task of paraphrase generation.
2. Demonstrating consistent improvement over the Vanilla approach on two paraphrase datasets (**Quora** question pair dataset and **Opusparcus** [6]).
3. Determining the impact of varying the number of encoding layers which provides further insights on the actual contributions of the proposed joint attention strategies.

The remainder of the paper is organized as follows. Section 2 provides a background to generating paraphrases via neural attention mechanism. The proposed attention strategies are presented in Sect. 3. The experiments conducted are presented in Sect. 4, and the results are compared and discussed in Sect. 5. The conclusion is presented in Sect. 6.

2   Background

Given a source sentence $\mathbf{x}$, represented as a sequence of m tokens $(x_1, x_2, \cdots, x_m)$, where x_i is the ith source token, the aim is to generate a target sequence of tokens $\mathbf{y} = (y_1, y_2, \cdots, y_k)$ of length k. Under the task of sentence paraphrasing, the pair of sentences is denoted as $(\mathbf{x}, \mathbf{y})$ and the $\mathbf{y}$ and $\mathbf{x}$ shares the same meaning. Paraphrase generation is a seq2seq problem and as such an encoder-decoder framework is employed to generate the target $\mathbf{y}$ from a given $\mathbf{x}$.

The goal of the encoding sub-network is to generate a source hidden vector representation $\mathbf{h}^e$ from a given sentence $\mathbf{x}$. Each source token x_i is represented by its embedding (distributed representation) $\mathbf{e}_i \in \mathbb{R}^d$ generated by a word embedding layer $\mathbf{W}(x_i)$. From the embedding vectors $\mathbf{e} = [\mathbf{e}_1, \cdots, \mathbf{e}_m]$, the encoder computes source hidden annotation $\mathbf{h}^e = [\mathbf{h}_1^e, \cdots, \mathbf{h}_m^e]$ using a bi-directional RNN where each $\mathbf{h}_i^e = [\overrightarrow{\mathbf{h}_i^e} \oplus \overleftarrow{\mathbf{h}_i^e}]$, $\overrightarrow{\mathbf{h}_i^e}$ and $\overleftarrow{\mathbf{h}_i^e}$ are the hidden states generated by the forward RNN and backward RNN respectively for the token x_i.

The decoder generates the output sentence $\mathbf{y}$ based on the hidden annotation of the source sentence. During the decoding step t, the probability of the target token is computed based on the hidden state of the decoder $\mathbf{s}_t$, previously generated tokens, $y_1, y_2, \cdots, y_{t-1}$, and a context vector $\mathbf{c}_t$ generated via neural attention mechanism over the output of the encoder network. This as shown in Eq.(1):

$$p(y_t | y_1, \cdots, y_{t-1}, \mathbf{x}) = \sigma(\mathbf{W}_o[\mathbf{c}_t \oplus \mathbf{s}_t] + \mathbf{b}_o). \tag{1}$$

where $\mathbf{W}_o$ and $\mathbf{b}_o$ are trainable parameters. $\sigma(\cdot)$ is the softmax activation function and $\oplus$ is a concatenation operator. The context vector is computed as a weighted sum of the encoder output annotations $\mathbf{h}^e$ with attention weight vectors $\mathbf{a}_t$:

$$\mathbf{c}_t = \sum_{i=1}^{m} \mathbf{a}_{t,i} \cdot \mathbf{h}_i^e \ . \tag{2}$$

The attention weight for each $\mathbf{h}_i^e$ at the decoding step t is computed as:

$$\mathbf{a}_{t,i} = \frac{\exp\left(\mathrm{score}(\mathbf{s}_t, \mathbf{h}_i^e)\right)}{\sum_{i=1}^{m} \exp\left(\mathrm{score}(\mathbf{s}_t, \mathbf{h}_i^e)\right)} \ . \tag{3}$$

where $\mathrm{score}(\mathbf{v}_1, \mathbf{v}_2)$ is a function modeling the alignment between the $\mathbf{s}_t$ and $\mathbf{h}_i^e$ vectors. A number of alignment functions exist in literature including:

- *bilinear* [12]: $\mathrm{score}(\mathbf{v}_1, \mathbf{v}_2) = \mathbf{v}_1^\mathsf{T} \mathbf{W}_a \mathbf{v}_2$, where $\mathbf{W}_a$ is a trainable parameter.
- *dot product*: $\mathrm{score}(\mathbf{v}_1, \mathbf{v}_2) = \mathbf{v}_1^\mathsf{T} \mathbf{v}_2$

The *dot product* variant is used for all experiments in this work.

3   JASs: Joint Attention Strategies

Consider a model with L encoding layers, the outputs from all the layers $\mathbf{h} = [\mathbf{h}^1, \mathbf{h}^2, \cdots, \mathbf{h}^L]$, where $\mathbf{h}^l = [\mathbf{h}_1^l, \mathbf{h}_2^l, \cdots, \mathbf{h}_m^l]$ is the output from layer l, conventional models proposed by [1,3,7,12] computes the alignment weights and context vector (see Eqs. (3) and (2)) based on only the outputs of the last encoding layer $\mathbf{h}^L$ (i.e $\mathbf{h}^e = \mathbf{h}^L$).

As the depth of the encoder increases, these models fail to train successfully as a result of the vanishing gradient or exploding gradients [9]. To improve the flow of information between the encoder-decoder sub-networks during training, we present strategies to perform the neural attention computation jointly across the entire encoding network.

3.1   Naive Joint Attention Mechanism

Instead of performing the attention computation across only the output $\mathbf{h}^L$ of the top-most encoding layer, the decoder is allowed to attend over a joint encoder output vector $\mathbf{v} = [\mathbf{v}_1, \mathbf{v}_2, \cdots, \mathbf{v}_m]$ generated as a combination of the outputs of all the encoding layers via a merging module as shown in Fig. 1a. We refer to this approach as the "Naive joint attention" strategy. Two merging function are considered in this work:

- *concatenation*: $\mathbf{v_i} = [\mathbf{h}_i^1 \oplus \mathbf{h}_i^2 \oplus \cdots \oplus \mathbf{h}_i^L]$
- *summation*: $\mathbf{v_i} = \sum_{l=1}^{L} \mathbf{h}_i^l$

Model employing the *concatenation* approach is referred to as the *Naive-concat* model and the *summation* strategy as *Naive-sum*. The overall context vector $\mathbf{c}_t$ is then computed as:

$$\mathbf{c}_t = \sum_{i=1}^{m} \hat{\mathbf{a}}_{t,i} \cdot \mathbf{v_i} \ . \tag{4}$$

where $\hat{\mathbf{a}}_{t,i}$ is the joint attention weights computed similarly as shown in Eq. (3) across $\mathbf{v_i}$.

3.2   Multi-Attention Mechanism

The naive approaches provide a simplistic way of performing neural attention across the entire encoding network but we hypothesis that a more direct access to each encoding layer can further enhance gradient flow hence improving performance. This implies performing the neural alignment computation directly across each of the encoding layers via the *Multi-Attention Module* as shown in Fig. 1b. The attention weight $a^l_{t,i}$ across the encoding layer l is computed as:

$$a^l_{t,i} = \frac{\exp\left(\text{score}(\mathbf{s}_t, \mathbf{h}^l_i)\right)}{\sum_{i=1}^m \exp\left(\text{score}(\mathbf{s}_t, \mathbf{h}^l_i)\right)} . \tag{5}$$

Given the attention weights across all encoding layers $\mathbf{a}_t = [\mathbf{a}^1_t, \mathbf{a}^2_t, \cdots, \mathbf{a}^L_t]$, a joint attention weight is computed as the summation of all the $\mathbf{a}_t$ vectors:

$$\hat{\mathbf{a}}_{t,i} = \sum_{l=1}^L \mathbf{a}^l_{t,i} . \tag{6}$$

The operation of the *Multi-Attention Module* is controlled by the value of the binary vector $\hat{\mathbf{Z}} = [\hat{Z}_1, \hat{Z}_2]$, where $\hat{Z}_i \in \{0,1\}$. The $\hat{Z}_1$ controls the choice of attention vector (either $\hat{\mathbf{a}}_t$ or $\mathbf{a}^l_t$) employed to compute the context vector $\mathbf{c}^l_t$ across the encoding layer l:

$$\mathbf{c}^l_t = \begin{cases} \sum_{i=1}^m \hat{\mathbf{a}}_{t,i} \cdot \mathbf{h}^l_i & \hat{Z}_1 = 0 \\ \sum_{i=1}^m \mathbf{a}^l_{t,i} \cdot \mathbf{h}^l_i & \hat{Z}_1 = 1 \end{cases} \tag{7}$$

The overall context vector $\mathbf{c}_t$ is then generated as the combination of all the $\mathbf{c}^l_t$. The choice of the combination function (either summation or concatenation) employed by the *Multi-Attention Module* is also controlled by the value of $\hat{Z}_2$ as shown in Eq. (8).

$$\mathbf{c}_t = \begin{cases} \sum_{l=1}^L \mathbf{c}^l_t & \hat{Z}_2 = 0 \\ [\mathbf{c}^1_t \oplus \mathbf{c}^2_t \oplus \cdots \oplus \mathbf{c}^L_t] & \hat{Z}_2 = 1 \end{cases} \tag{8}$$

Depending on the values of $\hat{\mathbf{Z}}$, the *Multi-Attention Module* is termed as operating in a particular mode (either 0, 1, 2 or 3) as summarized in Table.1.

Table 1. Mode of operation of the *Multi-Attention Module* as determined by the values of $\hat{Z}_1$ and $\hat{Z}_2$. $\mathbf{c}^l_t$ is the context vector computed across the encoding layer l and $\mathbf{c}_t$ is the overall context vector across the entire encoding sub-network.

Mode	$\hat{Z}_1$	$\hat{Z}_2$	$\mathbf{c}^l_t$	$\mathbf{c}_t$
0	0	0	$\sum_{i=1}^m \hat{\mathbf{a}}_{t,i} \cdot \mathbf{h}^l_i$	$\sum_{l=1}^L \mathbf{c}^l_t$
1	0	1	$\sum_{i=1}^m \hat{\mathbf{a}}_{t,i} \cdot \mathbf{h}^l_i$	$[\mathbf{c}^1_t \oplus \mathbf{c}^2_t \oplus \cdots \oplus \mathbf{c}^L_t]$
2	1	0	$\sum_{i=1}^m \mathbf{a}^l_{t,i} \cdot \mathbf{h}^l_i$	$\sum_{l=1}^L \mathbf{c}^l_t$
3	1	1	$\sum_{i=1}^m \mathbf{a}^l_{t,i} \cdot \mathbf{h}^l_i$	$[\mathbf{c}^1_t \oplus \mathbf{c}^2_t \oplus \cdots \oplus \mathbf{c}^L_t]$

4   Experimental Setup

4.1   Dataset and Preprocessing

Quora dataset[1] consists of question-pairs labeled as either paraphrase or non-paraphrase pair. Only the positive pairs are extracted. Sentences with length greater than 30 words were removed resulting in about 148 K question pairs. The models are evaluated on 143 K training and 5 K test set splits.

Opusparcus[2] is a large corpus consisting of sentential paraphrase pairs of extracted translation of subtitles from movies/TV shows. The dataset comes with separate training, validation and test set splits. The test and validation set splits (each consisting of about 1 K subtitle pairs) are manually annotated by two annotators and verified as acceptable paraphrase pairs. The training set consists of over 40M "potential" paraphrase pairs automatically ranked based on multiple probability ranking functions. The size of the training set comes at the cost of quality and also the sentence pairs have not been manually checked as paraphrase pairs. As a result, the training set is assumed to contain noise to some extent. To reduce the training time, about 1.16M highly ranked pairs from the training set were selected for evaluation with about 1.14M pairs for training and 20 K for validation. The manually annotated test and validation splits (clean pairs) are combined to create the evaluation test set (approximately 2 K paraphrase pairs).

NLTK is employed to tokenize the sentences. The vocabulary is limited to the top 28 K and 29.8 K frequent word (including the special BOS, EOS symbols) for experiments on the Quora and Opusparcus dataset respectively.

4.2   Model Setup and Hyperparameters

The focus of this study is to explore strategies for computing the overall context vector across all the layers within the encoder sub-network. The encoder consists of a 3-layers Bi-LSTM with the size of the LSTM's hidden units set as 256 across all the layers and a single layer LSTM for the decoder. The outputs from all the encoding layers are of the same size. The size of the decoder's hidden unit is set as 512 to be consistent with the *dot product* attention score function.

The hyperparameters were tuned in preliminary experiments conducted on a subset of each dataset. The word embedding layer is initialized with 300-D pre-trained fastText vectors [2] for experiments on the Quora dataset and on the Opusparcus dataset, 300-D pre-trained ConceptNet vectors [17]. These embedding vectors are not updated during training. The dropout with probability 0.1 is applied across the embedding, the LSTM layers for regularization. RMSProp optimizer was used to optimize the objective function for experiments on the

---

[1] https://www.kaggle.com/c/quora-question-pairs.
[2] http://urn.fi/urn:nbn:fi:lb-201804191.

Quora corpus and the Nadam optimizer was employed in the case of Opusparcus dataset. The initial learning rate and the batch size are set as 0.001 and 256
respectively. All models are implemented in Keras (version 2.2.4 with tensorflow
backend).

The performance of a sequence generation model is affected by several parameters such as dropout rate, choice of embedding vectors and the RNN cell variants. We mainly focus on the performance impact base on the attention computation strategy (Vanilla attention, "naive joint attention" and *Multi-Attention*)
and the number of encoding layers (depth of the encoder). The models presented in this work are evaluated based on the paraphrase sentences generated
via greedy search. The greedy search algorithm selects the most likely word at
each decoding time step in the output/target sequence.

4.3 Evaluation Metric

Following previous works [8,10], we evaluate the performance of our models
using well known evaluation metrics[3] for comparing multiple corpora: TER [16],
BLEU [14] and METEOR [11]. The TER measures the number of edits required
to transform a system generated paraphrase into the reference paraphrase. The
BLEU score considers the exact matching between the system generated paraphrases and reference paraphrases by considering the n-gram overlaps. METEOR
employs stemming and synonymy in WordNet to improve this measure. These
metrics perform well for the paraphrase identification task [13] and in most
instances correlates well with human judgement for evaluating the performance
of sentence generation models [19]. The results are reported with the p-values at
95% confidence interval. Higher BLEU and METEOR score is better. But for
the TER, the lower the score the better the model.

5 Results

In this section, we discuss the performance impact of the attention computation
and aggregation strategies presented in this paper. In the result tables, the number of trainable parameters per model is reported under the *Params* column and
in Tables 4 and 5, *Depth* refers to the number of encoding layers including the
word embedding layer. As shown in Tables 2, 3, 4 and 5, exposing the decoder to
all the encoding layers can significantly enhance the performance of the sequence
generation model. In most cases (*Naive-sum*, Mode 0 and Mode 2), the performance gain from employing the joint attention approaches comes at no increase
in the number of trainable parameters. Compared to the "Naive joint attention"
(*Naive-concat* and *Naive-sum*) models, the *Multi-Attention* strategies achieved
the better performance with the Mode 1 ($\hat{Z}_1 = 0$ and $\hat{Z}_2 = 1$) strategy yielding
the overall best performance on both datasets.

---

[3] https://github.com/jhclark/multeval.

5.1   Attention Strategies

We first evaluate the performance of the attention strategies presented in this work. Tables 2 and 3 show that the *Multi-Attention* strategies consistently outperform the Vanilla attention and the "Naive joint attention" approaches with the Mode 1 model achieving the higher performance in terms of the BLEU, METEOR and TER metrics on both datasets. Each mode allows the decoding sub-network direct access to not only the last layer but the individual layers within the entire encoding sub-network. The Vanilla attention approach offers a simplistic approach but for the same number of trainable parameters, models trained via *Naive-sum*(for the Quora dataset), *Multi-Attention* modes 0 and 2 produced higher performance improvement at a small overhead of computational cost.

On the Quora dataset (Table 2), the "naive joint attention" approaches achieved similar performance gain over the Vanilla attention strategy in terms of both the METEOR and BLEU score but *Naive-concat* approach had the worse

Table 2. Performance on the **Quora** dataset.

Model	Params	METEOR ↑	BLEU ↑	TER ↓
Vanilla attention	36.2M	28.1	25.8	55.6
Naive joint attention				
Naive-sum	36.2M	29.5	27.5	54.4
Naive-concat	142.6M	29.5	27.3	55.8
Multi-attention				
Mode 0	36.2M	30.7	28.8	52.8
Mode 1	80.6M	30.9	29.0	52.4
Mode 2	36.2M	29.9	27.6	53.5
Mode 3	80.6M	29.7	27.3	53.7

Table 3. Performance on the **Opusparcus** paraphrase dataset.

Model	Params	METEOR ↑	BLEU ↑	TER ↓
Vanilla attention	37.1M	23.1	18.2	57.7
Naive joint attention				
Naive-sum	37.1M	21.3	15	61
Naive-concat	146.3M	23.6	19.1	57.7
Multi-attention				
Mode 0	37.1M	23.5	18.8	57.7
Mode 1	83M	23.9	19.8	56.8
Mode 2	37.1M	23.4	18.5	57.5
Mode 3	83M	23.9	19.1	56.9

Table 4. Performance impact of varying the depth the encoding sub-network on the **Quora** dataset.

Model	Depth	Params	METEOR ↑	BLEU ↑	TER ↓
Vanilla attention	3	34.6M	29.2	27.0	54.3
	4	36.2M	28.1	25.8	55.6
	5	37.8M	27.1	24.7	56.6
Naive joint attention					
Naive-sum	3	34.6M	28.2	25.4	55.8
	4	36.2M	29.5	27.6	54.4
	5	37.8M	29.8	27.7	53.7
Naive-concat	3	103.4M	29.5	27.1	55.3
	4	142.6M	29.5	27.3	55.8
	5	183.9M	29.5	27.2	56.0
Multi-attention					
Mode 0	3	34.6M	30.4	28.6	53.1
	4	36.2M	30.7	28.8	52.8
	5	37.8M	30.7	28.9	53.2
Mode 1	3	64.2M	30.8	28.9	52.2
	4	80.6M	30.9	29.0	52.4
	5	97M	31.0	29.0	51.9
Mode 2	3	34.6M	29.8	27.6	53.8
	4	36.2M	29.9	27.6	53.5
	5	37.8M	30.1	28.0	53.3
Mode 3	3	64.2M	30.3	28.2	53.1
	4	80.6M	29.7	27.3	53.7
	5	97M	29.8	27.6	53.3

TER score among all the models under consideration. In the case of the *Multi-Attention* strategies, models trained via modes 0 and 1 obtained the best evaluation scores. For example, the Mode 1 strategy yields a performance improvement of about, -1.3 TER, 1.2 METEOR and 2.3 BLEU scores.

As shown in Table 3, among the joint attention strategies the *Naive-sum* model achieved the worse performance as it failed to match the performance of the Vanilla strategy on the Opusparcus dataset. The Opusparcus dataset as observed by [15] is quite noisy and as such the paraphrase generation models have to be more robust to noisy training data in order to achieve higher performance. The models trained via *Multi-Attention* modes 1 and 3 achieved the best performance. Clearly unlike the other joint attention approaches, the *Naive-sum* failed to deal efficiently with the noise within the training set resulting in the poor performance. On this dataset, the *Naive-concat* approach was

Table 5. Performance impact of varying the depth the encoding sub-network on the **Opusparcus** dataset.

Model	Depth	Params	METEOR ↑	BLEU ↑	TER ↓
Vanilla attention	3	35.6M	23.0	18.1	57.9
	4	37.1M	23.1	18.2	57.7
	5	38.7M	23.1	18.6	57.5
Naive joint attention					
Naive-sum	3	35.6M	21.2	14	62
	4	37.1M	21.3	15	61
	5	38.7M	21.4	14.2	61.8
Naive-concat	3	106.2M	23.4	19.5	57.7
	4	146.3M	23.6	19.2	57.7
	5	188.5M	23.7	19.8	56.8
Multi-attention					
Mode 0	3	35.6M	23.2	18.9	57.7
	4	37.1M	23.5	18.8	57.4
	5	38.7M	23.6	19.4	57.3
Mode 1	3	66.1M	23.7	20.1	56.9
	4	83M	23.9	19.8	56.8
	5	99.8M	24.1	20.1	56.7
Mode 2	3	35.6M	23.1	18.4	58.1
	4	37.1M	23.4	18.5	57.5
	5	38.7M	23.7	19.2	57
Mode 3	3	66.1M	23.6	19.4	57.1
	4	83M	23.9	19.1	56.9
	5	99.8M	23.7	19.6	57

able to match the performance of the *Multi-Attention* strategies but at the cost of a higher number of trainable model parameters of about 146.3M compared to that of Mode 1 and Mode 3 (83M parameters each).

Overall, the value of $\hat{Z}$ clearly contributes the most to the degree of performance gain obtained by the *Multi-Attention* strategies with $\hat{Z} = [0, 1]$ consistently providing the best evaluation scores.

5.2 Impact of Encoder Depth

Tables 4 and 5 show the impact of varying the depth of the encoder sub-network. Due to resource constraints, here we only explored the impact of the depth of the encoding sub-network up to 5 layers. The encoder depth count includes the embedding layer, therefore a model trained with encoder depth of 3 is equivalent to training the model with 2-layers BiLSTM on top of the word embedding layer.

The depth of the encoding sub-network clearly affects the overall performance of the models. For the Vanilla attention model, propagating information beyond a depth of 3 did not lead to any significant gain in performance especially on the Quora dataset. But on the Opusparcus dataset (as shown in Table 5), the Vanilla approach consistently outperforms only *Naive-sum* strategy as the depth increases but on the other hand the *Naive-concat* approach achieved a gain of about 0.6 METEOR, 1.2 BLEU and -0.7 TER over the Vanilla attention mechanism.

All the *Multi-Attention* strategies outperformed the Vanilla attention and the naive joint attention strategies (both *Naive-sum* and *Naive-concat* on the Quora dataset and *Naive-sum* on the Opusparcus dataset). Across the different operational modes, there is not much difference in performance as the depth increases from 3 to 5. Overall, the results obtained by the multi-attention and naive joint attention strategies prove that exposing the decoder to all the encoding layers can significantly enhance the performance of the sequence generation model.

6 Conclusion

In this work, we investigate different attention computation strategies which allow the decoder direct access to all layers within a multi-layer encoding sub-network. The experimental results showed the drawback of the Vanilla attention mechanism performed based on only the final layer of the encoding sub-network. The joint attention mechanisms ("naive joint attention" and *Multi-Attention*) consistently outperforms the Vanilla attention strategy under all the training instances.

As future work, we plan to further explore the impact of hyperparameters such as the choice of RNN cell variants, as well as the depth of decoding sub-network.

References

1. Bahdanau, D., Cho, K., Bengio, Y.: Neural machine translation by jointly learning to align and translate. arXiv:1409.0473 (2014)
2. Bojanowski, P., Grave, E., Joulin, A., Mikolov, T.: Enriching word vectors with subword information. Trans. Assoc. Comput. Linguist. **5**, 135–146 (2017)
3. Britz, D., Goldie, A., Luong, M.T., Le, Q.: Massive exploration of neural machine translation architectures. In: Proceedings of the EMNLP, pp. 1442–1451 (2017)
4. Cheng, J., Dong, L., Lapata, M.: Long short-term memory-networks for machine reading. In: Proceedings of the Conference on EMNLP, pp. 551–561 (2016)
5. Chollampatt, S., Ng, H.T.: A multilayer convolutional encoder-decoder neural network for grammatical error correction. In: Thirty-Second AAAI Conference on Artificial Intelligence (2018)
6. Creutz, M.: Open subtitles paraphrase corpus for six languages. In: Proceedings of the LREC (2018)
7. Gehring, J., Auli, M., Grangier, D., Yarats, D., Dauphin, Y.N.: Convolutional sequence to sequence learning. In: Proceedings of the 34th International Conference on Machine Learning, vol. 70, pp. 1243–1252. JMLR. org (2017)

8. Hasan, S.A., et al.: Neural paraphrase generation with stacked residual lstm networks. In: Proceedings of COLING: Technical Papers, pp. 2923–2934 (2016)
9. He, K., Zhang, X., Ren, S., Sun, J.: Deep residual learning for image recognition. In: Proceedings of the IEEE Conference on Computer Vision and Pattern Recognition, pp. 770–778 (2016)
10. Huang, S., Wu, Y., Wei, F., Zhou, M.: Dictionary-guided editing networks for paraphrase generation. arXiv:1806.08077 (2018)
11. Lavie, A., Agarwal, A.: Meteor: an automatic metric for mt evaluation with high levels of correlation with human judgments. In: Proceedings of the Second Workshop on Statistical Machine Translation, pp. 228–231. ACL (2007)
12. Luong, T., Pham, H., Manning, C.D.: Effective approaches to attention-based neural machine translation. In: Proceedings of EMNLP, pp. 1412–1421 (2015)
13. Madnani, N., Tetreault, J., Chodorow, M.: Re-examining machine translation metrics for paraphrase identification. In: Proceedings of the Conference of the NAACL:HLT, pp. 182–190. ACL (2012)
14. Papineni, K., Roukos, S., Ward, T., Zhu, W.J.: Bleu: a method for automatic evaluation of machine translation. In: Proceedings of the ACL, pp. 311–318 (2002)
15. Sjöblom, E., Creutz, M., Aulamo, M.: Paraphrase detection on noisy subtitles in six languages. W-NUT **2018**, 64 (2018)
16. Snover, M., Dorr, B., Schwartz, R., Micciulla, L., Makhoul, J.: A study of translation edit rate with targeted human annotation. In: Proceedings of Association for Machine Translation in the Americas. vol. 200 (2006)
17. Speer, R., Chin, J., Havasi, C.: Conceptnet 5.5: an open multilingual graph of general knowledge. In: AAAI Conference (2017)
18. Wu, Y., et al.: Google's neural machine translation system: Bridging the gap between human and machine translation. arXiv:1609.08144 (2016)
19. Wubben, S., Van Den Bosch, A., Krahmer, E.: Paraphrase generation as monolingual translation: data and evaluation. In: Proceedings of the NLG, pp. 203–207. ACL (2010)
20. Zhou, J., Cao, Y., Wang, X., Li, P., Xu, W.: Deep recurrent models with fast-forward connections for neural machine translation. Trans. ACL **4**, 371–383 (2016)
21. Zhu, J., Yang, M., Li, S., Zhao, T.: Sentence-level paraphrasing for machine translation system combination. In: Che, W., et al. (eds.) ICYCSEE 2016. CCIS, vol. 623, pp. 612–620. Springer, Singapore (2016). https://doi.org/10.1007/978-981-10-2053-7_54

Structure-Based Supervised Term Weighting and Regularization for Text Classification

Niloofer Shanavas[✉], Hui Wang, Zhiwei Lin, and Glenn Hawe

School of Computing, Ulster University, Jordanstown, UK
`shanavas-n@ulster.ac.uk`

Abstract. Text documents have rich information that can be useful for different tasks. How to utilise the rich information in texts effectively and efficiently for tasks such as text classification is still an active research topic. One approach is to weight the terms in a text document based on their relevance to the classification task at hand. Another approach is to utilise structural information in a text document to regularize learning so that the learned model is more accurate. An important question is, can we combine the two approaches to achieve better performance? This paper presents a novel method for utilising the rich information in texts. We use supervised term weighting, which utilises the class information in a set of pre-classified training documents, thus the resulting term weighting is class specific. We also use structured regularization, which incorporates structural information into the learning process. A graph is built for each class from the pre-classified training documents and structural information in the graphs is used to calculate the supervised term weights and to define the groups for structured regularization. Experimental results for six text classification tasks show the increase in text classification accuracy with the utilisation of structural information in text for both weighting and regularization. Using graph-based text representation for supervised term weighting and structured regularization can build a compact model with considerable improvement in the performance of text classification.

Keywords: Text mining · Classification ·
Graph-based text representation · Supervised term weighting ·
Node centrality · Structured regularization

1 Introduction

With the amount of text data increasing day by day, the ability to process this information effectively and efficiently also needs to increase [4]. Text mining helps to discover useful patterns in unstructured text data. An important step in text mining task is the effective representation of documents which involves the identification of relevant features that are useful for the task. Extracting a

© Springer Nature Switzerland AG 2019
E. Métais et al. (Eds.): NLDB 2019, LNCS 11608, pp. 105–117, 2019.
https://doi.org/10.1007/978-3-030-23281-8_9

relevant subset of features efficiently from the large pool of text data available is becoming challenging.

A natural language document has a syntactic and semantic structure which is implicit and hidden [4]. The features that represent the document should be effective to improve the performance of the text mining system. The bag-of-words model is simple, fast and most commonly used for document representation. It is based on a term independence assumption and represents a document as a set of weights that usually corresponds to the frequency of the terms in the document. However, it does not consider the structure of terms in the document, such as the order of terms or the syntactic and the semantic information. Since the structure of terms in the document is important to portray the meaning of the document, eliminating the structural information in the representation of a natural language document negatively affects the performance of a text mining system. A graph-based representation of text can capture important information in text, such as term order, term co-occurrence, and term context, that is lost in the bag-of-words model. Text modelled as graphs have been used for several text processing applications including classification, information retrieval, word-sense disambiguation, keyword extraction, sentence extraction and summarization.

Automatic text classification has gained more importance with the increased availability of text data. Text classification assigns natural language texts with class labels from a predefined set. Machine learning is the most commonly used approach for classifying texts automatically as it is effective and reduces expert labour [9]. The performance of text classification is strongly influenced by the features used to represent a text document [7]. Graphs help to encode the relationship between different text units that contribute to the meaning of text. The graph-based representation of text enhances the performance of text classification as it considers the rich structural information ignored by the bag-of-words model.

Overfitting is a problem in machine learning that results in poor generalization due to the model fitting the noise in the data. Regularization is a technique generally used to reduce overfitting. There are recent works on structured regularization based on the structural information in text [8,11,12,14,15], which serves two purposes: (i) to reduce overfitting and (ii) to consider the prior knowledge on the terms in the documents. It applies structured regularization to improve machine learning to classify text documents with a term frequency based representation. In this paper, we present a structure-based text classification system, that uses graphs to represent each document and class, and the structural information in the graphs is used not only for weighting the terms but also for regularization. Encoding the structural information in the regularization term in addition to the supervised graph-based term weights boosts the performance of text classification. To the best of our knowledge, this paper is the first work that applies structured regularization to classify text documents represented as graphs and explores the effect of term weighting methods on the performance of structured regularization.

The rest of the paper is organised as follows. Section 2 explains the proposed structure-based text classification system. Section 3 describes the experiments performed. The results of the experiments and the evaluation of the proposed system are discussed in Sect. 4. Finally, Sect. 5 summarises the paper and discusses the possible future work.

2   Structure-Based Approach for Text Classification

Graphs are becoming an alternative text representation as they have the ability to capture important information in text, such as term order, term co-occurrence and term relationships, that is not considered by the bag-of-words model. The nodes correspond to textual units and the links denote a relationship between the nodes it connects in a graph-based representation of text. The links can be directed or undirected. The directed links can capture information such as the dependency or the order of the textual units. In a weighted graph, the weights denote the strength of the relationship. The relationships encoded in a graph-based representation of text can be word co-occurrences, syntactic dependency, semantic information, cosine similarity, etc. The centrality measure helps to determine the importance of a node in a graph. The node centrality values help to convert a graph-based representation to a vector space representation.

2.1   Structure-Based Supervised Term Weighting

The performance of a text mining task depends on the weights of the features i.e. the values by which the features in the documents are represented. To improve the performance of text classification, terms should be assigned weights based on their relevance to the text classification task. In a supervised learning task, we can utilise the information in the labelled training documents to weight the terms. Term weighting schemes can be divided into unsupervised and supervised term weighting schemes based on whether we utilise the class-specific information in the training documents. The supervised term weighting schemes improve the performance of text classification compared to unsupervised term weighting schemes as each term is weighted based on its ability to place the documents in the right class.

tw-crc is an effective centrality-based supervised term weighting method that we developed and is explained in [10] how it utilises the rich information in text and the relationship of the terms to the predefined classes. Each document to be classified is represented as an undirected co-occurrence graph (called the document graph) where each node represents a unique term and the edges link terms that co-occur within a predefined sliding window (of size 2). We build a similar graph (called the class graph) for each class from the labelled training documents. The weight of the term in the document depends on the centrality of the node corresponding to the term in the document graph and the supervised term weight factor - class relevance centrality (crc). crc gives more weight to terms that are relevant to the text classification task. The term's relevance to the text classification task is based on the variation of its centrality in the class graphs. This results in higher weighting of class specific terms compared to common terms.

The computation of crc, the supervised term weight factor, is given in Eq. (1) where $M(t)$ is the maximum of the centralities of the term t in the class graphs, $\lambda(t)$ is the sum of the centralities of the term t in the class graphs, $L(t)$ is the number of classes in which the term t exists, $|S|$ is the total number of classes, $minc$ is the minimum of centralities in the class graphs and $A(t) = \frac{\lambda(t)-M(t)}{L(t)-1}$.

$$\text{crc} = \log_2\left(2 + \frac{M(t)}{\max(minc, A(t))} \times \frac{|S|}{L(t)}\right) \qquad (1)$$

2.2 Structured Regularization

Machine learning helps to build models that are learned from data. One common problem in machine learning is overfitting, resulting in a complex model having a large number of parameters that has good performance on training data and poor performance on predicting the labels of unseen data. This happens when the model learns the training data too well that it even captures the noise in it.

Regularization is a technique that reduces overfitting by penalizing models that are too complex. It adds a penalty term to the loss function. The loss function defines the cost associated with the error in prediction and hence measures the performance or predictive accuracy of the model. The loss functions can be log loss, hinge loss, square loss, etc. The regularization strength is controlled by the regularization parameter λ. The learning process finds the vector of optimal weight coefficients (or feature coefficients), $\mathbf{w}^*$, by minimizing the combination of loss function and penalty term as shown in Eq. (2) where $L(x_i, \mathbf{w}, y_i)$ is the loss function, N is the number of training documents, x_i is the feature vector representation of ith document, y_i corresponds to the class of ith document and $\Omega(\mathbf{w})$ is the regularizer. As the regularization strength increases, the coefficients of the weight vector $\mathbf{w}$ decrease.

$$\mathbf{w}^* = \arg\min_{\mathbf{w}} \sum_{i=1}^{N} L(x_i, \mathbf{w}, y_i) + \lambda\Omega(\mathbf{w}) \qquad (2)$$

The most common regularization methods to build a compact model are L1 regularization (aka lasso), L2 regularization (aka ridge) and elastic net regularization. L1 regularization, also called Least Absolute Shrinkage and Selection Operator (LASSO), adds the L1 norm of the coefficients of $\mathbf{w}$ to the loss function [13]. When the coefficients in $\mathbf{w}$ are set to 0, the corresponding features are removed during learning. Hence, it results in a model with a small set of features. L2 regularization, also called ridge, adds the square of the L2 norm of the coefficients of $\mathbf{w}$ to the loss function [6]. It has the effect of shrinking the magnitude of the weight coefficients. Elastic net is a linear combination of L1 and L2 regularization [17]. The L1, L2 and elastic net regularizers are unstructured regularizers as they do not consider the structure of the features and penalize the coefficients of weight vector in isolation. Structured regularizers have been introduced that allow groups of weight coefficients to be penalized together. Group lasso, based on mixed-norm, is a structured regularizer that sets the coefficients in a group to zero together resulting in group sparsity [2,16]. Sparse group lasso is a combination of lasso and group lasso [5]. It has the benefits of lasso and group lasso and brings sparsity at both feature and group level. Hence in sparse group lasso, unlike group lasso, all the features in the selected group need not be selected. The difference between lasso, group lasso and sparse group lasso is illustrated in Fig. 1 [1].

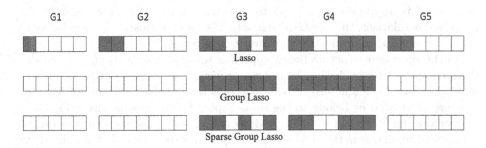

Fig. 1. Example of Lasso, Group Lasso, Sparse Group Lasso. The features are grouped into five non-overlapping groups - G1, G2, G3, G4 and G5. The dark block corresponds to the features selected and the light block denotes the discarded features.

Text contains features that are correlated. The prior knowledge on the groups of correlated features can be included in the penalty term. Recent works have incorporated linguistic structure information in statistical models as a regularization term that enhanced the effectiveness of text classification. Different kinds of linguistic information such as sentences, parse trees, word clusters, topics have been used to construct the sparse group lasso variants that encourages group behaviour of words [8,11,14,15]. The greedy variable selection algorithms such as Orthogonal Matching Pursuit (OMP) and group Orthogonal Matching Pursuit (GOMP) have been applied for regularization for text classification [12]. Even though the OMP and GOMP regularizers produce sparser models compared to sparse group lasso variants such as sentence, lda, lsi, graph-of-words (GoW) and word2vec regularizers, they do not consistently outperform these sparse group lasso variants in terms of text classification accuracy.

In a sentence regularizer, a group is defined for every sentence and contains the coefficients of words in a sentence [14,15]. Hence, it results in many overlapping groups. When there are overlapping groups, many coefficients are not set to zero. So a linear combination of lasso and sentence regularizer has been used to set the coefficients of irrelevant words to zero. In the parse tree regularizer, the sentences in training documents are parsed and a group is defined for every constituent in the parse [15]. The lasso-like penalty occurs naturally in the parse tree regularizer.

The unsupervised methods, Latent Dirichlet Allocation (LDA) and Latent Semantic Indexing (LSI), have been used to identify topics in the training documents and construct groups for each topic by selecting the top n words in it [11,15]. The lasso-like penalty is added to the LDA and LSI regularizers to penalize features at group and feature level. Brown clustering is also used for group identification where each node in the brown cluster is used to create a group, and like the parse tree regularizer, it includes the lasso-like penalty naturally [15].

Structured regularizers have been introduced based on graph-of-words and word2vec [11]. A single graph-of-words is created from the entire set of training documents, where each node corresponds to a term and the edges link terms that co-occur within a predefined sliding window. The Louvain community detection method [3] is applied on the graph to identify communities or groups. In the word2vec regularizer, the word2vec semantic vectors are clustered using k-means clustering algorithm to identify the clusters or groups.

In the proposed structure-based approach to text classification, we consider the structure of terms in the documents to weight the terms during document representation stage and also to penalize collectively the coefficients of features that are correlated or belong to the same topic. The commonly used regularizers penalize the terms in isolation and do not consider the structure of terms. Structured regularizers that encourage group behaviour of words have shown to improve the text classification performance. Studies show that the way by which the correlated terms are identified to build the groups for a structured regularizer affects the text classification performance [14,15]. The structured regularizers in the literature add linguistic bias to the bag-of-words model. Our work applies a graph-based regularizer to a linear model with graph-based representation of text where the terms are weighted by tw-crc, an effective graph-based supervised term weighting scheme [10], to boost the text classification performance.

Class Graph Regularizer. The class graphs that utilise the information in the training documents are used for building the proposed graph-based regularizer which is a variant of graph-of-words regularizer. We add the semantic similarity information to the class graphs to define accurate groups of correlated features. The semantic similarity between terms is determined using a word2vec model built from training documents. The terms that have semantic similarity greater than a particular threshold are linked by edges. The community detection algorithm, the Louvain method [3], is then applied to the class graphs to obtain the communities or groups. In this way, the relevant topics in each class are identified and the correlated terms in a topic define a group. Hence, we utilise the co-occurrence information and the semantic similarity information in the class graphs to create the proposed structured regularizer called the class graph regularizer as shown in Eq. (3) where $\Omega_{CG}(\mathbf{w})$ is the group lasso regularization term, s corresponds to the communities identified in the class graphs and $\mathbf{w_s}$ is the sub-vector of the weight vector $\mathbf{w}$ that contains the coefficients of the features in the community s.

$$\Omega_{CG}(\mathbf{w}) = \sum_{s=1}^{S} \lambda_s \|\mathbf{w_s}\|_2 \qquad (3)$$

The group identification is a step before learning the classification model. Each training document is represented as (x_i, y_i) where $y\epsilon(-1, +1)$ is the class label and x_i is the feature vector of document i.

The loss function used in our work is the log loss function as shown in Eq. (4).

$$L(x_i, \mathbf{w}, y_i) = \log(1 + \exp(-y_i \mathbf{w}^\mathrm{T} x_i)) \qquad (4)$$

We add the sparse group lasso regularization to the loss function. The optimal set of weight coefficients is estimated by minimizing the regularized training data loss shown in Eq. (5) where $\Omega_L(\mathbf{w})$ is the lasso penalty for each feature, $\mathbf{w}$ is the weight vector and N is the total number of training documents. This gives the following learning objective:

$$\min_{\mathbf{w}} \Omega_L(\mathbf{w}) + \Omega_{CG}(\mathbf{w}) + \sum_{i=1}^{N} L(x_i, \mathbf{w}, y_i) \qquad (5)$$

We used the optimization method based on alternating direction method of multipliers (ADMM) defined in [14] which has been proved to obtain a good solution for overlapping sparse group lasso.

2.3   Structure-Based Text Classification Pipeline

The goal of our proposed structure-based text classification pipeline is to take advantage of the structural information in text in order to build a model that is both accurate and compact. The first step in the pipeline is the pre-processing of the documents where the terms are stemmed by the Porter stemming algorithm. The pre-processing of documents plays an important role in the performance of the system as the representation, term weighting and the topics identified are dependent on it. Each pre-processed document is then converted to an undirected co-occurrence graph (document graph). An undirected co-occurrence graph (class graph) is created for each class from the pre-processed labelled training documents. The importance (or weight) of a term in a document is determined by the centrality (tw) of the node that corresponds to the term in the document graph and the supervised term weight factor (crc) that calculates the term's relevance to the text classification task using the centrality of the node that corresponds to the term in the class graphs. The graph-based representations of documents are converted to a document term matrix using the centrality-based weights of terms in the documents.

There are often many irrelevant parts in text that do not contribute to the classification task and can be eliminated. Discarding the irrelevant features can help in building a compact model that generalizes well on unseen data, improving its performance. The proposed structured regularizer, an instance of sparse group lasso, helps to incorporate the co-occurrence and semantic similarity information available in the class graphs for penalizing weight coefficients jointly. Hence, weight coefficients in the groups containing words belonging to irrelevant topics are driven to zero resulting in group sparsity. This leads to a compact model as words not contributing to the prediction are eliminated with this method.

3 Experiments

The performance of the proposed text classification approach is evaluated for six binary text classification tasks. The tasks involve classifying documents with closely related topics as it is more challenging than documents having distinct topics. The documents belong to subtopics within the area of computers, sports, science, religion, finance and diseases as given below.

- Computers: 'comp.sys.ibm.pc.hardware' vs 'comp.sys.mac.hardware'
- Sports: 'rec.sport.baseball' vs 'rec.sport.hockey'
- Science: 'sci.med' vs 'sci.space'
- Religion: 'alt.atheism' vs 'soc.religion.christian'
- Finance: 'oilseed' vs 'grain'
- Diseases: 'virus diseases' vs 'hemic & lymphatic diseases'

The documents in the topics of computers, sports, science and religion are obtained from the 20 newsgroup dataset (in scikit-learn). The documents related to finance and diseases are collected from reuters corpus (in NLTK) and ohsumed dataset[1] respectively. The number of training and testing documents in each of the six tasks is shown in Table 1. We split 20% of the training data for development dataset.

In our experiments, we show the effectiveness and the model compactness with the structure-based approach to text classification that uses structural information for both weighting the terms and regularization. The class graph regularizer that considers the semantic and co-occurrence information in text is used in our graph-based text classification system where the terms are weighted by the supervised graph-based term weighting scheme - tw-crc. tw is computed using the degree centrality of the node in the document graph and crc is calculated based on the variation of the degree centrality of the node in the class graphs. The semantic information is obtained by building a word2vec model from the training documents to link terms that have similarity greater than 0.9.

In the baseline systems, the features in the document representation are weighted by tf (term frequency) or tf-idf (term frequency-inverse document frequency) and the model is regularized using different standard regularizers and structured regularizers including the proposed class graph regularizer. The results of the experiments on the comparison of proposed text classification system with the baseline systems are shown in Tables 2, 3, 4, 5, 6 and 7. Tables 2, 4 and 6 show the classification results (in terms of accuracy and F1 score) of using regularizers in the baseline systems and the proposed system. Tables 3, 5 and 7

---

[1] http://disi.unitn.it/moschitti/corpora.htm.

present the percentage of non-zero features in the baseline models and the proposed model. The unstructured regularizers used are the standard regularizers - lasso, ridge and elastic net that penalize the weight coefficients independently. The structured regularizers compared are sentence regularizer, latent semantic indexing regularizer, graph-of-words regularizer, word2vec and the class graph regularizer which is the proposed regularizer based on class graphs. Regularization hyperparameters are tuned on the development dataset by performing a

Table 1. No of training and testing documents

Dataset	No of training documents	No of testing documents
Computers	1168	777
Sports	1197	796
Science	1187	790
Religion	1079	717
Finance	437	140
Diseases	1889	473

Table 2. Accuracy (in %) and F1 score (in %) of baseline system with tf weighting

Dataset	Accuracy & F1 Score															
	tf															
	Unstructured regularizer						Structured regularizer									
	L1		L2		Elastic		Sentence		Lsi		GoW		Word2vec		Class graph	
	Acc	F1	Acc	F1	Acc	F1	Acc	F1	Acc	F1	Acc	F1	Acc	F1	Acc	F1
Computers	85.97	85.52	84.94	84.54	85.84	84.89	**90.35**	**89.99**	87.64	87.43	89.45	89.43	89.58	89.55	89.45	89.43
Sports	93.47	93.42	93.47	93.48	93.47	93.42	**96.61**	**96.60**	96.11	96.12	96.11	96.12	96.23	96.25	95.85	95.85
Science	94.30	94.41	95.95	95.90	95.95	95.96	96.96	96.92	97.47	97.46	97.72	97.70	97.72	97.71	**97.85**	**97.84**
Religion	86.19	87.79	90.52	91.85	85.63	86.54	91.91	93.11	91.91	93.13	91.77	93.00	91.35	92.64	**92.19**	**93.35**
Finance	97.14	98.37	95.71	97.54	97.14	98.37	97.14	98.36	95.71	97.52	**97.86**	**98.78**	96.43	97.98	**97.86**	**98.78**
Diseases	89.43	90.27	91.33	91.98	89.22	90.10	**93.66**	**94.14**	91.97	92.55	92.81	93.36	92.39	92.94	92.18	92.73

Table 3. Percentage of non-zero features in baseline system with tf weighting

Dataset	Model size							
	tf							
	Unstructured regularizer			Structured regularizer				
	L1	L2	Elastic	Sentence	Lsi	GoW	Word2vec	Class graph
Computers	86.46	99.99	89.34	12.05	98.08	97.71	93.20	85.92
Sports	64.01	100.00	63.99	25.59	11.15	10.39	9.93	8.89
Science	59.11	100.00	84.83	15.49	41.10	33.44	41.04	42.93
Religion	93.05	100.00	99.99	80.15	73.37	61.84	70.52	91.46
Finance	4.98	100.00	3.83	2.92	6.08	2.57	4.82	2.57
Diseases	77.65	100.00	75.87	15.43	14.16	69.26	37.81	28.08

grid search on the values 0.01, 0.1, 1, 10, 100 for each of the hyperparameters, with accuracy as the evaluation criterion. Since there are more hyperparameters for structured regularizers compared to unstructured regularizers, the computational complexity is increased.

Table 4. Accuracy (in %) and F1 score (in %) of baseline system with tf-idf weighting

Dataset	Unstructured regularizer						Structured regularizer									
	L1		L2		Elastic		Sentence		Lsi		GoW		Word2vec		Class graph	
	Acc	F1	Acc	F1	Acc	F1	Acc	F1	Acc	F1	Acc	F1	Acc	F1	Acc	F1
Computers	88.03	87.78	**90.09**	**90.06**	86.87	85.39	87.77	86.68	86.23	85.75	86.23	84.95	86.36	85.11	86.36	85.07
Sports	96.61	96.60	95.85	95.80	95.35	95.30	95.60	95.61	95.98	95.99	96.11	96.12	**97.49**	**97.50**	**97.49**	**97.50**
Science	96.08	95.97	**97.59**	**97.58**	96.08	95.98	95.32	95.33	95.95	95.94	95.70	95.69	95.70	95.67	94.81	94.80
Religion	89.26	90.99	89.96	91.59	89.12	90.89	**91.91**	**93.13**	**91.91**	**93.13**	91.49	92.76	**91.91**	93.11	91.63	92.84
Finance	**97.14**	**98.36**	94.29	96.72	95.71	97.54	95.00	97.10	95.71	97.52	96.43	97.94	95.00	97.14	96.43	97.94
Diseases	92.60	93.07	91.97	92.46	92.39	92.89	**94.29**	**94.52**	93.87	94.26	93.87	94.26	93.87	94.26	93.87	94.26

Table 5. Percentage of non-zero features in baseline system with tf-idf weighting

Dataset	Unstructured regularizer			Structured regularizer				
	L1	L2	Elastic	Sentence	Lsi	GoW	Word2vec	Class graph
Computers	7.85	99.88	1.10	1.12	1.24	0.96	1.02	0.97
Sports	12.86	99.89	9.38	5.08	5.71	6.03	10.62	10.62
Science	2.79	99.89	2.82	3.29	4.20	4.15	4.10	3.36
Religion	17.70	99.79	17.37	91.89	81.11	44.56	55.52	24.95
Finance	6.42	99.87	1.50	5.41	6.08	6.32	8.89	6.26
Diseases	35.00	99.92	31.80	7.52	11.43	11.43	11.43	11.44

Table 6. Accuracy (in %) and F1 score (in %) of proposed structure-based text classification system

Dataset	Unstructured regularizer						Structured regularizer									
	L1		L2		Elastic		Sentence		Lsi		GoW		Word2vec		Class graph	
	Acc	F1	Acc	F1	Acc	F1	Acc	F1	Acc	F1	Acc	F1	Acc	F1	Acc	F1
Computers	90.22	89.73	89.58	89.16	89.83	89.51	92.02	91.67	92.02	91.80	92.28	92.00	**92.41**	**92.14**	**92.41**	**92.14**
Sports	97.36	97.35	97.61	97.60	97.61	97.60	**99.00**	**99.00**	98.37	98.39	98.37	98.39	98.37	98.39	98.74	98.75
Science	97.22	97.14	97.85	97.81	97.22	97.14	98.10	98.10	**98.35**	**98.35**	**98.35**	**98.35**	97.85	97.86	**98.35**	**98.35**
Religion	88.15	90.06	90.38	91.80	88.15	90.06	92.05	93.19	91.63	92.92	90.10	91.72	92.89	93.79	**93.03**	**93.95**
Finance	97.86	98.78	96.43	97.98	97.86	98.78	97.14	98.37	97.14	98.37	97.14	98.37	97.14	98.37	**99.29**	**99.59**
Diseases	93.45	93.84	94.08	94.47	93.23	93.63	**95.14**	**95.43**	94.71	95.05	94.29	94.61	94.29	94.55	94.93	95.24

Table 7. Percentage of non-zero features in proposed structure-based text classification system

Dataset	Model size							
	tw-crc							
	Unstructured regularizer			Structured regularizer				
	L1	L2	Elastic	Sentence	Lsi	GoW	Word2vec	Class graph
Computers	66.60	99.99	68.54	12.71	21.66	91.36	92.45	91.07
Sports	40.67	100.00	37.63	15.66	21.24	21.24	21.28	17.62
Science	5.97	100.00	6.05	43.22	13.37	11.03	15.59	12.84
Religion	59.97	100.00	58.08	78.96	77.16	17.67	18.23	18.11
Finance	2.41	100.00	2.92	1.23	1.18	1.18	1.18	2.38
Diseases	63.64	100.00	64.04	19.80	25.56	32.89	95.83	22.79

4   Results and Discussion

The effectiveness of text classification depends on the representation of text and the way in which the terms in the text are weighted. Tables 2, 4 and 6 show the classification accuracy obtained for the tf weighting, tf-idf weighting and tw-crc weighting with different regularizers. Regularization reduces overfitting by improving its generalization capability. Tables 3, 5 and 7 show the percentage of non-zero features in the baseline systems and the proposed text classification system. L2 regularization results in shrinkage of the weight coefficients and hence, it does not lead to a sparse model. L1 regularization increases sparsity resulting in a compact model. The structured regularizers increase sparsity and accuracy simultaneously. Hence, structured regularizers help in building a compact and accurate model. As there are more parameters to tune in structural regularizers, the computational complexity is high for structured regularizers compared to unstructured regularizers. In the structured regularizers compared, the sentence regularizer takes the maximum time to converge as there are a large number of overlapping groups, whereas the graph-based regularizers i.e. GoW regularizer and class graph regularizer, have a considerably faster convergence. The structured regularizers reach convergence with tw-crc term weighting method significantly faster than tf and tf-idf term weighting methods.

A structured regularizer encourages group behaviour of words and promotes group sparsity, thus performing better than unstructured regularizers as text has a structure and contains many irrelevant words not useful for the text classification task. The proposed text classification system with the class graph regularizer outperforms the baseline systems as shown in Table 6. The performance of text classification is dependent on both the term weighting scheme and the regularization method. The combination of the graph based term weighting scheme, tw-crc, and class graph regularizer improves the accuracy of text classification.

5 Conclusion

Text has an implicit syntactic and semantic structure that needs to be utilised to make the text processing more effective. We proposed a structure-based text classification pipeline where each document is represented by a graph and each class is represented by a class graph built from the labelled training documents. The information in the class graphs is used to weight the terms based on their relevance to the text classification task and also to build the structured regularizer. A graph is a powerful data structure where structural information can be encoded and utilised for different text processing applications. The outperformance of the proposed structure-based text classification framework over the baseline systems is due to the utilisation of structural information for both term weighting and regularization. In the future, the proposed system can be further improved by encoding more information in the graph and utilising it for weighting and regularization.

References

1. Aggarwal, C.C.: Data Classification: Algorithms and Applications. CRC Press, Boca Raton (2014)
2. Bakin, S., et al.: Adaptive regression and model selection in data mining problems (1999)
3. Blondel, V.D., Guillaume, J.L., Lambiotte, R., Lefebvre, E.: Fast unfolding of communities in large networks. J. Stat. Mech. Theory Exp. **2008**, P10008 (2008)
4. Feldman, R., Sanger, J.: The Text Mining Handbook: Advanced Approaches in Analyzing Unstructured Data. Cambridge University Press, New York (2007)
5. Friedman, J., Hastie, T., Tibshirani, R.: A note on the group lasso and a sparse group lasso. arXiv preprint arXiv:1001.0736 (2010)
6. Hoerl, A.E., Kennard, R.W.: Ridge regression: biased estimation for nonorthogonal problems. Technometrics **12**, 55–67 (1970)
7. Lewis, D.D.: Representation quality in text classification: an introduction and experiment. In: Speech and Natural Language: Proceedings of a Workshop Held at Hidden Valley, Pennsylvania, 24–27 June 1990 (1990)
8. Martins, A.F., Smith, N.A., Aguiar, P.M., Figueiredo, M.A.: Structured sparsity in structured prediction. In: Proceedings of the Conference on Empirical Methods in Natural Language Processing, pp. 1500–1511 (2011)
9. Sebastiani, F.: Machine learning in automated text categorization. ACM Comput. Surv. (CSUR) **34**, 1–47 (2002)
10. Shanavas, N., Wang, H., Lin, Z., Hawe, G.: Centrality-based approach for supervised term weighting. In: 2016 IEEE 16th International Conference on Data Mining Workshops (ICDMW), pp. 1261–1268. IEEE (2016)
11. Skianis, K., Rousseau, F., Vazirgiannis, M.: Regularizing text categorization with clusters of words. In: Proceedings of the 2016 Conference on Empirical Methods in Natural Language Processing, pp. 1827–1837 (2016)
12. Skianis, K., Tziortziotis, N., Vazirgiannis, M.: Orthogonal matching pursuit for text classification. In: Proceedings of the 2018 EMNLP Workshop W-NUT: The 4th Workshop on Noisy User-generated Text, pp. 93–103 (2018)

13. Tibshirani, R.: Regression shrinkage and selection via the lasso. J. R. Stat. Soc. Ser. B **58**, 267–288 (1996)
14. Yogatama, D., Smith, N.: Making the most of bag of words: sentence regularization with alternating direction method of multipliers. In: International Conference on Machine Learning, pp. 656–664 (2014)
15. Yogatama, D., Smith, N.A.: Linguistic structured sparsity in text categorization. In: Proceedings of the 52nd Annual Meeting of the Association for Computational Linguistics (Volume 1: Long Papers), pp. 786–796 (2014)
16. Yuan, M., Lin, Y.: Model selection and estimation in regression with grouped variables. J. R. Stat. Soc. Ser. B **68**, 49–67 (2006)
17. Zou, H., Hastie, T.: Regularization and variable selection via the elastic net. J. R. Stat. Soc. Ser. B **67**, 301–320 (2005)

Gated Convolutional Neural Networks for Domain Adaptation

Avinash Madasu and Vijjini Anvesh Rao$^{(\boxtimes)}$

Samsung R&D Institute, Bangalore, India
{m.avinash,a.vijjini}@samsung.com

Abstract. Domain Adaptation explores the idea of how to maximize performance on a target domain, distinct from source domain, upon which the model was trained. This idea has been explored for the task of sentiment analysis extensively. The training of reviews pertaining to one domain and evaluation on another domain is widely studied for modeling a domain independent algorithm. This further helps in understanding corelation of information between domains. In this paper, we show that Gated Convolutional Neural Networks (GCN) perform effectively at learning sentiment analysis in a manner where domain dependant knowledge is filtered out using its gates. We perform our experiments on multiple gate architectures: Gated Tanh ReLU Unit (GTRU), Gated Tanh Unit (GTU) and Gated Linear Unit (GLU). Extensive experimentation on two standard datasets relevant to the task, reveal that training with Gated Convolutional Neural Networks give significantly better performance on target domains than regular convolution and recurrent based architectures. While complex architectures like attention, filter domain specific knowledge as well, their complexity order is remarkably high as compared to gated architectures. GCNs rely on convolution hence gaining an upper hand through parallelization.

Keywords: Gated convolutional neural networks ·
Domain adaptation · Sentiment analysis

1 Introduction

With the advancement in technology and invention of modern web applications like Facebook and Twitter, users started expressing their opinions and ideologies at a scale unseen before. The growth of e-commerce companies like Amazon, Walmart have created a revolutionary impact in the field of consumer business. People buy products online through these companies and write reviews for their products. These consumer reviews act as a bridge between consumers and companies. Through these reviews, companies polish the quality of their services. Sentiment Classification (SC) is one of the major applications of Natural Language Processing (NLP) which aims to find the polarity of text. In the early stages [1] of text classification, sentiment classification was performed using traditional feature selection techniques like Bag-of-Words (BoW) [2] or TF-IDF.

E. Métais et al. (Eds.): NLDB 2019, LNCS 11608, pp. 118–130, 2019.
https://doi.org/10.1007/978-3-030-23281-8_10

These features were further used to train machine learning classifiers like Naive Bayes (NB) [3] and Support Vector Machines (SVM) [4]. They are shown to act as strong baselines for text classification [5]. However, these models ignore word level semantic knowledge and sequential nature of text. Neural networks were proposed to learn distributed representations of words [6]. Skip-gram and CBOW architectures [7] were introduced to learn high quality word representations which constituted a major breakthrough in NLP. Several neural network architectures like recursive neural networks [8] and convolutional neural networks [9] achieved excellent results in text classification. Recurrent neural networks which were proposed for dealing sequential inputs suffer from vanishing [10] and exploding gradient problems [11]. To overcome this problem, Long Short Term Memory (LSTM) was introduced [12].

All these architectures have been successful in performing sentiment classification for a specific domain utilizing large amounts of labelled data. However, there exists insufficient labelled data for a target domain of interest. Therefore, Domain Adaptation (DA) exploits knowledge from a relevant domain with abundant labeled data to perform sentiment classification on an unseen target domain. However, expressions of sentiment vary in each domain. For example, in *Books* domain, words *thoughtful* and *comprehensive* are used to express sentiment whereas *cheap* and *costly* are used in *Electronics* domain. Hence, models should generalize well for all domains. Several methods have been introduced for performing Domain Adaptation. Blitzer [13] proposed Structural Correspondence Learning (SCL) which relies on pivot features between source and target domains. Pan [14] performed Domain Adaptation using Spectral Feature Alignment (SFA) that aligns features across different domains. Glorot [15] proposed Stacked Denoising Autoencoder (SDA) that learns generalized feature representations across domains. Zheng [16] proposed end-to-end adversarial network for Domain Adaptation. Qi [17] proposed a memory network for Domain Adaptation. Zheng [18] proposed a Hierarchical transfer network relying on attention for Domain Adaptation.

However, all the above architectures use a different sub-network altogether to incorporate domain agnostic knowledge and is combined with main network in the final layers. This makes these architectures computationally intensive. To address this issue, we propose a Gated Convolutional Neural Network (GCN) model that learns domain agnostic knowledge using gated mechanism [19]. Convolution layers learns the higher level representations for source domain and gated layer selects domain agnostic representations. Unlike other models, GCN doesn't rely on a special sub-network for learning domain agnostic representations. As, gated mechanism is applied on Convolution layers, GCN is computationally efficient.

2   Related Work

Traditionally methods for tackling Domain Adaptation are lexicon based. Blitzer [20] used a pivot method to select features that occur frequently in both domains.

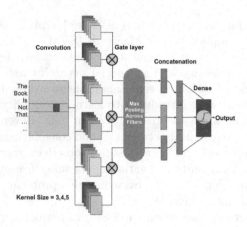

Fig. 1. Architecture of the proposed model

It assumes that the selected pivot features can reliably represent the source domain. The pivots are selected using mutual information between selected features and the source domain labels. SFA [14] method argues that pivot features selected from source domain cannot attest a representation of target domain. Hence, SFA tries to exploit the relationship between domain-specific and domain independent words via simultaneously co-clustering them in a common latent space. SDA [15] performs Domain Adaptation by learning intermediate representations through auto-encoders. Yu [21] used two auxiliary tasks to help induce sentence embeddings that work well across different domains. These embeddings are trained using Convolutional Neural Networks (CNN).

Gated convolutional neural networks have achieved state-of-art results in language modelling [19]. Since then, they have been used in different areas of natural language processing (NLP) like sentence similarity [22] and aspect based sentiment analysis [23].

3 Gated Convolutional Neural Networks

In this section, we introduce a model based on Gated Convolutional Neural Networks for Domain Adaptation. We present the problem definition of Domain Adaptation, followed by the architecture of the proposed model.

3.1 Problem Definition

Given a source domain D_S represented as $D_S = \{(x_{s_1}, y_{s_1}), (x_{s_2}, y_{s_2})....$ $(x_{s_n}, y_{s_n})\}$ where $x_{s_i} \in \mathbb{R}$ represents the vector of ith source text and y_{s_i} represents the corresponding source domain label. Let T_S represent the task in source domain. Given a target domain D_T represented as $D_T = \{(x_{t_1}, y_{t_1}),$ $(x_{t_2}, y_{t_2})....(x_{t_n}, y_{t_n})\}$, where $x_{t_i} \in \mathbb{R}$ represents the vector of ith target text and y_{t_i} represents corresponding target domain label. Let T_T represent the task in target domain. Domain Adaptation (DA) is defined by the target predictive function $f_T(D_T)$ calculated using the knowledge of D_S and T_S where $D_S \neq D_T$ but $T_S = T_T$. It is imperative to note that the domains are different but only a single task. In this paper, the task is sentiment classification.

3.2 Model Architecture

The proposed model architecture is shown in the Fig. 1. Recurrent Neural Networks like LSTM, GRU update their weights at every timestep sequentially and hence lack parallelization over inputs in training. In case of attention based models, the attention layer has to wait for outputs from all timesteps. Hence, these models fail to take the advantage of parallelism either. Since, proposed model is based on convolution layers and gated mechanism, it can be parallelized efficiently. The convolution layers learn higher level representations for the source domain. The gated mechanism learn the domain agnostic representations. They together control the information that has to flow through further fully connected output layer after max pooling.

Let I denote the input sentence represented as $I = \{w_1 w_2 w_3 ... w_N\}$ where w_i represents the i_{th} word in I and N is the maximum sentence length considered. Let B be the vocabulary size for each dataset and $X \in \mathbb{R}^{B \times d}$ denote the word embedding matrix where each X_i is a d dimensional vector. Input sentences whose length is less than N are padded with 0s to reach maximum sentence length. Words absent in the pretrained word embeddings[1] are initialized to 0s. Therefore each input sentence I is converted to $P \in \mathbb{R}^{N \times d}$ dimensional vector. Convolution operation is applied on P with kernel $K \in \mathbb{R}^{h \times d}$. The convolution operation is one-dimensional, applied with a fixed window size across words. We consider kernel size of 3, 4 and 5. The weight initialization of these kernels is done using glorot uniform [24]. Each kernel is a feature detector which extracts patterns from n-grams. After convolution we obtain a new feature map $C = [c_1 c_2 .. c_N]$ for each kernel K.

$$C_i = f(P_{i:i+h} * W_a + b_a) \tag{1}$$

where f represents the activation function in convolution layer. The gated mechanism is applied on each convolution layer. Each gated layer learns to filter domain agnostic representations for every time step i.

$$S_i = g(P_{i:i+h} * W_s + b_s) \tag{2}$$

where g is the activation function used in gated convolution layer. The outputs from convolution layer and gated convolution layer are element wise multiplied to compute a new feature representation G_i

$$G_i = C_i \times S_i \tag{3}$$

---

[1] https://nlp.stanford.edu/data/glove.840B.300d.zip.

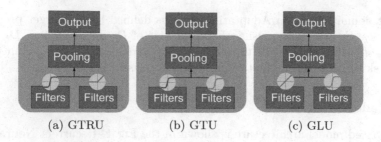

(a) GTRU   (b) GTU   (c) GLU

Fig. 2. Variations in gates of the proposed GCN architecture.

Maxpooling operation is applied across each filter in this new feature representation to get the most important features [9]. As shown in Fig. 1 the outputs from maxpooling layer across all filters are concatenated. The concatenated layer is fully connected to output layer. Sigmoid is used as the activation function in the output layer.

3.3   Gating Mechanisms

Gating mechanisms have been effective in Recurrent Neural Networks like GRU and LSTM. They control the information flow through their recurrent cells. In case of GCN, these gated units control the domain information that flows to pooling layers. The model must be robust to change in domain knowledge and should be able to generalize well across different domains. We use the gated mechanisms Gated Tanh Unit (GTU) and Gated Linear Unit (GLU) and Gated Tanh ReLU Unit (GTRU) [23] in proposed model. The gated architectures are shown in Fig. 2. The outputs from Gated Tanh Unit is calculated as $tanh(P * W + c) \times \sigma(P * V + c)$. In case of Gated Linear Unit, it is calculated as $(P * W + c) \times \sigma(P * V + c)$ where $tanh$ and σ denotes Tanh and Sigmoid activation functions respectively. In case of Gated Tanh ReLU Unit, output is calculated as $tanh(P * W + c) \times relu(P * V + c)$.

4   Experiments

4.1   Datasets

Multi Domain Dataset (MDD). Multi Domain Dataset [20] is a short dataset with reviews from distinct domains namely Books(B), DVD(D), Electronics(E) and Kitchen(K). Each domain consists of 2000 reviews equally divided among positive and negative sentiment. We consider 1280 reviews for training, 320 reviews for validation and 400 reviews for testing from each domain.

Amazon Reviews Dataset (ARD). Amazon Reviews Dataset [25] is a large dataset with millions of reviews from different product categories. For our experiments, we consider a subset of 20000 reviews from the domains Cell Phones and

Accessories(C), Clothing and Shoes(S), Home and Kitchen(H) and Tools and Home Improvement(T). Out of 20000 reviews, 10000 are positive and 10000 are negative. We use 12800 reviews for training, 3200 reviews for validation and 4000 reviews for testing from each domain.

4.2 Baselines

To evaluate the performance of proposed model, we consider various baselines like traditional lexicon approaches, CNN models without gating mechanisms and LSTM models.

BoW+LR. Bag-of-words (BoW) is one of the strongest baselines in text classification [5]. We consider all the words as features with a minimum frequency of 5. These features are trained using Logistic Regression (LR).

TF-IDF+LR. TF-IDF is a feature selection technique built upon Bag-of-Words. We consider all the words with a minimum frequency of 5. The features selected are trained using Logistic Regression (LR).

PV+FNN. Paragraph2vec or doc2vec [26] is a strong and popularly used baseline for text classification. Paragraph2Vec represents each sentence or paragraph in the form of a distributed representation. We trained our own doc2vec model using DBOW model. The paragraph vectors obtained are trained using Feed Forward Neural Network (FNN).

CNN. To show the effectiveness of gated layer, we consider a CNN model which does not contain gated layers. Hence, we consider Static CNN model, a popular CNN architecture proposed in Kim [9] as a baseline.

CRNN. Wang [27] proposed a combination of Convolutional and Recurrent Neural Network for sentiment Analysis of short texts. This model takes the advantages of features learned by CNN and long-distance dependencies learned by RNN. It achieved remarkable results on benchmark datasets. We report the results using code published by the authors[2].

LSTM. We offer a comparison with LSTM model with a single hidden layer. This model is trained with equivalent experimental settings as proposed model.

LSTM+Attention. In this baseline, attention mechanism [28] is applied on the top of LSTM outputs across different timesteps.

---

[2] https://github.com/ultimate010/crnn.

Table 1. Average training time for all the models on ARD

Model	Batchsize	Time for 1 epoch (in Sec)
CRNN	50	50
LSTM	50	70
LSTM+Attention	50	150
GLU	50	10
GRU	50	10
GTRU	50	10

4.3   Implementation Details

All the models are experimented with approximately matching number of parameters for a solid comparison using a Tesla K80 GPU.

Input. Each word in the input sentence is converted to a 300 dimensional vector using GloVe pretrained vectors [29]. A maximum sentence length 100 is considered for all the datasets. Sentences with length less than 100 are padded with 0s.

Architecture Details: The model is implemented using keras. We considered 100 convolution filters for each of the kernels of sizes 3, 4 and 5. To get the same sentence length after convolution operation zero padding is done on the input.

Training. Each sentence or paragraph is converted to lower case. Stopword removal is not done. A vocabulary size of 20000 is considered for all the datasets. We apply a dropout layer [30] with a probability of 0.5, on the embedding layer and probability 0.2, on the dense layer that connects the output layer. Adadelta [31] is used as the optimizer for training with gradient descent updates. Batchsize of 16 is taken for MDD and 50 for ARD. The model is trained for 50 epochs. We employ an early stopping mechanism based on validation loss for a patience of 10 epochs. The models are trained on source domain and tested on unseen target domain in all experiments.

5   Results and Discussion

5.1   Results

The performance of all models on MDD is shown in Tables 2 and 3 while for ARD, in Tables 4 and 5. All values are shown in accuracy percentage. Furthermore time complexity of each model is presented in Table 1.

Table 2. Accuracy scores on Multi Domain Dataset.

Source->Target	BoW	TFIDF	PV	CNN	CRNN	LSTM
B->D	72.5	73.75	63.749	57.75	68.75	69.5
B->E	67.5	68.5	53.25	53.5	63.249	58.75
B->K	69.25	72.5	57.75	56.25	66.5	64.75
D->B	66	68.5	64.75	54.25	66.75	74.75
D->E	71	69.5	56.75	57.25	69.25	64.25
D->K	68	69.75	60	58.25	67.5	70
E->B	63.249	64	54	57.25	69.5	67.75
E->D	65	66	47.25	56.499	64.5	67
E->K	76.25	76.75	59.25	63.249	76	76
K->B	61.5	67.75	50	57.75	69.25	66.25
K->D	68	70.5	52.25	60	64.75	71
K->E	81	80	50	59.25	69	76.75

Table 3. Accuracy scores on Multi Domain Dataset.

Source->Target	LSTM.Attention	GLU	GTU	GTRU
B->D	76.75	**79.5**	79.25	77.5
B->E	70	**71.75**	71.25	71.25
B->K	74.75	73	72.5	**74.25**
D->B	72.5	78	**80.25**	77.25
D->E	71	73	**74.5**	69.25
D->K	72.75	**77**	76	74.75
E->B	64.75	**71.75**	68.75	67.25
E->D	62.749	**71.75**	69	68.25
E->K	72	**82.25**	80.5	79
K->B	64.75	**70**	67.75	63.249
K->D	**75**	73.75	73.5	69.25
K->E	75.5	**82**	**82**	81.25

5.2 Discussion

Gated Outperform Regular Convolution. We find that gated architectures vastly outperform non gated CNN model. The effectiveness of gated architectures rely on the idea of training a gate with sole purpose of identifying a weightage. In the task of sentiment analysis this weightage corresponds to what weights will lead to a decrement in final loss or in other words, most accurate prediction of sentiment. In doing so, the gate architecture learns which words or n-grams contribute to the sentiment the most, these words or n-grams often co-relate with

Table 4. Accuracy scores on Amazon Reviews Dataset.

Source− >Target	BoW	TFIDF	PV	CNN	CRNN	LSTM
C− >S	79.3	81.175	69.625	62.324	84.95	83.7
C− >H	81.6	82.875	70.775	59.35	81.8	81.175
C− >T	76.25	77.475	66.4	54.5	79.025	77.175
S− >C	76.925	76.525	69.425	55.375	79.975	79.85
S− >T	80.125	81.575	74.524	62.7	81.45	82.925
S− >H	74.275	75.175	67.274	61.925	76.05	77.7
H− >S	76.149	73.575	65.3	53.55	79.074	78.574
H− >C	81.225	80.925	70.7	58.25	74.275	81.95
H− >T	79.175	75.449	69.425	59.4	76.325	76.725
T− >C	75.1	73.875	56.85	56	80.25	76.9
T− >S	78.875	80.5	59.199	60	**85.824**	81.8
T− >H	81.325	81.875	66.8	61.25	83.35	81

Table 5. Accuracy scores on Amazon Reviews Dataset.

Source− >Target	LSTM.Attention	GLU	GTU	GTRU
C− >S	84.15	**85.125**	84.95	84.8
C− >H	82.6	**84.85**	84.2	84.55
C− >T	77.9	79.5	79.274	**80.225**
S− >C	78.075	80.925	80.25	**83.1**
S− >H	82.325	83.95	83.399	**84.025**
S− >T	78.425	**79.475**	77.85	79.375
H− >C	81.375	**83.175**	81.85	82.1
H− >S	81.975	82.75	84.1	**85.425**
H− >T	80.95	**82.55**	81.774	81.825
T− >C	75.55	**82.125**	80.805	81.825
T− >S	82.375	82.625	83.975	84.775
T− >H	80.5	84.7	83.95	**85.275**

domain independent words. On the other hand the gate gives less weightage to n-grams which are largely either specific to domain or function word chunks which contribute negligible to the overall sentiment. This is what makes gated architectures effective at Domain Adaptation.

In Fig. 3, we have illustrated the visualization of convolution outputs (kernel size = 3) from the sigmoid gate in GLU across domains. As the kernel size is 3, each row in the output corresponds to a trigram from input sentence. This heat map visualizes values of all 100 filters and their average for every input trigram. These examples demonstrate what the convolution gate learns. Trigrams with domain independent but heavy polarity like "_ _ good" and "_ costly would" have higher weightage. Meanwhile, Trigrams with domain specific terms like "quality functional case" and "sell entire kitchen" get some of the least weights. In Fig. 3(b) example, the trigram "would have to" just consists of function words, hence gets the least weight. While "sell entire kitchen" gets more weight comparatively. This might be because while function words are merely grammatical units which contribute minimal to overall sentiment, domain specific terms like "sell" may contain sentiment level knowledge only relevant within the domain. In such a case it is possible that the filters effectively propagate sentiment level knowledge from domain specific terms as well.

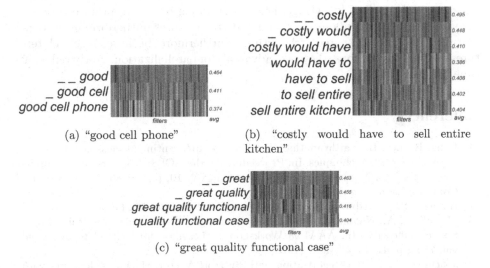

(a) "good cell phone"

(b) "costly would have to sell entire kitchen"

(c) "great quality functional case"

Fig. 3. Visualizing outputs from gated convolutions (filter size = 3) of GLU for example sentences, darker indicates higher weightage

Gated Outperform Attention and Linear. We see that gated architectures almost always outperform recurrent, attention and linear models BoW, TFIDF, PV. This is largely because while training and testing on same domains, these models especially recurrent and attention based may perform better. However, for Domain Adaptation, as they lack gated structure which is trained in parallel to learn importance, their performance on target domain is poor as compared to gated architectures. As gated architectures are based on convolutions, they exploit parallelization to give significant boost in time complexity as compared to other models. This is depicted in Table 1.

Comparison Among Gates. While the gated architectures outperform other baselines, within them as well we make observations. Gated Linear Unit (GLU) performs the best often over other gated architectures. In case of GTU, outputs from Sigmoid and Tanh are multiplied together, this may result in small gradients, and hence resulting in the vanishing gradient problem. However, this will not be the in the case of GLU, as the activation is linear. In case of GTRU, outputs from Tanh and ReLU are multiplied. In ReLU, because of absence of negative activations, corresponding Tanh outputs will be completely ignored, resulting in loss of some domain independent knowledge.

6 Conclusion

In this paper, we proposed Gated Convolutional Neural Network (GCN) model for Domain Adaptation in Sentiment Analysis. We show that gates in GCN, filter out domain dependant knowledge, hence performing better at an unseen target domain. Our experiments reveal that gated architectures outperform other popular recurrent and non-gated architectures. Furthermore, because these architectures rely on convolutions, they take advantage of parellalization, vastly reducing time complexity.

References

1. Pang, B., Lee, L., Vaithyanathan, S.: Thumbs up?: sentiment classification using machine learning techniques. In: Proceedings of the ACL-02 Conference on Empirical Methods in Natural Language Processing, vol. 10, pp. 79–86. Association for Computational Linguistics (2002)
2. Harris, Z.S.: Distributional structure. Word $10(2–3)$, 146–162 (1954)
3. McCallum, A., Nigam, K., et al.: A comparison of event models for Naive Bayes text classification. In: AAAI-98 Workshop on Learning for Text Categorization, vol. 752, pp. 41–48. Citeseer (1998)
4. Joachims, T.: Text categorization with Support Vector Machines: learning with many relevant features. In: Nédellec, C., Rouveirol, C. (eds.) ECML 1998. LNCS, vol. 1398, pp. 137–142. Springer, Heidelberg (1998). https://doi.org/10.1007/BFb0026683
5. Wang, S., Manning, C.D.: Baselines and bigrams: simple, good sentiment and topic classification. In: Proceedings of the 50th Annual Meeting of the Association for Computational Linguistics: Short Papers, vol. 2, pp. 90–94. Association for Computational Linguistics (2012)
6. Bengio, Y., Ducharme, R., Vincent, P., Jauvin, C.: A neural probabilistic language model. J. Mach. Learn. Res. 3, 1137–1155 (2003)
7. Mikolov, T., Sutskever, I., Chen, K., Corrado, G.S., Dean, J.: Distributed representations of words and phrases and their compositionality. In: Advances in Neural Information Processing Systems, pp. 3111–3119 (2013)
8. Socher, R., Lin, C.C., Manning, C., Ng, A.Y.: Parsing natural scenes and natural language with recursive neural networks. In: Proceedings of the 28th International Conference on Machine Learning (ICML 2011), pp. 129–136 (2011)
9. Kim, Y.: Convolutional neural networks for sentence classification. arXiv preprint arXiv:1408.5882 (2014)
10. Bengio, Y., Simard, P., Frasconi, P., et al.: Learning long-term dependencies with gradient descent is difficult. IEEE Trans. Neural Netw. $5(2)$, 157–166 (1994)

11. Pascanu, R., Mikolov, T., Bengio, Y.: On the difficulty of training recurrent neural networks. In: International Conference on Machine Learning, pp. 1310–1318 (2013)
12. Hochreiter, S., Schmidhuber, J.: Long short-term memory. Neural Comput. **9**(8), 1735–1780 (1997)
13. Blitzer, J., McDonald, R., Pereira, F.: Domain adaptation with structural correspondence learning. In: Proceedings of the 2006 Conference on Empirical Methods in Natural Language Processing, pp. 120–128. Association for Computational Linguistics (2006)
14. Pan, S.J., Ni, X., Sun, J.-T., Yang, Q., Chen, Z.: Cross-domain sentiment classification via spectral feature alignment. In: Proceedings of the 19th International Conference on World Wide Web, pp. 751–760. ACM (2010)
15. Glorot, X., Bordes, A., Bengio, Y.: Domain adaptation for large-scale sentiment classification: a deep learning approach. In: Proceedings of the 28th International Conference on Machine Learning (ICML 2011), pp. 513–520 (2011)
16. Li, Z., Zhang, Y., Wei, Y., Wu, Y., Yang, Q.: End-to-end adversarial memory network for cross-domain sentiment classification. In: IJCAI, pp. 2237–2243 (2017)
17. Liu, Q., Zhang, Y., Liu, J.: Learning domain representation for multi-domain sentiment classification. In: Proceedings of the 2018 Conference of the North American Chapter of the Association for Computational Linguistics: Human Language Technologies, Volume 1 (Long Papers), vol. 1, pp. 541–550 (2018)
18. Li, Z., Wei, Y., Zhang, Y., Yang, Q.: Hierarchical attention transfer network for cross-domain sentiment classification. In: Thirty-Second AAAI Conference on Artificial Intelligence (2018)
19. Dauphin, Y.N., Fan, A., Auli, M., Grangier, D.: Language modeling with gated convolutional networks. In: Proceedings of the 34th International Conference on Machine Learning, vol. 70, pp. 933–941. JMLR.org (2017)
20. Blitzer, J., Dredze, M., Pereira, F.: Biographies, bollywood, boom-boxes and blenders: domain adaptation for sentiment classification. In: Proceedings of the 45th Annual Meeting of the Association of Computational Linguistics, pp. 440–447 (2007)
21. Yu, J., Jiang, J.: Learning sentence embeddings with auxiliary tasks for cross-domain sentiment classification. In: Proceedings of the 2016 Conference on Empirical Methods in Natural Language Processing, pp. 236–246 (2016)
22. Chen, P., Guo, W., Chen, Z., Sun, J., You, L.: Gated convolutional neural network for sentence matching. In: Interspeech (2018)
23. Xue, W., Li, T.: Aspect based sentiment analysis with gated convolutional networks. arXiv preprint arXiv:1805.07043 (2018)
24. Glorot, X., Bengio, Y.: Understanding the difficulty of training deep feedforward neural networks. In: Proceedings of the Thirteenth International Conference on Artificial Intelligence and Statistics, pp. 249–256 (2010)
25. He, R., McAuley, J.: Ups and downs: modeling the visual evolution of fashion trends with one-class collaborative filtering. In: Proceedings of the 25th International Conference on World Wide Web, pp. 507–517. International World Wide Web Conferences Steering Committee (2016)
26. Le, Q., Mikolov, T.: Distributed representations of sentences and documents. In: International Conference on Machine Learning, pp. 1188–1196 (2014)
27. Wang, X., Jiang, W., Luo, Z.: Combination of convolutional and recurrent neural network for sentiment analysis of short texts. In: Proceedings of COLING 2016, the 26th International Conference on Computational Linguistics: Technical Papers, pp. 2428–2437 (2016)

28. Bahdanau, D., Cho, K., Bengio, Y.: Neural machine translation by jointly learning to align and translate. arXiv preprint arXiv:1409.0473 (2014)
29. Pennington, J., Socher, R., Manning, C.: GloVe: global vectors for word representation. In: Proceedings of the 2014 Conference on Empirical Methods in Natural Language Processing (EMNLP), pp. 1532–1543 (2014)
30. Srivastava, N., Hinton, G., Krizhevsky, A., Sutskever, I., Salakhutdinov, R.: Dropout: a simple way to prevent neural networks from overfitting. J. Mach. Learn. Res. **15**(1), 1929–1958 (2014)
31. Zeiler, M.D.: ADADELTA: an adaptive learning rate method. arXiv preprint arXiv:1212.5701 (2012)

A Keyword Search Approach
for Semantic Web Data

Mohamad Rihany[(✉)], Zoubida Kedad, and Stéphane Lopes

DAVID Lab, University of Versailles Saint-Quentin-en-Yvelines, Versailles, France
{Mohamad.rihany,zoubida.kedad,Stephane.lopes}@uvsq.fr

Abstract. More and more RDF datasets are available on the web. These datasets can be queried using the SPARQL language; to do so, one must be familiar with the query language itself, but also with the content of the dataset in terms of resources and properties in order to formulate the queries. Keyword search is an alternative way to query RDF data. In this paper, we present a keyword search approach which uses online lexical databases to bridge the terminological gap between the keywords and the dataset when searching for matching elements in the dataset. We formulate the problem of aggregating the matching elements as a Steiner tree problem and we adapt Kruskal's algorithm to provide a solution. We also propose a ranking approach if several answers are found for a given query. We have performed some experiments on the DBpedia and the AIFB datasets to illustrate the effectiveness of our approach.

Keywords: Keyword search · RDF graph · Steiner tree

1 Introduction

The rapid growth and huge amount of data published on the web in recent years has led to many challenges in searching and gathering meaningful information. These data are described by languages proposed by the W3C, such as the Resource Description Framework (RDF) [8]; the building block in RDF is a triple *(subject, predicate, object)*. An RDF dataset can be viewed as a labeled directed graph where nodes are resources or literals and where labeled edges represent properties. Figure 1 shows an example of an RDF graph. The SPARQL query language is used to query RDF datasets. A SPARQL query consists of a set of triples where the subject, predicate and/or object can be variables. The idea is to match the triple patterns of the query with the triples of the dataset, and to find the possible instances for the variables. Considering the dataset of

---

This work was funded by the National Council for Scientific Research of Lebanon (CNRS-L) and the French National Research Agency through the CAIR ANR-14-CE23-0006 project.

E. Métais et al. (Eds.): NLDB 2019, LNCS 11608, pp. 131–143, 2019.
https://doi.org/10.1007/978-3-030-23281-8_11

Fig. 1, finding the director and released date of the movie "Man on Fire" is expressed using the following query:

```
SELECT ?x ?y
WHERE {Man_on_Fire director ?x. Man_on_Fire Released_date ?y.}
```

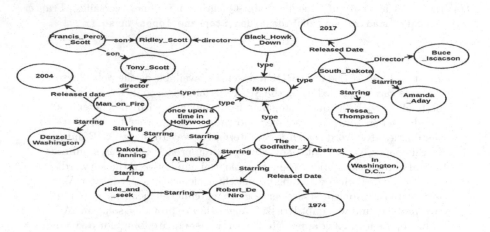

Fig. 1. An example of RDF dataset about movies

In order to write such queries, the user should be familiar with SPARQL, and should have some knowledge about the schema of the RDF data. An alternative way of querying RDF datasets is keyword search, in which a query is formulated as a set of keywords. This approach raises several challenges. One of them is finding the relevant elements by matching the keyword query with the elements of the dataset, taking into account the differences of terminology which may exist between them. In previous works, we have proposed a matching approach capable of handling these terminological differences [12]. In this paper, we will present some improvements of this approach

Another challenge addressed in this paper is aggregating the relevant elements, building the subgraphs which represent possible answers to the initial query, and ranking these results.

In this paper, we describe a keyword search approach for RDF datasets and present a novel method to aggregate the matching elements and to find the best paths between them in order to extract the subgraph corresponding to the keyword query. Our solution is an adaptation of an algorithm used to solve the Steiner tree problem and the minimum spanning tree problem; we also propose a method to rank the set of possible answers.

The rest of this paper is organized as follows. An overview of the approach is provided in Sect. 2. We present our solution for matching keywords with graph elements using external knowledge in Sect. 3. Section 4 presents the process of building the end results from the matching elements. The ranking method is discussed in Sect. 5. Section 6 presents our experiments and Sect. 7 reviews the related works. Finally, we conclude the paper and present some future works in Sect. 8.

2   Approach Overview

Our framework for keyword search, presented in Fig. 2, comprises three components: a matching component, which searches the graph elements corresponding to the keywords, a component aggregating the graph elements into a subgraph, and a ranking element to deal with multiple results for a given query.

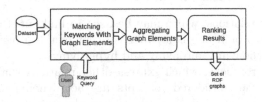

Fig. 2. Approach overview

The matching component takes as input the keyword query and searches for the matching elements in the dataset. Each keyword is compared to the graph elements (resource, class or property) and the matching elements are identified. In some cases, the user may enter a keyword for which an exact match can not be found in the dataset, but some graph elements could be close to the keyword such as a synonym or a close concept. The problem is to identify the equivalent concepts and the close concepts to a keyword in the dataset. To do so, we propose the use of external knowledge sources such as online linguistic dictionaries.

Once the matching elements in the RDF graph are identified for each keyword, the problem is to build the final result from these elements, and to aggregate them into a connected subgraph representing an answer to the query. Each keyword can be associated to more than one element in the RDF graph; we consider that each combination of matching elements containing exactly one element for each keyword is a possible answer to the query. The problem is to build the subgraph containing the elements of the considered combination.

As each keyword may have more than one matching element in the dataset, there may be several possible results to the query. The problem is to rank the different results and to find a ranking method capable of determining if there are better results than others.

3   Matching Keywords with the Dataset

One of the problems raised by keyword search in RDF datasets is matching the query keywords with the elements of the dataset. Let $Q = \{k_1, k_2, ..., k_n\}$ be the keyword query and let $G = (V, E)$ be the data graph, where V is the set of vertices and E the set of edges. The goal of the matching process is to find for each k_i the set of matching elements from the data graph G. A matching element can either be a node or an edge of G. To this end, we need to solve the terminological gap which may exist between the keywords and the data graph elements. Indeed, the keyword k_i itself may not be found in the dataset, but an equivalent or a close element could be found. We have proposed the use of some external knowledge stored in online linguistic resources such as observe from[1], which is a large lexical database of English providing numerous semantic relations among concepts. In our work, we have used WordNet in order to find the possible matchings between k_i and G by taking into account the provided semantic relations. We have first divide the semantic relations provided by WordNet into two sets:

- Exact semantic relations, which may be used to find out if two concepts are either equivalent, or if one is a super concept of the other; these semantic relations are: synonymy, antonymy and hypernymy;
- Close semantic relations, which express other relations than equivalence or generalization; the considered concepts are not synonyms, but are linked by some other semantic relation, which could be hyponymy, meronymy and holonymy.

An inverted index is used to improve the efficiency of the process of extracting the matching elements. Our index contains a set of documents, each one representing a graph element (resource, literal or property). Our matching process operates in two phases; firstly, all the exact matching relations for the keyword k_i are retrieved; secondly, the close semantic relations for k_i are retrieved. For each k_i in Q we start by querying WordNet to extract the exact semantic relations involving k_i; as a result, we obtain a set ES_i such that: $ES_i = \{c_{ij}| \exists \, sem - rel(k_i, c_{ij})\}$ where sem-rel(k_i, c_{ij}) is one of the following semantic relations: synonymy, antonymy or hypernymy between k_i and c_{ij}. For each concept c_{ij} in ES_i, and for k_i we check the index table to find matching elements in G and add them to a set of matching elements $ME_i(k_i)$.

At the end of this phase, if $ME_i(k_i)$ is empty, then we search for close semantic relations. This consists in querying WordNet to extract the close semantic relations involving k_i; as a result, we obtain a set CS_i such that: $CS_i = \{c_{ij}| \exists \, sem - rel(k_i, c_{ij})\}$ where sem-rel(k_i, c_{ij}) is one of the following semantic relations: hyponymy, meronymy or holonymy between k_i and c_{ij}. This whole process is repeated for each concept c_{ij} in CS_i.

The matching process is presented in Algorithm 1; let us start by presenting the notations used in the algorithm. Let $K = \{k_1, k_2, k_3...k_n\}$ be the keyword query, and ME(k_i) a function to extract the matching elements for k_i in the dataset (these elements can be literals, instances, classes, properties). The semantic relations between the keyword query k_i and WordNet are extracted by using two functions, SemanticRelationsExact(k_i) which returns the

---

[1] https://wordnet.princeton.edu/.

synonymy, antonymy and hypernymy relations and SemanticRelationsClose(k_i) which returns the hyponymy, meronymy and holonymy relations.

For each keyword k_i in the query, the matching elements ME(k_i) are extracted (line 4–5). Then WordNet is queried to extract the set SemanticRelationsExact of synonyms, antonyms and hyponyms. For each element of SemanticRelationsExact, the dataset is accessed to check if there is a matching element (line 6–9). If no matching element has been found, then a search for close matching elements is performed (line 12). In this phase, WordNet is queried to find close matching elements by searching for the hypernymy, holonymy and meronymy semantic relations involving k_i. For each concept c related to k_i by one of these relations, we search for elements labeled c in the dataset; these elements are added to the set of matching elements for the keyword k_i (line 13–20).

Algorithm 1. Matching the Query Keywords with the RDF Graph

1: $keywordQuery = \{k_1, k_2, k_3, ...k_i\}$
2: **procedure** MATCHING(keywordQuery)
3:     $hashmap(keyword, MatchingElements)$
4:     **for** each keyword k_i in $keywordQuery$ **do**
5:         ME(k_i)
6:         $SemanticRelationsExact \leftarrow SemanticRelationsExact(k_i)$
7:         **for** each s_j in $SemanticRelations$ **do**
8:             $ME(k_i) \leftarrow ME(k_i) + ME(s_j)$
9:         **end for**
10:        $hashmap.add(k_i, ME(k_i))$
11:        **if** ME(k_i) is empty **then**
12:            $SemanticRelationsClose \leftarrow SemanticRelationsClose(k_i)$
13:            **for** each s_k in $SemanticRelationsClose$ **do**
14:                $ME(k_i) \leftarrow ME(k_i) + ME(s_k)$
15:            **end for**
16:            $hashmap.add(k_i, ME(k_i))$
17:        **end if**
18:     **end for**
19:     Return $hashmap$
20: **end procedure**

Let us consider Q = {brother, Dakota, Washington, Hollywood} be a keyword query, if the user issues this keyword query on the data graph of Fig. 1 we will get a set of matching elements for each keyword.

As we can observe from the data graph, there is no graph element corresponding to the keyword "brother"; in previous work [14], we have introduced the notion of pattern, representing equivalences between a property and a path. These equivalences enable us to infer that "brother" is equivalent to the graph shown in Fig. 3, which can therefore be selected as a matching element for the keyword "brother". For the other keywords, the matching elements are presented in Fig. 4.

Fig. 3. Example of equivalent path for the keyword brother

Fig. 4. Matching elements for each keyword in the query K

4   Aggregating Matching Elements

The matching process produces for each keyword k_i in a query Q a set $S_i = \{s_{i1}, s_{i2}, ..., s_{in}\}$ of matching elements. From the sets of matching elements corresponding to each keyword, all the possible combinations are built by selecting one element from each set S_i. A combination C is a set $C = \{s_1, s_2, s_3..s_n\}$ where each s_i in the set S_i is a matching element for the keyword k_i. Our challenge is to aggregate the elements in the combination and produce the minimal connected subgraph containing the matching elements of the considered combinations. Each subgraph is a possible answer to the initial query. To derive all the possible combinations we compute the cartesian product of the different sets of matching elements. Let us consider the keyword query Q = {brother, Dakota, Washington, Hollywood}; the set of matching elements corresponding to each keyword is given in Fig. 4.

In Fig. 5, we can see all the possible combinations, obtained by performing the cartesian product of the sets of matching elements shown in Fig. 4. For each combination, a connected subgraph will be extracted.

Our problem can be stated as a Steiner tree problem [6]: given a graph G = (V, E), a subset $T \subseteq V$ of vertices called terminals, and a weighted function $d : E \to \mathbb{R}$ on the edges, the goal is to find a subgraph S of minimal weight in G containing all the terminals. Other nodes than the terminals can be added

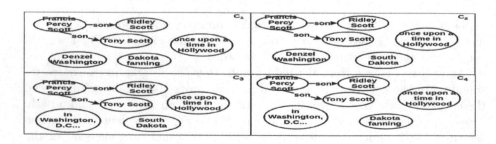

Fig. 5. Combinations of matching elements

to S; they are called Steiner nodes. S should be a tree, which means that from every node s and t $\in$ V there should exist exactly one path between s and t. The Steiner tree problem is NP-hard, and there are many research works on finding approximate solutions to this problem. The quality of the approximation algorithm is measured by calculating the ratio between the weight of the resulting tree and the optimal Steiner tree.

If we compare the Steiner tree problem with the aggregation of matching elements in our context, the set of terminals represents a combinations of matching elements. But in our case, the terminals can be either nodes, or edges, or subgraphs. We assume that all the nodes have a weight equal to one.

We have adapted the distance network heuristic (DNH) [9] with some modifications to solve our problem. The distance network heuristic has an approximation ratio equal to $2 - \frac{2}{p}$ where p is the total number of the terminal nodes. The distance network heuristic comprises the following steps:

1. Compute the complete distance graph DG = (T, E, d) induced by T.
2. Compute the minimum spanning tree T for the distance graph DG.
3. Construct a subgraph G' of G by replacing each edge in T by a corresponding minimum cost path in G; if several paths are found, one of them is randomly selected.
4. Compute a minimum spanning tree T' for the subgraph G'.
5. Delete from T' all non-terminals of degree 1.
6. The resulting tree T'' is the solution.

As the DNH requires some modifications to fit our needs, we first translate all the matching elements (nodes, edges and subgraphs) into terminal nodes.

A matching element m_i is translated into a node as follows:

- If m_i is a node then the resulting node is m_i
- If m_i is a edge then the edge and the 2 connected nodes are replaced by a single node m_i
- If m_i is a subgraph then the subgraph is replaced by one node.

Another adaptation of DNH is related to the selection of the most relevant path. In our approach, we have introduced the notion of centrality degree weight instead of selecting an arbitrary path if there are several paths between a pair of terminals. The centrality degree weight is defined for a node as the number of both incoming and outgoing edges. The centrality degree of a path is calculated as follows. Let p = $\{(v_1, e_1, v_2)(v_2, e_2, v_3)....(v_{n-1}, e_{n-1}, v_n)\}$ be the path connecting v_1 to v_n; the centrality degree weight of p $CDW(p) = A(p)$, where $A(p) = \sum_{i=1}^{n} \frac{deg(v_i)}{n}$ is the average degree of the nodes in the path. We can also limit the computation of $A(p)$ to the top k nodes having the highest centrality.

The distance graph DG is computed from G by using the shortest path between all the matching elements. Let me_i and me_j be two matching elements, then w(me_i, me_j) is the number of edges in the shortest path connecting me_i and me_j. We compute the MST of DG by using the modified Kruskal's algorithm presented in Algorithm 2.

Algorithm 2. Modified Kruskal's algorithm

```
1: A = φ
2: procedure KRUSKAL(G)
3:     Make − set(V)
4:     for each v ∈ G.V do
5:         Make-set(v)
6:     end for
7:     for each (u, v) in G.E ordered by weight(u, v) and CDW, increasing do
8:         if FIND − SET(u) ≠ FIND − SET(v) then
9:             A = A ∪ {(u, v)}
10:            UNION(u, v)
11:        end if
12:    end for
13:    return A
14: end procedure
```

The Kruskal's algorithm consists in creating a forest F (a set of trees) where each vertex in the graph is a separate tree, and creating a set S containing all the edges in the graph. The algorithm finds an edge with minimal weight connecting any pair of trees in the forest and without forming a cycle. If several edges match these criteria, then one is chosen arbitrarily. The edge is added to the spanning tree, and this step is repeated until there are |V|-1 edges in the spanning tree (where V is the number of vertices). We have adapted Kruskal's algorithm by modifying the weights of the edges and taking into account the centrality degree weight. If $w(i, j)$ in DG represents the shortest path connecting i with j, CDW is calculated for each path. The edges are first sorted according to w and then if we have two or more edges having the same w, we sort them according to CDW.

Let us take the combination C_3 as an example. We first translate the matching elements into nodes, and create the distance graph shown in Fig. 6a.

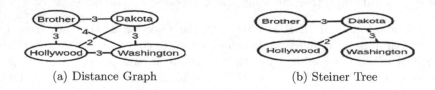

(a) Distance Graph   (b) Steiner Tree

Fig. 6. From distance graph to Steiner tree

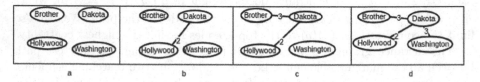

a   b   c   d

Fig. 7. Steiner tree using Kruskal's algorithm

Figure 6b shows the Steiner tree constructed using the adapted Kruskal's algorithm. The steps are shown in Fig. 7. After selecting the first edge which connects "Hollywood" with "Dakota" in Fig. 7a, we select the edge with the minimal weight. But as we can see from Fig. 6a, there are four edges with the same weight equals to three. We then compute the CDW for these edges; the results are shown in the table below (Table 1). We select the edge with the highest CDW as shown in Fig. 7c and then select the edge connecting "Washington" and "Dakota" as presented in Fig. 7d. If the number of edges equals |V|-1, then we can deduce that this is the Steiner tree for the distance graph (Fig. 6a). After replacing the nodes and the edges in ST by the matching elements and the paths in G then we obtain the subgraph of Fig. 8.

Table 1. CDW for edges in distance graph

Path	A(p)	Atop2(p)	CDW
Brother-Hollywood	3.75	5	3.95
Brother-Dakota	4.25	5	4.45
Dakota-washington	4	5	4.2
Washington-Hollywood	3.5	5	3.7

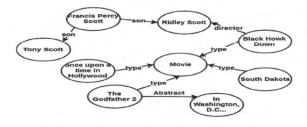

Fig. 8. Solution for the keyword query

5   Ranking the Results

The elicitation of all the combinations of graph elements and the aggregation of graph elements for each combination lead to several subgraphs, each one being a possible answer for the query. One problem is to rank these answers, and to determine if there are better results than others. In our approach, we have ranked the results according to the matching process. We calculate the ranking score as follows:

$$\text{Score} = 1 - \frac{[w_a * A + (1 - w_a) * L]}{N}$$

where A is the number of approximate matching elements, L is the number of linking elements, N is the total number of nodes and edges in the subgraph and

w_a is the weight for A. Intuitively, the above score expresses that the less linking elements in a subgraph, the better the solution. It also expresses that the more exact matching elements in a subgraph, the better the solution.

6   Experimental Evaluation

Our approach is implemented in Java, we have used the Jena API for the manipulation of RDF data. For indexing and searching the keyword query, we have used the Lucene API. The Jung API is used for graph manipulation and visualization.

In the rest of this section, we describe our experiments to validate the performances of our approach. Our goal is to study the impact of using WordNet as an external knowledge source to fill the gap between the keywords and the dataset terminologies, as well as to evaluate the ranking model with various keyword queries. All the experiments have been performed on Intel Core i7 with 32 GB RAM. We have used two datasets: AIFB and DBpedia. AIFB contains 8281 entities and 29 233 triples. The subset of DBpedia we have used is related to movies, their title, stares, director, released data and other properties. This dataset contains 30 793 triples. The size of the keyword queries was between 3 and 7 keywords. The total number of queries was 30 queries (15 for each dataset).

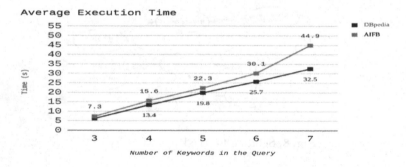

Fig. 9. Average execution time according to the size of the query

Figure 9 shows the execution time with respect to the size of the query. The graph shows that the execution time increases when the number of keywords increases for both datasets. We can also see that the execution time for AIFB is greater than the execution time of DBpedia because the size of data in AIFB is greater than the size of the dataset extracted from DBpedia.

We have tested and compared our keyword search approach both with and without the use of the adapted Kruskal's algorithm to solve the Steiner tree problem. We have compared our approach (referred to as ST) to an approach consisting in picking a random matching element and computing the shortest path between this element an all the other ones in a combination. We refer to this approach as the basic approach. As we can observe from Table 2, the number of results decreases when the adapted Kruskal's algorithm is used during aggregation process, because using this method decreases the number of inaccurate results as shown in Table 3.

Table 2. Number of results for each keyword query (DBpedia)

Query	Q1	Q2	Q3	Q4	Q5	Q6	Q7	Q8	Q9	Q10
Number of results (ST approach)	5	17	7	5	32	48	4	10	2	9
Number of results (basic approach)	7	32	9	5	53	62	6	17	3	11

To check the effectiveness of our approach we have used 10 queries from Table 2 and asked five users to check the top-k results for each query and give the number of relevant results. We have computed the Top-K precision as follows:
$Top - kPrecision = \frac{NumberOfRelevantResults}{K}$.

All the results were above 0.92 as shown in Table 3; this means that the results were accurate according to the users.

Table 3. Top-K precision

Data	AIFB	
K	5	10
Top-K ST approach	0.98	0.94
Top-K basic approach	0.89	0.87

7   Related Words

Keyword search and the translation of a keyword query into a formal query have been the topic of several research works. The early research works were on keyword query over relational databases [5], XML data [2] and then RDF data [4,10,15,17].

The SPARK approach [17] consists in finding the corresponding ontology for each term in the keyword query and try to map and find a relation between the ontology and the keyword query; other approaches use external knowledge or resources such as in the Q2semantic approach [15], where Wikipedia is used to extract related keywords; for each keyword in the dataset, a document is created containing features that are matched to the keyword query. The approach described in [18] also uses some external knowledge; it uses the supporting entity pairs in order to paraphrase dictionary records. Each record represents the semantic equivalence between an entity in the phrase and the dataset; but the supporting entity pairs are specific to Wikipedia and the New York Times [13].

In order to aggregate the matching elements in a dataset, SPARK [17] uses an ontology to discover the relations between the keywords and the dataset and uses a minimal spanning tree algorithm to create a possible query graph. The approaches described in [1,4,10,11,15] transform the data graph into a summarized graph; some of them start from the leaf nodes containing the keyword query and perform a traversal until all the paths converge to the same node; the other works use a summarized graph and try to extract a SPARQL query by finding relationships between the nodes. In [3], keywords are classified into two sets: the first one contains the vertices and the second one contains the edges; then the final possible solutions are computed. In [16] and [7], the keywords are translated into SPARQL by using schema-related declarations in the dataset.

In all these works, the keywords in the query are matched with the nodes of the considered graph, unlike our approach which considers semantic relations and searches for matching elements in both the nodes and the edges in order to build the final result. For the aggregation of matching elements, some approaches have formulated the problem as a Steiner tree problem, like our approach. Our approach differs in the selection of the edges, which is done using the concept of node centrality.

8 Conclusion and Future Works

In this paper we have provided an approach for keyword search over RDF datasets. We have focused on two problems, the first one is bridging the terminological gap between the keyword query and the terms in the dataset by using an online linguistic dictionary, the second problem is aggregating the matching elements to find the best way to connect them; we have adapted the Kruskal's algorithm after stating our problem as a Steiner tree problem. We have conducted some experiments that shows a good quality and the efficiency of our approach.

In future works, we will study the scalability issues and enable efficient keyword search for massive datasets.

References

1. Ayvaz, S., Aydar, M.: Using RDF summary graph for keyword-based semantic searches. arXiv preprint arXiv:1707.03602 (2017)
2. Guo, L., Shao, F., Botev, C., Shanmugasundaram, J.: XRANK: ranked keyword search over xml documents. In: Proceedings of the 2003 ACM SIGMOD International Conference on Management of Data, pp. 16–27. ACM (2003)
3. Han, S., Zou, L., Yu, J.X., Zhao, D.: Keyword search on RDF graphs-a query graph assembly approach. In: Proceedings of the 2017 ACM on Conference on Information and Knowledge Management, pp. 227–236. ACM (2017)
4. He, H., Wang, H., Yang, J., Yu, P.S.: BLINKS: ranked keyword searches on graphs. In: Proceedings of the 2007 ACM SIGMOD International Conference on Management of Data, pp. 305–316. ACM (2007)

5. Hristidis, V., Papakonstantinou, Y.: DISCOVER: keyword search in relational databases. In: VLDB 2002: Proceedings of the 28th International Conference on Very Large Databases, pp. 670–681. Elsevier (2002)
6. Hwang, F.K., Richards, D.S.: Steiner tree problems. Networks **22**(1), 55–89 (1992)
7. Izquierdo, Y.T., García, G.M., Menendez, E.S., Casanova, M.A., Dartayre, F., Levy, C.H.: *QUIOW*: a keyword-based query processing tool for RDF datasets and relational databases. In: Hartmann, S., Ma, H., Hameurlain, A., Pernul, G., Wagner, R.R. (eds.) DEXA 2018. LNCS, vol. 11030, pp. 259–269. Springer, Cham (2018). https://doi.org/10.1007/978-3-319-98812-2_22
8. Klyne, G., Carroll, J.J.: Resource description framework (RDF): concepts and abstract syntax. W3C Recommendation (2004). http://www.w3.org/TR/2004/REC-rdf-concepts-20040210/
9. Kou, L., Markowsky, G., Berman, L.: A fast algorithm for steiner trees. Acta Informatica **15**(2), 141–145 (1981)
10. Le, W., Li, F., Kementsietsidis, A., Duan, S.: Scalable keyword search on large RDF data. IEEE Trans. Knowl. Data Eng. **26**(11), 2774–2788 (2014)
11. Lin, X.q., Ma, Z.M., Yan, L.: RDF keyword search using a type-based summary. J. Inf. Sci. Eng. **34**(2), 489–504 (2018)
12. Rihany, M., Kedad, Z., Lopes, S.: Keyword search over RDF graphs using WordNet. In: Big Data and Cyber-Security Intelligence (2018)
13. Nakashole, N., Weikum, G., Suchanek, F.: PATTY: a taxonomy of relational patterns with semantic types. In: Proceedings of the 2012 Joint Conference on Empirical Methods in Natural Language Processing and Computational Natural Language Learning, pp. 1135–1145. Association for Computational Linguistics (2012)
14. Ouksili, H., Kedad, Z., Lopes, S., Nugier, S.: Using patterns for keyword search in RDF graphs. In: EDBT/ICDT Workshops (2017)
15. Wang, H., Zhang, K., Liu, Q., Tran, T., Yu, Y.: Q2Semantic: a lightweight keyword interface to semantic search. In: Bechhofer, S., Hauswirth, M., Hoffmann, J., Koubarakis, M. (eds.) ESWC 2008. LNCS, vol. 5021, pp. 584–598. Springer, Heidelberg (2008). https://doi.org/10.1007/978-3-540-68234-9_43
16. Wen, Y., Jin, Y., Yuan, X.: KAT: keywords-to-SPARQL translation over RDF graphs. In: Pei, J., Manolopoulos, Y., Sadiq, S., Li, J. (eds.) DASFAA 2018. LNCS, vol. 10827, pp. 802–810. Springer, Cham (2018). https://doi.org/10.1007/978-3-319-91452-7_51
17. Zhou, Q., Wang, C., Xiong, M., Wang, H., Yu, Y.: SPARK: adapting keyword query to semantic search. In: Aberer, K., et al. (eds.) ASWC/ISWC -2007. LNCS, vol. 4825, pp. 694–707. Springer, Heidelberg (2007). https://doi.org/10.1007/978-3-540-76298-0_50
18. Zou, L., Huang, R., Wang, H., Yu, J.X., He, W., Zhao, D.: Natural language question answering over RDF: a graph data driven approach. In: Proceedings of the 2014 ACM SIGMOD International Conference on Management of Data, pp. 313–324. ACM (2014)

Intent Based Association Modeling
for E-commerce

Sailesh Kumar Sathish[(✉)] and Anish Patankar

Samsung R&D Institute Bangalore – India, Bengaluru, India
{sailesh.sk,anish.p}@samsung.com

Abstract. Online e-commerce sites track user behavior through use of in-house analytics or by integrating with third party platforms such as Google Analytics. Understanding user behavioral data assists with strategies for user retention, buy-in loyalty and optimizing objective completions. One of the more difficult problems though is understanding user intent that can be dynamic or built over time. Knowing user intent is key to enabling user conversions - the term used to denote completion of a particular goal. Current industry approaches for intent inference have an inherent disadvantage of having the need for embedded tracking code per site-sections as well as the inability to track user's intent over longer periods. In this paper, we present our work on mining dynamic as well as evolving user's intents, using a latent multi-topic estimation approach over user's web browsing activity. Further, based on the intent patterns, we look at generating association rules that model purchasing behavior. Our studies show that users typically go through multiple states of intent behavior, dependent on key features of products under consideration. We test the behavioral model by coupling it with Google Analytics platform to augment a re-marketing campaign, analyzing purchasing behavior changes. We prove statistically that user conversions are possible, provided purchase category dependent associations are effectively used.

Keywords: Intent mining · Latent models · Association modeling

1 Introduction

The market for global online shopping, as of 2018, is estimated at US$ 1.9 trillion [1]. A foothold into this ever-competitive market requires a thorough understanding of one's user base. E-commerce sites are now capable of monitoring every single of their user's online activities and are coming up with ingenious ways to make use of such information in order to translate them into valued transactions. The term used is "user conversions", where conversion means the execution of a series of steps that result in a goal completion. Using third party analytics services such as Google Analytics (GA) [2] and Microsoft Azure Analytics [3], services can now go beyond their own domains to track user activity through cross-domain linking. Such services can track and identify user events, building higher-level inferences, and simultaneously track the number of goal completions resulting from individual marketing campaigns. In the case of Google Analytics, by linking with Google Ads, entities can track full customer cycle

© Springer Nature Switzerland AG 2019
E. Métais et al. (Eds.): NLDB 2019, LNCS 11608, pp. 144–156, 2019.
https://doi.org/10.1007/978-3-030-23281-8_12

through ad impressions and enable better remarketing campaigns. Such cycles translate to short-term intent of users, which being generally dynamic, is difficult to predict based on past behavior. The difference between intent and long term profiling is that intent behavior does not conform to consistent preferences that are categorical and contextual as captured by a profile. From an industry perspective, services such as GA have made a big impact on the way user behavior and conversions are affected. The disadvantage is that such services require embedded scripts that address each page sections and events requiring tracking in order to do backend analysis. External marketing channels require proprietary campaign tagging with explicit tracking that makes scaling over other sites difficult. Furthermore, such systems only capture short-term intents, relying more on syntactic matches than a semantic behavioral understanding behind user intents.

The paper describes our intent mining framework that captures short and long-term user intent. The aim is to aid third party services (subject to user permission) with remarketing campaigns. By using web browser as application platform, we are able to model an original intent that may have evolved over multiple sessions and across content categories. Our goal is to recognize, understand and model accurately the evolution of user intent over time and study behavioral aspects on the different content associations that user makes based on their intent.

2 Prior Art

The study and analysis of online consumer behavior is a well-researched subject that can be traced back to the onset of e-commerce on web. The fields of research in this area mainly focus on profile-based behavior, prediction models based on user purchase expectations, social media based behavior correlations and works that model user intent itself. Work done by Kumar et al. [4] looks at how profile based behaviors including demography affects purchase behavior. Other works look at predicting purchase behavior. Several diverse approaches have been proposed such as predictions based on statistical purchase probability based on pre-modeled scales [5], visual feature based matching and recommendation [6], modeling customer attitudes towards products & companies [7], machine learning based approaches on customer segmentation with predictive analysis [8], looking at repeat behaviors to predict future purchases [9, 10] and using web search data [11]. Work done by Ioanas et al. [12] looks at social media behavior and its correlation with online purchase behavior. A study on user behavior over Pinterest [13] looks at predicting intent spread between applications through use of a cross application model. Our work differs in the way intents are inferred and modeled. We account for reinforcement and decay of intent and associate intents with a step transition model based on pre-determined threshold values. Our study goes a step further by generating unsupervised association models for user behavior based on intents inferred.

3 Intent Capture and Analysis Framework

We have implemented the intent mining software framework as a hybrid client-server model, with the client part provided as an extension to Samsung Internet Browser v6.2. To address the issue of cross-content analysis, we analyze all web content that are article qualified (over 200 words) that contribute to user intent. The framework creates intent objects for the user based on the topics user consumes through the web browser. An intent object expires when that intent results in a successful purchase.

3.1 Intent States

We look at only those intents that may result in purchase of a particular product or service. All other categories, which are unrelated to any commercial transactions are ignored as they are outside the scope of our study. We observed that users go through at least three phases between recognition of a first intent and a consequent purchase decision. When user browse a new product related topic, a "weak intent" is generated for that topic. The product or category is referenced through our ontology and intent recognition can be seen as the predominant topic identification for that page. We use a refined latent multi-topic classification that follows from our earlier work [14], which can perform multi-categorical inference within a single page.

The second phase of intent transition is "in-progress" that represents a stronger conviction on the intent state. In this state, we observed that typically, the user's browsing habits indicated a more narrowed focus on a specific range of products as opposed to a topic related to that product. Example, when user starts browsing on "home theatre audio reviews", the system generates a "weak intent" for "Technology => home audio". The "weak intent" transitions into an "in-progress" state once user reinforces the intent through consistent browsing on that topic. It is also typical that user may have narrowed on particular brands at this stage.

The final intent phase is "strong" phase. This is the typical phase where users may make a purchase (conversion) decision. The modeling function takes into account reinforcement and decay of intent with respect to other active intents within a user dependent activity period. In calculating reinforcement and decay, we do not take into account the correlation or independence between the products and categories. Instead, we factor in topic dependence. This is because relational intent is hard to model, as product relations, even if modeled within the ontology, often results in erroneous inference.

Figure 1 shows the normalized intent progression time against intent level values for two product's price range: less than and greater than 100 USD. We normalized progression time, as users have different time range for intent realization but on average, apart from outliers, tend to follow a similar transition pattern. We observed that the intent state times are highly dependent on price of the product under consideration. We observed variance between product classes but for brevity, we factor for price, which is the more significant influencer.

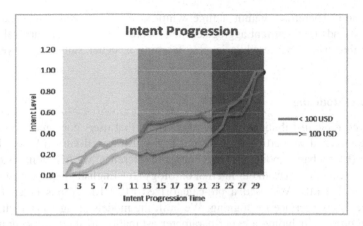

Fig. 1. User intent state progression. The three regions denote states of "weak", "in-progress" and "strong" respectively. The two curves indicate, progression for low value (<100 USD) and high value priced (>=100 USD) products respectively

3.2 Modeling Intent

Figure 1 shows the progression of user intent states. The three regions denote states of "weak", "in-progress" and "strong" respectively. The two curves indicate progression for low value (<100 USD) and high value priced (>=100 USD) products respectively. We chose to make our analysis over these two price points as we were able to cluster behavior patterns to within these brackets. Based on captured data points, we present an empirical model for user intent. We use co-occurrence of categories as well as specific product models within each category. Some categories have well defined product names while other product types have generic names. We observed that having specific product names resulted in stronger intent.

To normalize across such categories, we use a factor called *Product Specificity* for a category (C_{PS}). The categories having specific product names get lower weightage while other categories get higher weightage. In addition, extracting product category and product names is important for estimating price. It so happens that products with less monetary value move from weak to strong intent faster than products with high monetary value. We factor in the inverse relation with monetary value using parameter $Z = k/V$, where, V is price of product and k is a constant empirically taken as 100. Occurrences of category and products in the user topic vector are referred as C_C and P_C respectively. From e-commerce data [4], we understand that certain categories have more customers as compared to others. This variation across categories is captured as C_W. The intent capture model is given in (1).

$$I = f(C_W, C_C, P_C, Z) = \varphi(C_W(C_C C_{PS} + P_C)Z) \qquad (1)$$

Where, φ is a constant, and I is the intent score. It should be noted that this intent score is for a category for a given time window and needs to be adapted as new data becomes available. We applied temporal learning to learn the intent score for category over a period. Therefore, within a time window if intent score for a category is improved, it adds up to current aggregate intent score. For high monetary value items, linear learning works better while inverse learning is better suited for lower valued items.

3.3   Topic Modeling

We analyzed topic distributions within web content that users consumed to determine intent. We modeled web article content into topics using well-known Latent Dirichlet Allocation (LDA) based modeling [15]. For topic inference, we first built a supervised model for 36 categories (from our internal ontology) for English language resulting in a model size of 2 MB. We used a pre-categorized 6K URL corpus taken from our internal web proxy service for training. We built the models using a proprietary batch integration process including a hyper-parameter estimation method for accurate model convergence. This derives from our earlier work on building semantic indices [18], described briefly in following sub-sections.

Determination of Hyper-parameters
For LDA model, hyper-parameter α indicates the distribution of topics over a document and β indicates the distribution of words over a given topic. We need optimal α, and β values that give best converged model for a given set of topics. For this reason, we use an internally developed metric called Averaged Normalized Mode (ANM), given by Eq. 2 as a score to compare the purity of the mixture produced by LDA clustering.

$$ANM = \frac{1}{n}\sum \frac{\max\{T_i\}}{C_i} \tag{2}$$

Where C_i denotes the ith cluster out of n and T_i represents the ith topic. The ANM score is computed for $\forall\ \alpha \in [0.1, 3.0]$ and $\forall\ \beta \in [0.1, 1.0]$, with incremental steps of 0.1 each. In repeated runs with fixed α and β, if the ANM score of 1.0 repeats consistently, then the model is considered stable.

Incremental LDA
Incremental LDA (iLDA) performs supervised inference against a set of pre-built LDA batch models. Inference is performed thorough an incremental Gibbs sampler by applying sampling process to a pre-set of the sampled distribution and sampling for particular topic to which word i belongs conditioning on the previous word model $(i - n)$ (as shown in (3)).

$$P(Z_i|Z_{i/j}w_i)\alpha \frac{n_{Zi,i|j}^{(Wj)} + \beta n_{Zi,i|j}^{(dj)} + \alpha}{n_{Zi,i|j}^{(i)} + W\beta n_{i,i|j}^{(dj)} + K\alpha} \tag{3}$$

Here, K is the number of topics, W is vocabulary size, Z_i represents i^{th} topic assignment; $n_{Z_{i,i|j}}^{(W_i)}$ represents word – topic Z_j assignment, and $n_{Z_{i,i|j}}^{(d_j)}$ is document-to-topic Z_j assignment. After incremental inference, we perform cluster process by aligning all vectors together that fall within a set threshold.

3.4 Product Name Extraction

We further developed a product name extractor to detect product name from web content. The product name extractor is based on Stanford NLP's Named Entity Recognition (NER) extractor software [16]. A rule-based module augments the NER to capture n-gram tokens as probable candidate for product names. The rules were written specifically for every base category. The title and URL tokens extracted helped in boosting the confidence level in the product names from the web page content.

3.5 User Intent Structure

A data structure, called User Intent Structure (UIS) hosts the identified user intent. The client generates the UIS and depending on the application, the UIS may be used locally or sent to an external service. When there is a change in intent data, the client sends an updated UIS to server for use by validated services. The UIS covers intent, topic vector, state values, categories and optional fields for URL values.

3.6 Observations on User Intent

The cut-off threshold values for the intent states were determined based on the observed drop-off probability values and the amount of time users spent within these states. We observed that users spent least time within the "strong" intent state, often marked by increased activity and lowest time intervals between topic reads. When in weak intent state, user concentration on topic was lower and on average, users created more intent-objects during this time. When user intent was in "in-progress" state, the number of weak intents created (for other intents than the current one) were lower. The number of new intent creation was another way of observing whether other user intents were in weak or in-progress state. The observations made here were for majority users and there were few outlier cases where users did not fit within this model. As Fig. 1 shows, purchase intent for higher priced products follows a lower level of intent between weak and in-progress states. We also observed that activities on search and reviews pick up at a much higher rate during the strong intent phase. Such observations were not applicable across categories. One example was "Fashion Apparels" where the transition states were not applicable for majority users.

4 Association Mining

Our association-mining engine looked at intent-based purchase paths that led to actual purchases on partner web sites. The association-mining engine is part of the intent mining and reinforcement framework.

4.1 Unsupervised Association Rule Mining

The association rule-mining engine on server generated unsupervised rules pertaining to user journey on specific topics and associated topic features similar to "if-then" rules. A typical rule may encompass one of the many user journeys possible from the beginning of an identified intent to a goal completion (such as a purchase). Association rules or affinity rules look at identifying proper antecedents (X) and their corresponding consequents (Y). The antecedents and their corresponding consequents form item sets where the items within the sets are disjoint. We derived association rules from user behavioral patterns and thus, the probabilistic nature of behaviors also extend to association rules. The association rules were generated for each topic/product items identified as an intent (indicated via UIS).

Clients would send UIS updates on identified intent that contain fields for "recognized association journey" for that intent. Clients (in default mode) sent updates through UIS once a week to server and when connected over Wi-Fi. The association "journey" captured on the client side will identify the next "similar" topic (web page) user had consumed, either as search and read, direct read and/or bookmarked by the user. The capture and update to server continued until the intent (UIS) closed at user end.

The server combined all the UIS received to date (within a processing window) and computed association rules per intent. The first stage of the process was to group all similar intents together. This was done based on topics contained with the UIS that were used for grouping. Note that even though the topics were same, the underlying vectors that defined each topic could be different. At this stage though, the differences in topics were not considered. Topic based identification of frequent sets were done in second stage. The set of UIS with their unique IDs were stored within a table in the server database for a particular intent (topic). All subsequent updates received from clients for the same UIS were stored against the corresponding entry within the database. If a new UIS was received for an existing topic, a new entry for that UIS was made within the association content table for that topic. The further steps for identifying a set of unsupervised association rules per intent is described in subsection B. At present, we are only focusing on positive rule mining ($X \rightarrow Y$) and not considering negative rule mining. Considering that total product catalog can be huge, all types of negative rule associations ($\neg X \rightarrow Y$, $X \rightarrow \neg Y, \neg X \rightarrow \neg Y$) are computationally expensive and cumbersome to formulate. However, since, we have only 36 topics negative rule mining of type $\neg X \rightarrow Y$ is still possible to some extent with comparatively less computational complexity. That is absence of topic from browsing history can be a cue for intent. E.g., Person not interested in Science might be interested in fashion accessories.

4.2   Identifying Frequent Sets

After grouping of intent sets were made in first stage, the following stage looked to identify frequent item sets within the intent and its associations. The second stage for identifying frequent items consisted of two sub stages. The first sub-stage performed a time-windowed alignment of page jumps by a single user based on the unique UIS. What this did was, within a time window (2 days), all the UIS association updates received were considered as a single journey path so as to build a simple and reasonable corpus for rule mining. Any sites re-visited by the user within this window was ignored and the first visit time (or bookmark) of that URL (within the window) was used in building the disjoint path set. After this process, we ended up with a corpus for a topic with sets ranging from single antecedent – single consequent to single ante-cedent – multi consequents.

For the second sub-stage, we used the Apriori algorithm [17] for reducing set complexity for generating frequent item sets. If we use all antecedent-consequents – combinations of single items, paired items, and triples, it would require high compu-tational resources that would grow exponentially with each combination addressed. With the Apriori algorithm, we initially generated frequent item sets with just one item. The frequency of occurrence formed the support for that set. The one-item lists where the support was below our set threshold were dropped and only those above the threshold were chosen. The next step was to identify the most frequent sets containing at least two consequents by using the frequent one-item sets that were identified in the first iteration. We used the same one item set for identifying the next two-item set that contained the same antecedent and consequent sets and so on. Through this iterative process, we identified all k-item sets based on the frequent (k − 1) item sets identified through the preceding step with each step calculation requiring only a single database query. The second stage of association mining still produced a large set of association rules that did not indicate the level of bond between the antecedents and the conse-quents. In order to filter out the weak associations and to end up with a convincing rule set, we used two additional metrics: confidence score and lift ratio, described in sub-section C.

4.3   Determining Association Strength

Confidence score measures the degree of uncertainty amongst the identified association set. Confidence score compares the item co-occurrences (transactions with both ante-cedent and consequent sets) to the total antecedent occurrences. Confidence is given by (4).

$$C_f = \frac{P(antecedent\ AND\ consequent)}{P(antecedent)} \tag{4}$$

Where C_f, denotes the conditional probability that a randomly selected rule cor-responding to an antecedent will contain all consequent transactions.

Once confidence score C_f, was calculated for an association rule, we further filtered the rule based on lift ratio. Confidence score is a conditional probability score with the consequent dependent on an antecedent occurrence. Lift gives us a benchmark ratio that compares an independent probability score against the conditional probability. Lift metric thus indicates how valuable the conditional clause is as compared to the case where the antecedent set and the consequent sets are independent of each other. Lift is interpreted as a benchmark score (5).

$$\text{Lift ratio, } L_r = \frac{C_f}{P(consequent)} \tag{5}$$

A value for $L_r > 1$ indicates that the level of antecedent → consequent association is higher than what would be expected if the two sets were independent. It gives the level of correlation between the two sets and is thus a useful metric for determining strength of associations. Table 1 shows a sample of the mined association rules. Note that some of the association rules gets repeated. For these rules, even though the topic association was the same, the underlying sub-topics or their corresponding token vectors would be different indicating the various user preferences on topic variations among users.

Table 1. Association rules example

Rule#	Antecedent (a)	Consequent (c)	Antecedent Vector (Va)	Consequent Vector (Vc)	Lift ratio
1	Photography	Camera, Smartphone	{mirrorless, speedlite, AI, BSI, image processing, autofocus, EV, AF, exposure, buffer, JPEG, log gamma, SD, UHS-C...}	{Google, clips, shutter, AI smarts, gesture, goofy, galaxy...}, {aperture, variable, dual lens, jack, snapdragon, android, exynos, IP68, emoji, iPhoneX...}	1.38
2	Photography	Camera, Smartphone	{macro, large aperture, landscape, focal points, horizon, sport metering, exposure lock...}	{dual pixel, cmos, eye detection, autofocus, dual lens, touch and drag, itu, cr3, digic, 4k...}, {OIS, lowlight, android, portrait, dualcamera, wireless, pixel, poled, amoled...}	1.265553

5 Evaluating Intent Behavior

We used the browser client extension running the intent mining engine for both user data collection and for validating behavioral activities. Users could invoke a mock screen which showed a dynamic selection UI with a purchase button next to the detected intent. Users, when they were ready to make a purchase, could invoke this UI and click the purchase icon. A modified version of this client was used for our evaluation phase. The association engine, user account management and recommender (rule based) is server based running on an AWS (Amazon Web Service) instance.

The trial and evaluation period was spread over 3.5 months. We collected 873 intents from the 53 users over approximately 11K URLs. Using this set, we built 903 association rules. The second activity was to validate our hypothesis that intent behavior could be affected through effective content recommendations. 88 users participated in this trial over a 45-day period. We benchmarked user click through rate (CTR) against Google AdWords in order to determine whether provisioning additional intent information can help with user conversions. We omit implementation details due to space constraints.

AdWords can track user activity through Google ID across domain and service provided ID linking if coupled with Google Analytics backend. Google uses a combination of search keywords, extracted keys, keyword bids, quality score (for ads) and cost-per-click (CPC) factors to determine what adverts to serve to user. Our end goal was to make this more relevant by feeding contextually, intent keywords at the right context through Google Ads or other services such as Yahoo Bing Network and Samsung's own Ad platform. The program scope was limited to evaluating effectiveness of intent based recommendations through a simulated UI in addition to Google Ads. We base our observations on relative CTR with respect to Google Ads. Our comparative analysis only looks at the time when we do an intent based recommendation to user. This happened for under 5% of total user session time. Ad word interactions (as a non-recommended set) during the overlap time per user was collected over GA data by using Google Ad id of user while intent recommendations were collected as a custom dimension over GA. Figure 2 shows the CTR differences between the two sets. A second collection looked at a specific category of purchase. We figured that decision time taken for a purchase differed between categories. Seasonal factors can also affect decision times. We found that mobile category had an overlap of 28% on the topics searched, and so we analyzed specifically the topic of mobile related purchases. Twenty-seven users from the content recommended set ended up purchasing within the mobile category while corresponding number of buyers from the

Fig. 2. (a) Purchase decision time plot: Set 1 represents recommended content group. Set 2 represents non-recommended group. (b) Purchase decision time plot for mobile category: Set 1 represents recommended content group. Set 2 represents non-recommended group.

non-recommended set was 22. We used the two different data sets to validate two hypotheses.

1. There is a significant difference on purchase decision time between the intent determined content recommendations set vs. the non-recommended ad-word set over all shopping categories.
2. There is a significant difference on purchase decision time between the intent determined content recommendations set vs. the non-recommended set on any particular shopping category (in this case mobile shopping category).

We used the two means t-test for statistically evaluating our hypothesis. The values within the two sets were independent and we assumed a normal distribution of the values. Set 1, in Fig. 2a, shows a sample plot for average CTRs for 44 users belonging to the recommended content group. Set 2 gives the corresponding plots (average CTR) for the users based on AdWords recommendation. Set 1 plots the purchase decision time, recorded as day counts, for the 44 user set who were provided with recommended content once the intent was captured based on the pre-modeled association rules. For Set 1, the mean value for the purchase decision time was 5.15 days with a standard deviation of 4.17. The average decision time for a purchase was 7.16 days with a standard deviation of 5.38 in Set 2. Our t-test evaluation over the two distributions, taking an un-paired two-tail analysis gave a p-value of 0.1002. Given that the p-value exceeds the alpha value of 0.05, we cannot claim statistical significance for our hypothesis 1. One reason for this may be that the data has a wide spread of category specific purchases. In addition, since the concentration of categories is non-uniform, this would affect the average purchase time as well as the deviation.

Figure 2b shows the plot for purchase decision time taken within the mobile shopping category by two different user sets. The mean value for the content recommended group (set 1) was 3.97 days as compared to decision time of 7.87 days for users within set 2. Set 1 also showed a smaller standard deviation of 3.44 as against 5.58 for users within set 2. The un-paired two-tailed p-value for our hypothesis 2 was 0.007268 indicating a high statistical significance. The value indicates a confidence > 99% that our hypothesis holds, meaning there is a significant difference between purchase decision times on mobile category between the two sets.

The average time and lower deviation also suggests acceleration in purchase decision times. We believe this validates our assumption that recommending latest content based on intent detection might work to reduce average purchase decision times. We inferred that keeping intent "alive" was key to boosting intent transitions achieved through our content recommendation service. The rate of keeping "alive" i.e. recommendation rate was however user specific (according to their preference), as was revealed through our user questionnaire. We acknowledge that this is still a small sample set and validated for a single category. It is difficult to validate hypothesis 1 given a mixture of purchases across categories. To do a full evaluation, we need to consider multiple category purchase timelines by users.

6   Conclusion

As part of our user insight activity, we have built an intent mining engine, as an experimental extension to our browser (Samsung Internet). The intents were created based on topic inference over web articles browsed by user and maintained within client till the user made a purchase decision based on the intent. Our experimental activity for intent behavioral inference, conducted using 53 volunteers gave interesting insights into each of these states, helping us to derive a user model that brought in a quantitative measure to intent calculations. Based on the behavioral data collection activity, we built a further unsupervised association rule set allowing us to test our hypothesis, using a further 88 volunteer set, as to whether targeted recommendations can affect intent behavior. We found that intent state transitions can be affected if done over specific categories. Our future task will to be increase our test user base through a beta release of the system with more user controls and permissions on mining intent categories. We aim to study correlations between demography and intent to check if intent dependent personalized content can further reinforce and assist in purchase decisions.

References

1. KPMG: Truth about online consumers, 2017 Global Online Consumer Report (2017)
2. Google Analytics. https://analytics.google.com/analytics/web
3. Microsoft Azure Analytics. https://azure.microsoft.com/en-us/product-categories/analytics/
4. Kumar, A.H., John, S.F., Senith, S.: A study on factors influencing consumer buying behavior in cosmetic products. Int. J. Sci. Res. Publ. 4(9), 6 (2014)
5. Day, D., Gan, B., Gendall, P., Esslemont, D.: Predicting purchase behavior. Market. Bull. 2, 18–30 (1991). Article 3
6. Bell, S., Bala, K.: Learning visual similarity for product design with convolutional neural networks. ACM Trans. Graph. (TOG) 34(4) (2015). SIGGRAPH 2015. Article no 98
7. Cesar, A.C.: Impact of Consumer Attitude in Predicting Purchasing Behaviour (2007)
8. Arulkumar, S., Kannaiah, D.: Predicting purchase intention of online consumers using discriminant analysis approach. Eur. J. Bus. Manag. 7(4), 319–324 (2015)
9. Pal, S.: Know your buyer: a predictive approach to understand online buyer's behavior', white paper, Happiest Minds
10. Banerjee, N., Chakraborty, D., Joshi, A., Mittal, S., Rai, A., Ravindran, B.: Towards analyzing micro-blogs for detection and classification of real-time intentions. In: International Conference on Web and Social Media (2012)
11. Rose, D.E., Levinson, D.: Understanding user goals in web search. In: Proceedings of the 13th Conference on World Wide Web (2004)
12. Ioanas, E., Stoica, I.: Social media and its impact on consumers behavior. Int. J. Econ. Pract. Theor. 4(2), 295–303 (2014)
13. Guo, S., Wang, M., Leskovec, J.: The role of social networks in online shopping: information passing, price of trust, and consumer choice. In: Proceedings of the 12th ACM Conference on Electronic Commerce, pp. 157–166 (2011)
14. Sathish, S., Patankar, A., Neema, N.: Semantics-based browsing using latent topic warped indexes. In: International Conference on Semantic Computing, ICSC 2016

15. Blei, D.M., Ng, A.Y., Jordan, M.I.: Latent Dirichlet allocation. J. Mach. Learn. Res. **3**, 993–1022 (2003)
16. Finkel, J.R., Manning, C.D.: Nested named entity recognition. In: Conference on Empirical Methods in Natural Language Processing, vol. 1, pp. 141–150 (2009)
17. Agrawal, R., Srikant, R.: Fast algorithms for mining association rules. In: Proceedings of the 20th VLDB Conference, Chile (1994)
18. Sathish, S., Patankar, A., Priyodit, N.: Enabling multi-topic and cross-language browsing using web-semantics service. In: International Conference on Web Services (ICWS) (2017)

From Web Crawled Text to Project Descriptions: Automatic Summarizing of Social Innovation Projects

Nikola Milošević[1]([✉])(iD), Dimitar Marinov[1](iD), Abdullah Gök[2](iD), and Goran Nenadić[1](iD)

[1] School of Computer Science, University of Manchester, Manchester M13 9PL, UK
nikola.milosevic@manchester.ac.uk,
dimitar.marinov@student.manchester.ac.uk
[2] Hunter Centre For Entrepreneurship, Strathclyde Business School, University of Stratclyde, Glasgow, UK

Abstract. In the past decade, social innovation projects have gained the attention of policy makers, as they address important social issues in an innovative manner. A database of social innovation is an important source of information that can expand collaboration between social innovators, drive policy and serve as an important resource for research. Such a database needs to have projects described and summarized. In this paper, we propose and compare several methods (e.g. SVM-based, recurrent neural network based, ensambled) for describing projects based on the text that is available on project websites. We also address and propose a new metric for automated evaluation of summaries based on topic modelling.

Keywords: Summarization · Evaluation metrics · Text mining · Natural language processing · Social innovation · SVM · Neural networks

1 Introduction

Social innovations are projects or initiatives that address social issues and needs in an innovative manner [3]. In the past decade, social innovation has gained significant attention from policy makers and funding agencies around the worlds, especially in the EU, USA, and Canada. Policy makers and researchers are particularly interested in monitoring social innovation projects, the effects of policies on these projects and the effects of these projects for the society.

In order to enable monitoring of social innovation projects a number of database creation projects were funded over time. In the KNOWMAK project, we aim to integrate and expand on previously collected information by utilizing automation approaches enabled by machine learning and natural language processing techniques.

© Springer Nature Switzerland AG 2019
E. Métais et al. (Eds.): NLDB 2019, LNCS 11608, pp. 157–169, 2019.
https://doi.org/10.1007/978-3-030-23281-8_13

The existing data sources for social innovation are varied in their levels of depth and detail. Therefore, in KNOWMAK we aim to normalize the information, providing the same wealth of information for each reported project. In order to do this, we utilize the data from original data sources, as well as the data from the projects' webpages and social media sites, such as Facebook and Twitter.

In order to provide relevant information to the researchers and policy makers, the projects in the database need to be described. Some of the original data sources have descriptions, but many data sources do not have. Additionally, some of the descriptions in existing data sources may be too long (e.g. over 500 words), or too short (1 sentence) and therefore need to be normalized.

Automated summarization can be used to automate and speed up the process of summarizing texts about a project in the database. Summarization is a well-known task in natural language processing, however solutions in literature do not address the domain specific issues. Project description building using summarization has challenges that may not be present with a usual text summarization task. In this task, it is necessary to generate short, cohesive description that best portrays the project, which may be described over several web pages, contain noisy text (pages or portions of pages with irrelevant text) and align project description to the theme of the database.

In this paper, we compare several methods for creating project descriptions and summaries in the semi-automated system that takes texts about social innovation projects from the web. We develop a method that makes human readable project descriptions from the scraped pages from the project web sources. This paper presents an automated project description method applied in the KNOW-MAK project that aims to create a tool for mapping knowledge creation in the European area. The project focuses on collecting information on publications, patents, EU projects and social innovation projects. As publications, patents and EU projects would have abstracts or short descriptions, this paper aims at the particular case of describing social innovation projects.

2   Background

Automatic summarization is a complex natural language processing task which has been approached from several perspectives. We will review the main approaches.

On the whole, it is challenging to evaluate automatic summarization. Summaries of text will look different depending on who is doing them and which approach is used. However, it has to be ensured that the main points of the text that is analysed have been retained. Over the years, there have been a couple of evaluation metrics proposed. In this section, we will also review the proposed metrics.

2.1 Summarization Approaches

Summarization approaches can be classified into two main categories: (1) extractive and (2) abstractive [11]. Extractive approaches try to find snippets, sentences and paragraphs that are important, while abstractive approaches attempt to paraphrase important information from the original text. The types of summarizers may also depend on how many documents are used as input (single-document or multi-document), on the languages of input and output (monolingual, multilingual or cross-lingual), or purpose factors (informative, indicative, user-oriented, generic or domain specific) [5].

Summarization approaches can be both supervised and unsupervised. Unsupervised methods usually use sentence or phrase scoring algorithms to extract the relevant parts of the original text [6,16]. Most of the extractive summarization approaches model the problem as a classification task, classifying whether certain sentences should be included in the summary or not [19]. These approaches usually use graphs, linguistic scoring or machine learning in order to classify sentences. Standard machine learning classifiers, such as Naive Bayes or Support Vector Machines (SVM) using features such as the frequency of words [1,14,18], as well as neural network-based classifiers [5,11,19] have been proposed. Traditional machine learning classifiers usually use features such as the frequency of phrases, relational ranks, positions of the sentences in the text, or overlapping rate with the text title. Neural network approaches utilize word, sentence and document representations as vectors, pre-trained on large corpora (word, document or sentence embeddings). Then these vectors are imputed into convolutional or recurrent neural networks for classification training.

Abstractive summarization is considered less traditional [21]. Approaches usually include neural network architectures trained on both original texts and human created summaries. Approaches using sequence-to-sequence neural architectures [12], but also attention mechanism have been proposed [17].

2.2 Evaluation Measures for Summarization

A good summary should be a short version of the original text, carrying the majority of relevant content and topics in condensed format. Summarization of a text is a subjective problem for humans and it is hard to define what a good summary would consist of. However, a number of quantitative metrics have been proposed, such as ROUGE or Pyramid.

Recall-Oriented Understudy for Gisting Evaluation (ROUGE) is a commonly used metric in summarization literature [5] that is based on overlapping n-grams in summary and original text. There are several variants of ROUGE, such as ROUGE-N (computing percentage of the overlapping n-grams), ROUGE-L (computing the longest overlapping N-gram), ROUGE-S (computing the overlapping skip-grams in the sentence) [7]. Since ROUGE takes into account only overlapping n-grams, it often favors the summaries that are long, where the summarizer did not sufficiently reduced the size of the original text.

Pyramid is another metric that is based on the assumption that there is no one best summary of the given original text [13]. Pyramid requires a number of human generated summaries for each text as well as human annotations for summarization content units (SCU). For each SCU a weight is assigned based on the number of human generated summaries containing it. Newly created summaries are evaluated based on the overlapping SCUs and their weights. This method is expensive, since it requires a lot of human labour for annotating and generating multiple summaries for evaluated texts [5].

While ROUGE and Pyramid metrics are the most used in current literature, other approaches have been proposed. A Latent Semantic Analysis-based metric was proposed based on the hypothesis that the analysis of semantic elements of the original text and summary will provide a better metric about the portion of important information that is represented in the summary [20]. As ROUGE metrics often do not correlate with human rankings, the evidence was provided that LSA based metric correlates better than ROUGE and cosine similarity metric based on the most significant terms or topics.

Human ranking and scoring is a measure that is often used for evaluation of summarization systems [20]. Human annotations are more expensive than automatic annotations, however, they provide a good metric that accounts for all elements of a good summary definition (main topics, condensed length, readability).

3   Method

3.1   Method Overview

We present a comparison and implementation of four summarization or description generation methods for social innovation. The input to all summarization methods is text crawled from the social innovation project websites, while the expected output is a short and condensed description of the project (summary).

The method consists of data collection, training data set generation, data cleaning, classification and evaluation steps. Figure 1 presents the methodology overview.

3.2   Data Collection and Data Set Generation

The initial set of social innovation projects was collected using existing databases of social innovation, such as MOPACT, Digital Social Innovation, InnovAge, SI-Drive, etc. The data was collected from a compiled list of about 40 data sources. Some of the data sources contained data that can be downloaded in CSV, JSON or XML format, however many data sources contained data accessible only through the website and therefore needed to be crawled. As these data sources contained structured data, with humanly created descriptions of the projects, websites and social media, a set of crawlers were created that were able to locate these structured data points on the page and store them in our database. Only

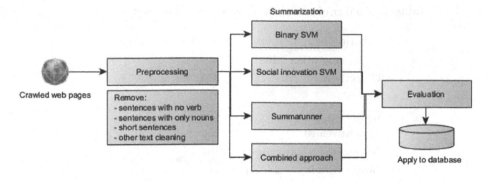

Fig. 1. Methodology overview

a small number of data sources already contained descriptions of the projects and they were used for the creation of the training set.

We collected 3560 projects. Out of these, 2893 project had identifiable websites. In order to provide data for describing the projects, we created a crawler that collects text from the websites.

We performed a set of annotation tasks in which annotators were annotating sentences that describe how each project satisfies some of the following social innovation criteria:

– *Social objective* - project addresses certain (often unmet) societal needs, including the needs of particular social groups; or aims at social value creation.
– *Social actors and actor interactions* - involves actors who would not normally engage in innovation as an economic activity, including formal (e.g. NGOs, public sector organisations etc.) and informal organisations (e.g. grassroots movements, citizen groups, etc.) or creates collaborations between "social actors", small and large businesses and the public sector in different combinations
– *Social outputs* - creates socially oriented outputs/outcomes. Often these outputs go beyond those created by conventional innovative activity (e.g. products, services, new technologies, patents, and publications), but conventional outputs/outcomes might also be present.
– *Innovativeness* - There should be a form of "implementation of a new or significantly improved product (good or service), or process, a new marketing method, or a new organisational method".

Data annotation is further explained in [10]. The data set contained 315 documents, 43 of which were annotated by 4 different annotators, while the rest were mainly single annotated. The distribution of annotated sentences is presented in Table 1. Annotated data, descriptions from the original data sources and crawled websites were used for training and evaluating summarization approaches.

Table 1. Number of sentences satisfying social innovation criteria

Criteria	Number of sentences
Social innovation criteria	
Objectives	374
Actors	217
Outputs	309
Innovativeness	256
Not satisfying any criteria	3167
Binary (inside/outside summary)	
Inside	2459
Outside	12962

3.3 Data Cleaning

The data from the websites may be quite noisy, as the crawler was collecting all textual information, including menus, footers of the pages and at times advertisements. Additionally, many pages contained events and blog posts that were not relevant for describing the core of the project. Therefore, we have performed some data cleaning before proceeding with training the summarizers.

In order to reduce the amount of irrelevant text in form of menus and footers, we have performed part of speech tagging and excluded sentences that did not contain verbs.

For further summarization, only main pages, about pages and project description pages were used. In case the page was not in English it was translated using Google Translate.

3.4 SVM Based Summarizer

The first summarization approach is based on the assumption that the task can be modelled as a classification task, where sentences would be classified as part of a summary or not. It was hypothesized that words in a sentence would indicate whether it describes the project (e.g. "project aims to...", "the goal of the project is to...", etc.).

In order to create a training data set, we utilized projects that had both project description in the original data sources and crawled websites. Since the descriptions were created by humans, they usually cannot be matched with the sentences from the website. In order to overcome this issue, we generated sent2vec embedding vectors of the sentences in both the description and the crawled text [15]. We then computed cosine similarities between the sentences from the description and the ones from the crawled text. If the cosine similarity is higher than 0.8, the sentence is labeled as part of the summary, otherwise it is labeled as a sentence that should not be part of the summary.

These sentences were used as training data for the SVM classifier. Before training we balanced the number of positive (sentences that should be part of the summary) and negative (sentences that should remain outside the summary) instances. The bag-of-words transformed to TF-IDF scores, the position of a sentence in the document (normalized to the score between 0–1) and keywords were used as features for the SVM classifier. The keywords are extracted using KNOWMAK ontology [8, 22] API that for the given text returns grand societal challenge topics and a set of keywords that were matched for the given topic and text[1].

3.5 Social Innovation Criteria Classifier

The social innovation criteria classifier utilized an annotated data set. In this data set, sentences that were marked as explaining why a project satisfies any of the social innovation criteria (objectives, actors, outputs, innovativeness), were used as positive training instances for the SVM classifier. The classifier used a bag-of-words transformed to TF-IDF scores.

3.6 Summarunner

Summarunner is an extractive summarization method developed by IBM Watson [11] that utilizes recurrent neural networks (GRU). If compared using ROUGE metrics, the algorithm outperforms state-of-the-art methods. The method visits sentences sequentially and classifies each sentence by whether or not it should be part of the summary. The method is using a 100-dimensional word2vec language model [9]. The model was originally trained on a CNN/DailyMail data set [4]. The social innovation data set that we have created was quite small and not sufficient for training a neural network model (about 350 texts compared to over 200,000 in DailyMail data). However, we performed a model fitting on our social innovation data set.

3.7 Stacked SVM-Based Summarizer and Summarunner

Our final summarization method was developed as a combination of SVM-based method and Summarunner. We have noticed that binary SVM model produces quite long summaries and may be efficient for initial cleaning of the text. Once the unimportant parts have been cleaned up by the SVM-based classifier, Summarunner shortens the text and generates the final summary.

4   Evaluation Methodology

In order to evaluate our methodologies and select the best performing model we used ROGUE metrics, human scoring and two topic-based evaluation methods.

---

[1] https://gate.ac.uk/projects/knowmak/.

ROUGE metrics are the most popular and widely used summarization scoring approaches which were presented back in 2004 [5, 7, 11]. As such, we are utilizing them as well.

Since a good summary should include the most important topics from the original text, topic-related metrics can be devised. We have used two topic based metrics: one was based on KNOWMAK ontology and the proportion of matched topics related to EU defined Grand Societal Challenges[2] and Key Enabling Technologies[3] in the original and summarized text. The other method was based on latent Dirichlet allocation (LDA) [2]. We have extracted 30 topics using LDA from merged corpus of original texts and summaries and then we have calculated the proportion of topics that match. In order to prevent favouring long summaries, we have normalized the scores, assuming that the perfect summary should be no longer than 25% of the length of the original text (longer texts were penalized).

5 Evaluation and Results

The evaluation of summarization techniques is a challenging process, therefore, we have employed several techniques.

Since SVMs classifiers are utilizing classification, we have calculated their precision, recall and F1-scores. These are measures commonly used for evaluating classification tasks. These metrics are calculated on a test (unseen) data set, containing 40 documents (286 sentences labeled as inside summary, 2014 sentences as outside). The results can be seen in Table 2.

Table 2. Evaluation based on classification metrics (precision, recall and F1-score) for classification-based summarizers (binary and social innovation criteria-based)

Classifier	Precision	Recall	F1-score
Binary SVM	0.8601	0.7130	0.7594
Objectives SVM	0.8423	0.5601	0.6226
Actors SVM	0.8821	0.4687	0.5659
Innovativeness SVM	0.8263	0.4456	0.5166
Outputs SVM	0.8636	0.6284	0.7089

The data set for training these classifiers is quite small, containing between 200–400 sentences. It is interesting to note that the criteria classifiers containing larger number of training sentences (compare Tables 1 and 2), perform with a better F1-score (Objectives and Outputs). This indicates that scores can be

---

[2] https://ec.europa.eu/programmes/horizon2020/en/h2020-section/societal-challenges.

[3] http://ec.europa.eu/growth/industry/policy/key-enabling-technologies_en.

improved by creating a larger data set. The classifiers perform with quite good precision, which means there are few false positive sentences (the majority of the sentences that end up in summary are correct).

Since ROUGE metrics are commonly used in summarization literature, we have evaluated all our summarization approaches with ROUGE 1, ROUGE 2 and ROUGE-L metrics. The evaluation was performed again on an unseen test set, containing 40 documents and their summaries. The results can be seen in Table 3.

Table 3. ROUGE scores for the developed summarization methodologies

Classifier	ROUGE 1	ROUGE 2	ROUGE-L
Binary SVM	0.6096	0.5544	0.5553
Social innovation SVM	0.6388	0.6140	0.5846
Summarunner	0.6426	0.5788	0.5762
Binary SVM + Summarunner	0.5947	0.5197	0.5279
Binary SVM + Summarunner relative length	0.5496	0.4731	0.4668

Summarunner has the best performance based on unigram ROUGE (ROUGE-1) score. However, the social innovation SVM-based summarizer performs better in terms of bigram ROUGE (ROUGE-2) and ROUGE-L score (measuring longest common token sequence). Based on these results, it is possible to conclude that a specifically crafted classifier for the problem will outperform a generic summarizer, even if it was trained only on a small data set. Stacked binary SVM and Summarunner perform worse than single summarizers on their own in terms of ROUGE.

In order to further evaluate the methodologies used, we have used an LDA-based metric. The assumption behind using this approach was that a good summarizer would have a high number of topics in the summary/description and the original text matching. The results of the LDA topic similarity evaluation can be seen in Table 4.

Table 4. LDA topic similarity scores for the developed summarization methodologies

Classifier	LDA topic similarity
Binary SVM	0.2703
Social innovation SVM	0.2485
Summarunner	0.2398
Binary SVM + Summarunner	0.2683

The most matching topics are found with the binary SVM classifier. However, this classifier is also producing the longest summaries. Stacked SVM and Summarunner are performing similar matches with much shorter summaries being generated.

The second topic-based approach utilizes topics about grand societal challenges and key-enabling technologies retrieved from the KNOWMAK topic-modelling tool. The results can be seen in Table 5.

Table 5. Topic similarity evaluation using KNOWMAK ontology topics

Classifier	KNOWMAK topic similarity
Binary SVM	0.3725
Social innovation SVM	0.3625
Summarunner	0.3025
Binary SVM + Summarunner	0.3025

The binary SVM summarizer, followed by the social innovation summarizer are the best methodologies according to this metric.

Finally, summaries were scored by human annotators. Human scorers were presented with an interface containing the original text and a summary for each of the three methods (binary SVM, social innovation SVM and Summarunner). For each of the summaries they could give a score between 0–5. In Table 6 are presented averaged scores made by the human scorers. We have also averaged the scores in order to account for document length. In order to do that we used the following formula:

$$LengthAveragedScore = \frac{docLen - summaryLen}{docLen} * human\_score$$

Table 6. Human scores for the developed summarization methodologies

Classifier	Number of ratings	Human score	Length averaged human score
Binary SVM	23	2.7391	0.8647
Social innovation SVM	20	2.4500	1.6862
Summarunner	22	2.0000	1.5110

The best human scores were for binary SVM. However, this classifier excluded only a few sentences from the original text, and it was generally creating longer summaries. If the scores are normalized for length, the best performing summarizer was the one based on social innovation criteria, followed by Summarunner.

At the time of the human scoring, the stacked approach consisting of binary SVM and Summarunner was not yet developed, so results for this approach are not available.

We have used stacked (SVM+Summarunner) and social innovation classifier in order to generate summaries for our database. Stacked model was used as fallback, in case summary based on social innovation model was empty or contained only one sentence. The approach was summarizing and generating project descriptions where either the description was too long (longer than 1000 words), or was missing. The summarizer generated new summaries for 2186 projects.

6   Conclusion

Making project descriptions and summaries based on the textual data available on the internet is a challenging task. The text from the websites may be noisy, different length, and important parts may be presented in different pages of the website. In this paper, we have presented and compared several approaches for a particular problem of summarizing social innovation projects based on the information that is available about them on the web. The presented approaches are part of a wider information system, including the ESID database[4] and the KNOWMAK[5] tool. Since these approaches make extractive summaries, they may not have connected sentences in the best manner, and therefore additional manual checks and corrections would be performed before final publication of the data. However, these approaches significantly speed up the process of generating project descriptions.

Evaluating automatically-generated summaries remains a challenge. A good summary should carry the most important content, but also significantly shorten the text. Finding a balance between the content and meaning that was carried from original text to the summary and final length can be quite challenging. Most of the currently used measures in the literature do not account for the summary length, which may lead to biases towards longer summaries. There are a number of measure that we have used and proposed in this work. Often, it is not easy to indicate strengths and weaknesses of summarization approaches using single measures and using multiple measures may be beneficial.

Most of the current research presents summarization approaches for general use. Even though, these approaches can be used in specific domains and for specific cases (such as social innovation), our evaluation shows that approaches developed for a particular purpose perform better overall.

Our evaluation indicated that it may be useful to combine multiple summarization approaches. Certain approaches can be used to clear the text, while the others may be used to further shorten the text by carrying the most important elements of the text. In the end, we used a combined approach for the production of the summaries in our system.

---

[4] https://esid.manchester.ac.uk/.
[5] https://www.knowmak.eu/.

Acknowledgments. The work presented in this paper is part of the KNOWMAK project that has received funding from the European Union's Horizon 2020 research and innovation programme under grant agreement No. 726992.

References

1. Bazrfkan, M., Radmanesh, M.: Using machine learning methods to summarize persian texts. Indian J. Sci. Res. **7**(1), 1325–1333 (2014)
2. Blei, D.M., Ng, A.Y., Jordan, M.I.: Latent dirichlet allocation. J. Mach. Learn. Res. **3**, 993–1022 (2003)
3. Bonifacio, M.: Social innovation: a novel policy stream or a policy compromise? An EU perspective. Eur. Rev. **22**(1), 145–169 (2014)
4. Cheng, J., Lapata, M.: Neural summarization by extracting sentences and words. arXiv preprint arXiv:1603.07252 (2016)
5. Dong, Y.: A survey on neural network-based summarization methods. arXiv preprint arXiv:1804.04589 (2018)
6. Fattah, M.A., Ren, F.: GA, MR, FFNN, PNN and GMM based models for automatic text summarization. Comput. Speech Lang. **23**(1), 126–144 (2009)
7. Lin, C.Y.: ROUGE: a package for automatic evaluation of summaries. Text Summarization Branches Out (2004)
8. Maynard, D., Lepori, B.: Ontologies as bridges between data sources and user queries: the KNOWMAK project experience. In: Proceedings of Science, Technology and Innovation Indicators 2017, STI 2017 (2017)
9. Mikolov, T., Sutskever, I., Chen, K., Corrado, G.S., Dean, J.: Distributed representations of words and phrases and their compositionality. In: Advances in Neural Information Processing Systems, pp. 3111–3119 (2013)
10. Milosevic, N., Gok, A., Nenadic, G.: Classification of intangible social innovation concepts. In: Silberztein, M., Atigui, F., Kornyshova, E., Métais, E., Meziane, F. (eds.) NLDB 2018. LNCS, vol. 10859, pp. 407–418. Springer, Cham (2018). https://doi.org/10.1007/978-3-319-91947-8_42
11. Nallapati, R., Zhai, F., Zhou, B.: SummaRuNNer: a recurrent neural network based sequence model for extractive summarization of documents. In: Thirty-First AAAI Conference on Artificial Intelligence (2017)
12. Nallapati, R., Zhou, B., Gulcehre, C., Xiang, B., et al.: Abstractive text summarization using sequence-to-sequence RNNs and beyond. arXiv preprint arXiv:1602.06023 (2016)
13. Nenkova, A., Passonneau, R.: Evaluating content selection in summarization: the pyramid method. In: Proceedings of the Human Language Technology Conference of the North American Chapter of the Association for Computational Linguistics: HLT-NAACL 2004 (2004)
14. Neto, J.L., Freitas, A.A., Kaestner, C.A.A.: Automatic text summarization using a machine learning approach. In: Bittencourt, G., Ramalho, G.L. (eds.) SBIA 2002. LNCS (LNAI), vol. 2507, pp. 205–215. Springer, Heidelberg (2002). https://doi.org/10.1007/3-540-36127-8_20
15. Pagliardini, M., Gupta, P., Jaggi, M.: Unsupervised learning of sentence embeddings using compositional n-gram features. In: Proceedings of the 2018 Conference of the North American Chapter of the Association for Computational Linguistics: Human Language Technologies, Volume 1 (Long Papers), vol. 1, pp. 528–540 (2018)

16. Riedhammer, K., Favre, B., Hakkani-Tür, D.: Long story short-global unsupervised models for keyphrase based meeting summarization. Speech Commun. **52**(10), 801–815 (2010)
17. Rush, A.M., Chopra, S., Weston, J.: A neural attention model for abstractive sentence summarization. arXiv preprint arXiv:1509.00685 (2015)
18. Sarkar, K., Nasipuri, M., Ghose, S.: Using machine learning for medical document summarization. Int. J. Database Theory Appl. **4**(1), 31–48 (2011)
19. Sinha, A., Yadav, A., Gahlot, A.: Extractive text summarization using neural networks. arXiv preprint arXiv:1802.10137 (2018)
20. Steinberger, J., Ježek, K.: Evaluation measures for text summarization. Comput. Inform. **28**(2), 251–275 (2012)
21. Young, T., Hazarika, D., Poria, S., Cambria, E.: Recent trends in deep learning based natural language processing. IEEE Comput. Intell. Mag. **13**(3), 55–75 (2018)
22. Zhang, Z., Petrak, J., Maynard, D.: Adapted textrank for term extraction: a generic method of improving automatic term extraction algorithms. Procedia Comput. Sci. **137**, 102–108 (2018)

Cross-Corpus Training with CNN to Classify Imbalanced Biomedical Relation Data

S. S. Deepika$^{(\boxtimes)}$, M. Saranya, and T. V. Geetha

CEG, Anna University, Chennai 600025, Tamilnadu, India
deepu.deepika26@gmail.com

Abstract. Information extraction from unstructured text is a challenging task which demands automation. Relation extraction is an important sub-task of information extraction and it is usually modeled as a classification problem. In the field of biomedicine, relation extraction helps in improving the health-care system and also in manufacturing safer drugs. But lack of a huge single annotated corpus for all biomedical relation types and the presence of class imbalance problem hinders most of the classifier's performance. For this reason, cross-corpus training of a deep-learning model namely the convolutional neural network (CNN) is carried out with annotated corpora developed for different biomedical relation types. Before modeling the CNN, SMOTE, an oversampling technique is used to balance the dataset. The input to the CNN is a concatenated feature embedding vector of a sentence obtained from six distinct features. From the six distinct features, word context and dependency context primarily contributed to the system's performance. Three types of biomedical relations namely drug-drug, drug-adverse effect and drug-disease are handled in this work. The experimental results showed that training an algorithm with a balanced dataset gives better results than using an imbalanced dataset. Additionally, use of cross-corpus training improves the relation classification task's performance for annotated datasets that are limited in size.

Keywords: Biomedical relation extraction · Cross-corpus training
Class-imbalance · CNN

1 Introduction

In the digital era, one of the biggest challenges is handling the vast amount of complex, unstructured text data. The biomedical domain is one such field where there is data explosion due to the advancement in drug discovery and development technologies [1]. The number of biomedical-related articles being published has seen an exponential growth in the last few years. It is important to mine these scientific articles to extract information submerged in it. Manual curation of the articles consumes a lot of time and it is a very expensive task [1]. So, automatic text mining methods are being used for information extraction. Biomedical relation extraction from text has gained a lot of attention in recent times because of the insights it gives about the different types of interactions that happens among the biomedical entities.

© Springer Nature Switzerland AG 2019
E. Métais et al. (Eds.): NLDB 2019, LNCS 11608, pp. 170–181, 2019.
https://doi.org/10.1007/978-3-030-23281-8_14

One of the crucial problems in the clinical care system is the adverse drug reaction. The US Food and Drug Administration (FDA) has reported a situation of polypharmacy (use of four or more drugs) in U.S patients especially among older patients [2]. Over 35.8% of adults in U.S aged over 65 are prescribed 5 or more drugs [2]. In a polypharmacy situation, there is a very high chance for the drugs to interact with each other. Drug-Drug Interaction (DDI), a type of biomedical relation happens when two drugs interact with each other causing a positive or negative effect on the patient [3]. Besides, DDI is an important factor contributing towards Adverse Drug Effects (ADE). Adverse drug effects lead to more deaths per year in most of the countries and increases the health care expenditure [4]. Hence, identifying drug-drug interaction and adverse drug effects from the scientific literature would help in improving the health-care system. When identifying the adverse drug effects, it is also important to distinguish it from the drug-disease treatment relations present in the documents. So, in this paper three types of biomedical relations are handled namely drug-drug, drug-disease and adverse drug effects.

Recently deep learning methods are used in the Natural Language Processing (NLP) problems and have achieved state-of art results [5]. Deep learning models learn the required high-level features from the given data on its own and this makes it more powerful. In this work, we design a Convolutional Neural Network (CNN) with cross-corpus training incorporating six features based on the word-context, entity-position, concept-type, part-of-speech tagging, parse-tree and dependency tree. These features are represented as vector embedding and given as input to the CNN. Many relation classification datasets do not contain equal number of instances in each class and it affects the classifier's performance. Consequently, before training the CNN, a random over sampling algorithm named SMOTE [6] is used to resolve the class imbalance problem.

2 Related Work

In the past decade, computational methods using machine-learning algorithms with different feature sets have been designed to address the biomedical relation extraction task [7]. Traditional machine learning methods needs a lot of feature engineering to model the relation extraction task. It is time-consuming, expensive and needs domain expertise. Deep neural networks (DNN) automatically learn the features necessary for a particular task and have demonstrated good performance for the relation extraction tasks [8]. Convolutional neural network (CNN), a deep learning model which has proved in the image processing field is being utilized for the NLP tasks in recent times [5]. Clinical records have a lot of medical concepts and relations. CNN model has been used by He et al. [8] to classify these relations and was able to extract rich and precise features. The system performed better compared to the supervised learning algorithms for the 2010 i2b2/VA relation corpus. CNN-based approach is also used for DDI extraction by Liu et al. [9] and it has achieved a better F_1-score compared to the other state-of-art methods that existed.

Supervised deep learning models require a lot of annotated training data to model a system that performs fairly well. Due to the limited availability of huge annotated corpora for each and every domain-specific task, alternate solutions like semi-supervised learning [10], multi-task learning [11] and transfer-learning [12] are being employed. Multi-task learning learns a shared representation for various related tasks and improves the model generalization. Collobert et al. [13], trained a single CNN model for various NLP tasks like named entity recognition, POS tagging, chunking and semantic role labelling. Each NLP task performed better by sharing the learned weights across the tasks. Multi-task learning and cross-corpus training is utilized to recognize the audio emotions by Zhang et al. [14]. The system through multi-task learning incorporates the domain, gender and corpus variability and found that it outperforms the independent task learning approach.

Most of the deep-learning approaches for NLP tasks utilize embedding methods to represent their input sentences. To mitigate the drawbacks of traditional feature representation (sparse, high dimension, varying length vectors), word embedding methods are being employed. Word embedding [15] is a language modelling technique used in NLP tasks to learn the features and represent it in a dense, low-dimensional vector representation of fixed length. In the biomedical domain, word embedding has been successfully applied for named entity recognition [16] and relation extraction tasks [9]. Along with word context features, lexical and syntactic features also contribute to relation classification task. Ghannay et al. [17] carried out experiments to evaluate the different word embedding models for four NLP tasks. The experiments showed that dependency-based embedding performs well and there would be significant improvement if combination of embeddings are used.

One of the major problems in any biomedical relation classification task is the uneven distribution of the data samples across the distinct classes causing a class imbalance problem. Experimental studies show that an unbalanced dataset degrades the classifier's performance and it should be handled [18]. Over-sampling and under sampling techniques are two techniques that are frequently used to handle the imbalances. Synthetic Minority Oversampling Technique (SMOTE) [6], is an oversampling method to generate new instances for the minority classes instead of just duplicating the existing instances. SMOTE has the potential to avoid overfitting to a certain extent. Zhao et al. [19] incorporated different rebalancing strategies to identify rare events in the health care data. The experimental study showed that SMOTE combined with logistic regression increased the recall by 45.3%. To resolve the issues arising out of an unbalanced, small-sized dataset, SMOTE algorithm is employed along with cross-corpus training to build a CNN model. The input to the CNN is the feature embedding vector obtained using the six features namely word-context, entity-position, concept-type, part-of-speech tagging, parse-tree and dependency tree.

3 Biomedical Relation Extraction

Biomedical relation extraction is modeled as a classification problem, where the classifier outputs the class label for each entity pair. The proposed system identifies relation between entity pairs that are present within a sentence. Depending on the type of biomedical relation, the classification can be binary or multi-class. A CNN model with cross-corpus training is built to classify the drug-drug pairs, drug-adverse effects pairs and drug-disease pairs present within a sentence. It is a modeled as a multi-class problem with six positive classes - four DDI types, one drug-adverse effect type and one drug-disease type. The system architecture for the proposed system is illustrated in Fig. 1. The system uses different types of feature embedding to represent the input sentence and employs SMOTE [6] algorithm to mitigate the class imbalance problem.

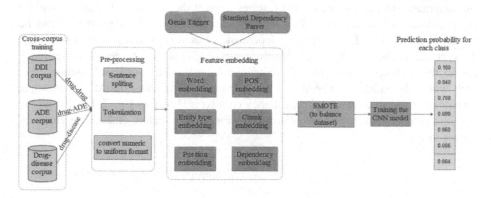

Fig. 1. System architecture

3.1   Pre-processing

Considering the different formats (xml, csv etc.,) in which the annotated corpus is available, sentence splitting and entity extraction is done in the pre-processing stage for both the positive and negative instances. Then, the sentences are tokenized using GENIA Tagger [20] and converted to lower-case. Float and integer numbers commonly occurs in the biomedical corpora as they show the results of various experiments. As these numbers are not important for the relation extraction task, they are converted to a common representation named 'number'. This would improve the result of the context-based embedding and reduce the vocabulary size.

3.2   Representation Learning

Data representation plays a vital role in any machine learning algorithm's performance. In case of unstructured data like text, it is even more essential to represent the data in a way that is useful for the learning algorithm. In the proposed work, each word in the sentence is represented using the concatenation of different feature embedding vectors and this concatenated vector is fed as input to the CNN. The beneficial features identified for relation extraction are (a) word-context (b) entity position (c) entity type, (d) part-of-speech tag (e) chunk tag and (f) dependency path. Based on the popular skip-gram model [15], feature embedding vector for each considered feature is generated.

In this work, the word context feature is represented using word embedding vector, w_i obtained by using the word2vec [15] on the extracted sentences. The two associated entities in the sentence are represent using BIO tags where B and I tagare used to

174   S. S. Deepika et al.

represent the entities and O for non-entity words. Based on these tags, entity type embedding, t_i is generated. P_1 and P_2 are positions of the entities that are present in the sentence and are related. The relative distance of each word from entity position, P_1 and P_2 is embedded as two position embedding vectors, p_{i1} and p_{i2} respectively. GENIA Tagger [20] is used to obtain the POS tag and chunk tag of the words in the sentence. After which, POS-based embedding, s_i and chunk-tag based embedding, c_i are generated. The above mentioned features exploit the linguistic features along the linear context of the sentence. Two words located far away in the sentence can be related to each other. To handle this, dependency based embedding is used to exploit the syntactic context of the words. Stanford parser [21] is used to get the dependency tree of the sentence, from which dependency context of the words are obtained. For each word, dependency embedding e_i is generated using this context. An example sentence with word context and dependency context is shown in Fig. 2. Finally, all the generated embeddings are concatenated to get the word's feature embedding vector as shown in Eq. 1.

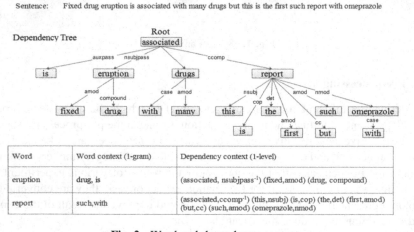

Fig. 2. Word and dependency context

$$f_i = w_i \oplus t_i \oplus p_{i1} \oplus p_{i2} \oplus s_i \oplus c_i \oplus e_i \tag{1}$$

where, f_i − concatenated feature embedding for word i;

w_i − word embedding; t_i − entity type embedding;

p_{i1} − position embedding w.r.t entity1;

p_{i2} − position embedding w.r.t entity2; s_i − POS embedding;

c_i − chunk embedding; e_i − dependency embedding

3.3 SMOTE

Before training the CNN model with the generated feature embedding, the imbalance present in the dataset has to be addressed. SMOTE, a synthetic sampling technique is used to balance the unbalanced dataset. The minority classes which have a smaller number of samples are first selected and the SMOTE algorithm is applied. For each minority class instance, k-nearest neighbors are found using kNN algorithm. For each nearest neighbor, the difference between the instance and the considered nearest neighbor is calculated as Z_i. Then this value is multiplied with a random vector R and added to the instance to generate the synthetic instance. The input representation is in the form of numeric vector with value range of -1 to $+1$, as sigmoid activation function is used in the skip-gram model. Hence, the random vector values are generated from -0.1 to $+0.1$. The number of neighbors that should be considered for each class instance depends on the degree of imbalance present in the class.

3.4 Convolutional Neural Network

Convolutional neural networks extract the informative features present in the text based on the set of convolution filters applied on n-grams of different length. To train the CNN, a sequence of embedding representation for each sentence is given as input. The output of CNN is a vector of length equaling the number of relation types (in this work including the negative type, there are 7 relation types) and the vector has the probability values corresponding to each relation type. The different layers in the CNN are convolution layer, max-pooling layer and feed-forward, fully-connected layer. The CNN architecture for the proposed biomedical relation extraction task is illustrated in Fig. 3.

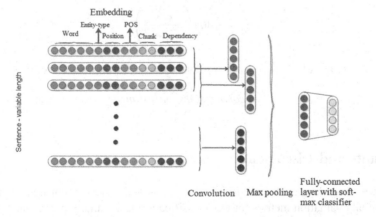

Fig. 3. Convolutional neural network with feature embedding

Convolution Layer. Convolution layer is the core component of CNN, where multiple filters of varying size are used to extract the local features present in the sentence. As mentioned in Eq. 1, f_i is the concatenated feature embedding for word i. $f_1 f_2 \ldots f_n$ is the sequence of feature embedding for a sentence of length n. In Fig. 3, for illustration filter of length three is used. Each convolution layer outputs c^i as shown in Eq. 2. Rectified linear unit (ReLU) is used as the activation function, w and b are the learning parameters.

$$c^i = g(w.f_{i:i+l-1} + b) \; for \; i = 1, 2, \ldots n - l + 1 \tag{2}$$

$$where, \; g - activation \, function; \; l - length \, of \, the \, filter;$$

$$w - weight \, vector; \; b - bias \, term; \; n - length \, of \, the \, sentence$$

Max-Pooling Layer. The output of the convolution layer i.e. convolved vectors will vary in length as different length filters are used. To obtain the global and the most relevant features from the sentence, max-pooling aggregation function is utilized. For each filter, max pooling is applied and vector of fixed size is generated using Eq. 3.

$$z = [c_1^{max}, c_2^{max}, \ldots c_p^{max}] \tag{3}$$

$$where, \; c_j^{max} = \max \, (c_j^1, c_j^2, \ldots c_j^{n-l+1});$$

Fully Connected Layer. The output of the max pooling layer is fed to the fully connected layer and soft-max classifier is used to predict the type of biomedical relation. Before giving the output of the max pooling layer to the fully connected layer, dropout, a regularization technique is used to prevent overfitting. Objective of the soft-max classifier is the minimization of the loss function given in the Eq. 4.

$$L_i = -\log \left(\frac{e^{x_{y_i}^{(i)}}}{\sum_k e_j^{x^{(i)}}} \right) \tag{4}$$

$$where, \; x^{(i)} - output \, of \, the \, fully \, connected \, layer \, for \, instance \, i;$$

$$y_i - correct \, class \, for \, instance \, i$$

4   Results and Discussion

The proposed work is evaluated using 5-fold cross validation of considered datasets. The essential evaluation metrics for any classification task namely precision, recall and F_1-score are used in this work and their equations are given below.

$$Precision = \frac{TruePositive}{TruePositive + FalsePositive}$$

$$Recall = \frac{TruePositive}{TruePositive + FalseNegative}$$

$$F_1 Score = \frac{2 * Precision * Recall}{Precision + Recall}$$

4.1   Dataset

The proposed system is experimentally evaluated using DDI extraction challenge 2013 dataset [22], ADE corpus [23] and EU-ADR corpus [24] for the drug-drug, drug-adverse effect and drug-disease relations respectively. DDI corpus has 233 MEDLINE abstracts and 792 DrugBank documents. DDI has four types of interactions namely (a) mechanism (b) effect (c) advice and (d) int. The ADE corpus contains 2,972 MEDLINE case reports annotated with drugs, adverse effects and dosages from which 20,967 sentences were generated. The corpus has 4,272 positive and 16,695 negative instances. EU-ADR corpus has annotations for relations among drug, gene and disease entities. Drug-disease relation has been annotated for 100 abstracts and contains three sub-types in which only positive and negative associations are considered. The dataset description is given in Table 1.

Table 1. Dataset description

Relation type	Positive instances	Negative instances
DDI-mechanism	1625	28,554
DDI-effect	2069	
DDI-advice	1050	
DDI-int	284	
Drug-adverse effect	4272	16,695
Drug-disease	162	68
Total:	9462	45,317

4.2   Convolutional Neural Network

CNN can be modeled using filters of varying size. For the considered corpora, the optimal filter length is chosen as two, four and six. In the convolution layer, 100 filters for each filter length (2, 4, and 6) are being used. Dropout rate probability is fixed as 0.2. The performance of the deep-learning CNN model is compared to best performing machine learning algorithm for the considered application. Support vector machine (SVM) algorithm with the same set of features is used as the baseline model. Table 2 shows the comparison between the precision, recall and F_1-score of the baseline and the CNN model using 5-fold cross validation. For both SVM and CNN, all three datasets are used for training and SMOTE is used to solve the imbalance present in the corpora. CNN has improved the system performance for all the considered corpora. This is because, CNN extracts both local and global features and also it uses filters of varying size. The highest F_1-score gain of 4.6% is achieved for the DDI-DrugBank corpora.

178   S. S. Deepika et al.

Table 2. Performance comparison between CNN and SVM

Learning model	Corpora	Precision	Recall	F_1-Score
CNN	ADE	0.772	0.716	0.743
	DDI-DrugBank	0.787	0.701	0.742
	DDI-Medline	0.775	0.686	0.728
	EU-ADR	0.726	0.679	0.702
SVM	ADE	0.719	0.697	0.707
	DDI-DrugBank	0.714	0.678	0.696
	DDI-Medline	0.711	0.661	0.685
	EU-ADR	0.699	0.664	0.681

4.3 Contribution of Different Features

A total of six features are considered to train the CNN with cross-corpus training. The optimal embedding vector size for word-embedding, entity-type embedding, position embedding, POS embedding, chunk embedding (parse tree) and dependency embedding (dependency tree) is fixed as 50, 10, 10, 10, 20. The contribution of each feature, X_i is calculated by assessing the drop in F_1-score of the model, after removing the feature, X_i. Among the various features considered, some features contributed more to the system's performance and it is illustrated in Fig. 4. The drop in F1-score from the original value (considering all features) is high for word context and dependency-tree feature. The ADE dataset without the word embedding showed the highest drop in F1-score of 9%. By removing the dependency embedding, DDI dataset showed the highest drop in F1score of 7%.

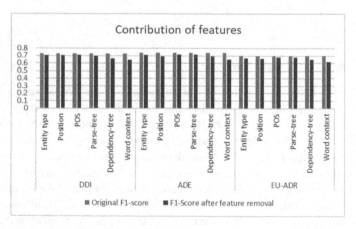

Fig. 4. Contribution of each feature to the system performance

4.4   SMOTE

The ratio of positive to negative instance in total is approximately 1:4.8. The ratio of the small positive relation type (drug-disease) to the predominant positive relation type (drug-adverse effect) is 1:26.3. The ratio of positive to negative instance separately for DDI, drug-ADE and drug-disease relation type are 1:5.7, 1:3.9 and 1:2.4 respectively. The imbalance present in the datasets are mitigated using the SMOTE algorithm. Table 3, shows the performance on different datasets when the CNN is trained with all the three considered datasets with and without the SMOTE algorithm. The results show that the CNN classifier performs better with balanced dataset generated from SMOTE than with the unbalanced data. The highest increase in F_1-score of 7.1% is achieved for the EU-ADR corpus. This is due to the highly imbalanced nature of the EU-ADR corpus with respect to the other corpora.

Table 3. Performance of SMOTE

Corpus	F_1-score (without SMOTE)	F_1-score (with SMOTE)
ADE	0.707	0.743
DDI-DrugBank	0.713	0.742
DDI-Medline	0.691	0.728
EU-ADR	0.631	0.702

4.5   Cross-Training

In this work, the CNN is trained with all the three datasets which has annotated instances for different biomedical relations. The advantage of cross-corpus training is that it increases the generalization of model and also it helps the relations with very minimal dataset instances to perform better. The performance of the cross-corpus training is evaluated by training the CNN with different combinations of the datasets and tested across all the three datasets as demonstrated in Table 4. EU-ADR corpus has comparatively very less positive and negative instances. By training the CNN with additional corpus, there is an improvement in the performance of drug-disease

Table 4. Performance of cross-corpus training

Training set	Testing set (F_1-score with SMOTE)		
	ADE	DDI	EU-ADR
ADE	**0.790**	0.561	0.562
DDI	0.554	0.712	0.579
EU-ADR	0.557	0.579	0.648
ADE+DDI	0.781	0.724	0.645
ADE+EU-ADR	0.713	0.654	0.666
DDI+EU-ADR	0.673	0.679	**0.753**
ADE+DDI+EU-ADR	0.743	**0.732**	0.702

classification task as seen in Table 4. EU-ADR corpus performs best with an F_1-score of 0.753 when trained with DDI and EU-ADR corpus. The DDI corpus has the best F_1-score of 0.732 when trained with all the three datasets. But ADE corpus is not benefitted by the cross-corpus training method. The F_1-score decreases when additional dataset is used to train the CNN. So, cross-corpus training is very beneficial for small-size corpora.

5 Conclusion

A CNN model with SMOTE and cross-corpus training has been developed to extract the biomedical relations from text. The CNN model trained with balanced corpora obtained by applying SMOTE algorithm gave better results. Besides, the experiments showed that relations with small annotated corpus can be benefitted from cross-corpus training. In this work, only a single deep-learning model (CNN) has been modeled with three biomedical relation types. In the future, the same framework can be modeled using various other types of biomedical relations and also can be experimented with different deep-learning techniques like recursive neural networks, auto-encoders etc.

References

1. Nagaraj, K., Sharvani, G.S., Sridhar, A.: Emerging trend of big data analytics in bioinformatics: a literature review. Int. J. Bioinform. Res. Appl. **14**(1–2), 144–205 (2018)
2. Qato, D.M., Wilder, J., Philip Schumm, L., Gillet, V., Caleb Alexander, G.: Changes in prescription and over-the-counter medication and dietary supplement use among older adults in the United States, 2005 vs 2011. JAMA Intern. Med. **176**(4), 473–482 (2016)
3. Sutherland, J.J., Daly, T.M., Liu, X., Goldstein, K., Johnston, J.A., Ryan, T.P.: Co-prescription trends in a large cohort of subjects predict substantial drug-drug interactions. PLoS ONE **10**(3), e0118991 (2015)
4. Giardina, C., et al.: Adverse drug reactions in hospitalized patients: results of the FORWARD (facilitation of reporting in hospital ward) Study. Front. Pharmacol. **9**, 350 (2018)
5. Sharma, R.D., Tripathi, S., Sahu, S.K., Mittal, S., Anand, A.: Predicting online doctor ratings from user reviews using convolutional neural networks. Int. J. Mach. Learn. Comput. **6**(2), 149 (2016)
6. Chawla, N.V., Bowyer, K.W., Hall, L.O., Philip Kegelmeyer, W.: SMOTE: synthetic minority over-sampling technique. J. Artif. Intell. Res. **16**, 321–357 (2002)
7. Kim, S., Liu, H., Yeganova, L., John Wilbur, W.: Extracting drug–drug interactions from literature using a rich feature-based linear kernel approach. J. Biomed. Inf. **55**, 23–30 (2015)
8. He, B., Guan, Y., Dai, R.: Classifying medical relations in clinical text via convolutional neural networks. Artif. Intell. Med. **93**, 43–49 (2019)
9. Liu, S., Tang, B., Chen, Q., Wang, X.: Drug-drug interaction extraction via convolutional neural networks. Comput. Math. Methods Med. **2016** (2016)
10. Krasakis, A.M., Kanoulas, E., Tsatsaronis, G.: Semi-supervised ensemble learning with weak supervision for biomedical relationship extraction (2018)
11. Caruana, R.: Multitask learning. Mach. Learn. **28**(1), 41–75 (1997)

12. Torrey, L., Shavlik, J.: Transfer learning. In: Handbook of Research on Machine Learning Applications and Trends: Algorithms, Methods, and Techniques, pp. 242–264. IGI Global (2010)
13. Collobert, R., Weston, J., Bottou, L., Karlen, M., Kavukcuoglu, K., Kuksa, P.: Natural language processing (almost) from scratch. J. Mach. Learn. Res. **12**, 2493–2537 (2011)
14. Zhang, B., Provost, E.M., Essl, G.: Cross-corpus acoustic emotion recognition with multitask learning: seeking common ground while preserving differences. IEEE Trans. Affect. Comput. (2017)
15. Mikolov, T., Sutskever, I., Chen, K., Corrado, G.S., Dean, J.: Distributed representations of words and phrases and their compositionality. In: Advances in Neural Information Processing Systems, pp. 3111–3119 (2013)
16. Habibi, M., Weber, L., Neves, M., Wiegandt, D.L., Leser, U.: Deep learning with word embeddings improves biomedical named entity recognition. Bioinformatics **33**(14), i37–i48 (2017)
17. Ghannay, S., Favre, B., Esteve, Y., Camelin, N.: Word embedding evaluation and combination. In: LREC, pp. 300–305 (2016)
18. Japkowicz, N., Stephen, S.: The class imbalance problem: a systematic study. Intell. Data Anal. **6**(5), 429–449 (2002)
19. Zhao, Y., Wong, Z.S.-Y., Tsui, K.L.: A framework of rebalancing imbalanced healthcare data for rare events' classification: a case of look-alike sound-alike mix-up incident detection. J. Healthc. Eng. **2018** (2018)
20. Tsuruoka, Y., Tateishi, Y., Kim, J.-D., Ohta, T., McNaught, J., Ananiadou, S., Tsujii, J.: Developing a robust part-of-speech tagger for biomedical text. In: Bozanis, P., Houstis, E.N. (eds.) PCI 2005. LNCS, vol. 3746, pp. 382–392. Springer, Heidelberg (2005). https://doi.org/10.1007/11573036_36
21. Manning, C., Surdeanu, M., Bauer, J., Finkel, J., Bethard, S., McClosky, D.: The stanford CoreNLP natural language processing toolkit. In: Proceedings of 52nd Annual Meeting of the Association for Computational Linguistics: System Demonstrations, pp. 55–60 (2014)
22. Segura-Bedmar, I., Martínez, P., Herrero Zazo, M.: Semeval-2013 task 9: extraction of drug-drug interactions from biomedical texts (ddiextraction 2013). In: Second Joint Conference on Lexical and Computational Semantics (* SEM), Volume 2: Proceedings of the Seventh International Workshop on Semantic Evaluation (SemEval 2013), vol. 2, pp. 341–350 (2013)
23. Gurulingappa, H., Rajput, A.M., Roberts, A., Fluck, J., Hofmann-Apitius, M., Toldo, L.: Development of a benchmark corpus to support the automatic extraction of drug-related adverse effects from medical case reports. J. Biomed. Inf. **45**(5), 885–892 (2012)
24. Van Mulligen, E.M., et al.: The EU-ADR corpus: annotated drugs, diseases, targets, and their relationships. J. Biomed. Inf. **45**(5), 879–884 (2012)

Discourse-Driven Argument Mining
in Scientific Abstracts

Pablo Accuosto$^{(\boxtimes)}$ and Horacio Saggion

LaSTUS/TALN Research Group, DTIC, Universitat Pompeu Fabra,
C/Tànger 122-140, 08018 Barcelona, Spain
{pablo.accuosto,horacio.saggion}@upf.edu

Abstract. Argument mining consists in the automatic identification of argumentative structures in texts. In this work we address the open question of whether discourse-level annotations can contribute to facilitate the identification of argumentative components and relations in scientific literature. We conduct a pilot study by enriching a corpus of computational linguistics abstracts that contains discourse annotations with a new argumentative annotation level. The results obtained from preliminary experiments confirm the potential value of the proposed approach.

Keywords: Argument mining · RST · Scientific corpus

1 Introduction

Argument mining [16,18]–the automatic identification of arguments, its components and relations in texts–, has recently gained increased interest in natural language processing and computational linguistics research both in the academia [16] and the industry [1]. Being able to automatically extract not only what is being stated by the authors of a text but also the evidence that they provide to support their claims would enable multiple applications, including argumentative summarization, computer-assisted text quality assessment, information retrieval systems, reasoning engines and fact-checking tools. The identification of argumentative units and relations in scientific texts, in particular, would enable tools that could contribute to alleviate the information overload experienced by researchers, editors and students as a consequence of the accelerated pace at which scientific knowledge is being produced [3]. Argument mining in scientific texts has, however, proven as a highly challenging task. This is, mainly, due to the complexity of the underlying argumentative structures of the scientific discourse [9]. These difficulties are not only faced when trying to develop automated systems but also in the production of gold standards with which to train those systems. It has been observed [7] that even humans with expert domain knowledge can find it difficult to unambiguously identify premises, conclusions and argumentation schemes in scientific articles. The lack of annotated corpora, in turn, represents a major barrier for advancing argumentation mining research in the scientific domain.

© Springer Nature Switzerland AG 2019
E. Métais et al. (Eds.): NLDB 2019, LNCS 11608, pp. 182–194, 2019.
https://doi.org/10.1007/978-3-030-23281-8_15

In this work we investigate the potential exploitation of existing linguistic resources in order to facilitate the annotation of argumentative components and relations in the domain of computational linguistics. We propose a fine-grained annotation schema particularly tailored at scientific texts which we use to enrich a subset of abstracts from the SciDTB corpus [33], which have been previously annotated with discourse relations from the Rhetorical Structure Theory (RST) [17]. RST provides a set of coherence relations with which adjacent spans in a text can be linked together in a discourse analysis, resulting in a tree structure that covers the whole text. The minimal units that are joined together in RST are called *elementary discourse units* (EDUs). Let us consider the following example from [32], included in the SciDTB corpus, in which EDUs are numbered and identified by square brackets:

[Text-based document geolocation is commonly rooted in language-based information retrieval techniques over geodesic grids.]$_1$ [These methods ignore the natural hierarchy of cells in such grids]$_2$ [and fall afoul of independence assumptions.]$_3$ [We demonstrate the effectiveness]$_4$ [of using logistic regression models on a hierarchy of nodes in the grid,]$_5$ [which improves upon the state of the art accuracy by several percent]$_6$ [and reduces mean error distances by hundreds of kilometers on data from Twitter, Wikipedia, and Flickr.]$_7$ [We also show]$_8$ [that logistic regression performs feature selection effectively,]$_9$ [assigning high weights to geocentric terms.]$_{10}$

From the argument mining perspective, we would like to identify, for instance, that the authors support their claim about the *effectiveness of using regression models* for text-based document geolocation (EDUs 4–5) by stating that this method *improves upon the state of the art accuracy* (EDU 6) and it *performs feature selection effectively* (EDU 9), which in turn is supported by the fact that it *assigns high weights to geocentric terms* (EDU 10).

In this work we aim at exploring if the information provided by the discourse layer of the corpus, which establishes that these elements are linked by chains of discourse relations[1] can contribute to facilitate this task. With this objective we conduct a set of experiments aimed at the identification of argumentative structures in the abstracts, including their argumentative components, functions and attachment. As described in Sect. 5, we propose to learn each of these subtasks separately as well as together, in a multi-task framework. Multi-task learning is a way of transferring information between machine learning processes, so they can positively influence each other. Caruana [5] describes multi-task learning as a way of improving generalization when training a machine learning model, by taking advantage of information contained in the training signals of related tasks. In order to do this, the tasks are trained in parallel while using a shared representation (such as the hidden layers of a neural newtork). We propose to

---

[1] For instance, EDUs 4 and 9 are linked by an *evaluation* relation, which can provide a clue for the identification of a *support* relation from the argumentative perspective.

compare the performance of training the argument mining subtasks in a multi-task architecture in order to explore to what degree the natural connections between them is actually captured in the training process and reflected in an improved performance of the resulting models. This idea is in line with current research across multiple natural language processing problems. In particular, those that include the identification of units and relations as related tasks, as is the case of argument mining. In [24], Ruder provides a thorough overview of the current state of multi-task learning in the context of deep learning architectures.

Contributions. Our main contributions can therefore be summarized as:

- The proposal of a new, fine-grained, schema for the annotation of arguments in scientific texts;
- The first iteration in the development of a corpus of computational linguistic abstracts containing a layer of argumentative annotations (in addition to a previously existing discourse annotation layer);
- New evidence for analyzing the interplay between argumentative and discursive components and relations and how existing tools and resources for discourse analysis can effectively be exploited in argument mining;
- Experimental results obtained by neural and non-neural architectures for mining arguments in scientific short texts;
- Additional elements to feed ongoing investigations on scenarios in which multi-task architectures have a positive effect over the independent learning of related tasks.

The rest of the paper is organized as follows: in Sect. 2 we briefly report previous work in the area. In Sect. 3 we describe the SciDTB corpus and in Sect. 4 our proposed annotation schema for new the argumentative layer. In Sect. 5 we describe our experimental settings and in Sect. 6 we report and analyze the results. Finally, in Sect. 7, we summarize our main contributions and propose additional research avenues as follow-up to the current work.

2   Related Work

The Argumentative Zoning (AZ) model [30,31] is a key antecedent in the identification of the discursive and rhetorical structure of scientific papers. It includes annotations for knowledge claims made by the authors of scientific articles. In turn, the CoreSC annotation scheme [14] adopts the view of a scientific paper as a readable representation of a scientific research by associating research components to the sentences describing them. Our proposal for annotating argumentative units (described in Sect. 4), lies between CoreSC and AZ: while the set of annotation labels resembles that of CoreSC, they are intended to express argumentative propositions, as in the case of AZ. Unlike our proposal, neither AZ nor CoreSC consider relations between rhetorical units.

Due to the challenges posed by the identification of arguments in scientific texts, most of the previous works in argument mining are targeted at other textual registers (news, product reviews, online discussions). Lippi and Torroni [16] provide a thorough summary of initiatives in these areas. The corpus created by Kirschner et al. [9] was one of the first intended for the analysis of the argumentative structure of scientific texts. The authors introduce an annotation schema that represents arguments as graph structures with two argumentative relations (*support, attack*) and two discourse relations (*detail, sequence*). Recently, Lauscher et al. [12] enriched a corpus of scientific articles with argumentative components and relations and analyzed the information shared by the rhetorical and argumentative structure of the documents by means of normalized mutual information (NMI) [29]. They then used the enriched corpus to train a tool (*ArguminSci*[2]) aimed at the automatic analysis of scientific publications, including the identification of claims, and citation contexts and the classification of sentences according to their rhetorical role, subjective information and summarization relevance [11]. Stab et al. [27] conducted preliminary annotation studies to analyze the relation between argument identification and discourse analysis in scientific texts and persuasive essays. In line with previous work [2,4], the authors acknowledge the differences between both tasks (in particular, as discourse schema are not specifically aimed at identifying argumentative relations), while they also affirm that work in automated discourse analysis is highly relevant for argumentation mining, leaving as an open question how can this relation be exploited in practice.

Our work is inspired by that of Peldszus and Stede [20]. In this work, an annotation study of 112 argumentatively rich short texts using RST and argumentation schemes is produced. The authors provide a qualitative analysis of commonalities and differences between the two levels of representation in the corpus and report on experiments in automatically mapping RST trees to argumentation structures. The argumentative components that they consider are argumentative discourse units (ADUs), which consist of one or more EDUs of the RST scheme. They propose two basic argumentative relations: *support* and *attack*, further dividing attacks between *rebutals* (denying the validity of a claim) and *undercuts* (denying the relevance of a premise for a claim). They also include a non-argumentative meta-relation (*join*) to link together EDUs that are part of the same argumentative unit. In their case the experiments are conducted at the discourse units level.[3] We, instead, propose our analysis at the level of the argumentative units (which can be formed by more than one EDU) and can therefore compare the results obtained with and without including explicit[4] discourse information in argument mining tasks.

---

[2] http://lelystad.informatik.uni-mannheim.de/.

[3] For instance, if given two EDUs they are connected by an argumentative relation.

[4] We have not generated annotations without previously segmented text, so the implicit effect of considering already available EDUs as building blocks is not analyzed in this work.

3   SciDTB Corpus

The Discourse Dependency TreeBank for Scientific Abstracts (SciDTB) [33] is a corpus containing 798 abstracts from the ACL Anthology [22] annotated with elementary discourse units and relations from the RST Framework with minor adaptations to the scientific domain.

The SciDTB annotations use 17 coarse-grained relation types and 26 fine-grained relations. Polynary discourse relations in RST are binarized in SciDTB following a criteria similar to the "right-heavy" transformation used in other works that represent discourse structures as dependency trees [13, 19, 28], which makes it particularly suitable as input of sequence tagging algorithms.

4   Argumentation Annotations

We propose an annotation schema for scientific argument mining and test it in a pilot study with 60 abstracts.[5] The annotation are made by means of an adapted version of the GraPAT [26][6] tool for graph annotations.

4.1   Relations

In line with [9], we adopt in our annotation scheme the classic *support* and *attack* argumentative relations and the two discourse relations *detail* and *sequence*. In order to simplify both the creation and processing of the annotations we restrict the accepted argumentative structures to dependency trees.[7] To account for cases in which two or more units are mutually needed to justify an argumentative relation, we introduce the *additional* meta-relation. In this case the annotator chooses one premise to explicitly link to the supported or attacked unit while the rest are chained together by *additional* links. We observed that this restriction does not limit the expressiveness of the schema but, on the contrary, contributes to hierarchically organize the arguments according to their relevance or logical sequence.

4.2   Argumentative Units

Previous works in argument mining [16] frequently use *claims* and *premises* as basic argumentative units. Due to the specificity of the scientific discourse in general [8], and abstracts, in particular, we consider this schema to be too limiting, as it does not account for essential aspects such as the degree of assertiveness and subjectivity of a given statement. We therefore propose a finer-grained annotation schema that includes the following set of classes for argumentative

---

[5] All of the abstracts are from papers included in the Proceedings of the 2014 Conference on Empirical Methods in Natural Language Processing (EMNLP).

[6] http://angcl.ling.uni-potsdam.de/resources/grapat.html.

[7] Each argumentative unit can only have one argumentative function and is attached to one parent.

components: **proposal** (problem or approach), **assertion** (conclusion or known fact), **result** (interpretation of data), **observation** (data), **means** (implementation), and **description** (definitions/other information). While *proposal* could broadly be associated with claims, *result* and *observation* are in general used to provide supporting evidence. The units labeled as *assertion* can have a dual role of claim and premise and *means* and *description* are, in general, used to provide non-argumentative information.

In line with [20], and unlike previous works that consider sentences as annotation units [15,30], we consider EDUs as the minimal spans that can be annotated, while there is not a pre-established maximum span. Argumentative units can cover multiple sentences.

Figure 1 shows a subset of the argumentative components and relations annotated in the abstract included in Sect. 1. The color of the units represent their type: yellow for units of type *result*, pink and red for *assertion*.[8]

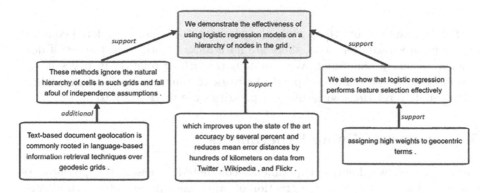

Fig. 1. Argumentative tree

The example shows how, in the case of scientific abstracts, claims and evidence provided to support them are frequently not stated explicitly in an argumentative writing style but are instead implicit.[9]

4.3   Argumentation Corpus Statistics

The corpus enriched with the argumentation level contains 60 documents with a total of 327 sentences, 8012 tokens, 862 discourse units and 352 argumentative units.[10] Even if not enforced by the annotation schema, argumentative unit boundaries coincide with sentences in 93% of the cases.

---

[8] Background assertions are displayed with a red border while assertions stated by the authors of the paper are displayed with a pink border.

[9] For instance, implicit claims in relation to the relevance of the problem at stake.

[10] The annotations are made available to download at http://scientmin.taln.upf.edu/argmin/scidtb_argmin_annotations.tgz.

Table 1 shows the distribution of the argumentative units in relation to their type, argumentative function and distance to their parents.[11]

Table 1. Statistics of the corpus enriched with the argumentative layer.

Type		Function		Distance to parent	
proposal	110	*support*	124	*adjacent*	167
assertion	88	*attack*	0	*1 arg. unit*	55
result	73	*detail*	130	*2 arg. units*	36
observation	11	*additional*	27	*3 arg. units*	17
means	63	*sequence*	11	*4 arg. units*	11
description	7			*5 arg. units*	5
				6 arg. units	1

It is relevant to note that, while almost every document considered contains one or more *support* relations, there are no *attacks* identified in the set of documents currently annotated. We maintain the *attack* relation in our schema, nevertheless, as we plan to expand our work to longer scientific texts, where argumentative relations with different polarities are more likely to occur.

5   Argument Mining Experiments

In this section we describe experiments conducted to assess the potential of discourse annotations for the extraction of argumentative structures (units and relations) in computational linguistics abstracts. We model all of these subtasks as sequence tagging problems, which allows us to compare the performance obtained by learning them separately as well as jointly, in a multi-task setting. We also compare the performance obtained when learning and evaluating these high-level tasks with neural and non-neural models.

5.1   Tasks

In order to capture the argumentative structure of a text it is necessary to identify its components and how they are linked to each other. The following set of interrelated tasks are aimed at this objective:

- **B (units)**: Identify the boundaries of the argumentative units.
- **Ty (types)**: Identify the types of the units (i.e.: *proposal*)
- **Fu (function)**: Identify the argumentative functions (i.e.: *support*).
- **Pa (attachment)**: Identify the position of the parent argumentative unit.

---

[11] According to the position of the parent unit, there are 200 relations pointing forward and 92 in which the parent appears before in the text.

5.2   Experimental Setups

We compare the results obtained for each of the tasks mentioned in Sect. 5.1 with and without considering rhetorical information available in the RST layer of the corpus. In each case, we run four different learning algorithms.

- **Baselines**: Basic classifiers, that exploit correlations between the abstracts' argumentative structure and rhetorical or syntactic information.
- **CRF**: Conditional random fields (CRF) tagger.
- **BiLSTM-ST**: A separate BiLSTM-CRF sequence tagger for each task.
- **BiLSTM-MT**: A BiLSTM-CRF sequence tagger with the four main tasks jointly trained in a mult-task setting.

In all of the settings the tasks are modeled as token classification problems where the argumentative units are encoded with the BIO tagging scheme. All the classifiers (including the baselines) are trained and evaluated in a 10-fold cross-validation setting.[12]

Baselines. The result of the annotation process shows a correspondence of 93% between argumentative units and sentence boundaries. We therefore consider sentences as argumentative components for the implementation of the baseline algorithms. In order to predict argumentative functions, types and parents we generate simple classifiers based on the values of syntactic and/or discourse-level features and the classes to be predicted. We do this by mapping each value of the considered feature to its most frequent class in the training set. When no rhetorical information is included, we observed that the concatenation of the *lemma* of the syntactic root of the sentence and the sentence position is the best predictor in average (when all the tasks are considered). When rhetorical information is available, instead, the concatenation of the *discourse function* in which the token participates and the sentence BIO tag is the most predictive feature. It is relevant to note that this is a strong baseline to beat. For instance, the discourse relation predicts correctly the argumentative parent (Pa) for 57% of the argumentative units.[13]

CRF. For the CRF classifier we used Stanford's CRFClassifier [6] with unweighted attributes, including positional, syntactic and discourse features for the current, previous and next tokens.

- **Positional features**: sentence and EDU information: their position in the text, their boundaries and position of the token within them;

---

[12] For this pilot study the algorithms' hyperparameters—including the number of training epochs in the case of BiLSTM networks—were not optimized, as the main goal of this work was not to produce the best possible argument mining system but to obtain elements that would allow us to establish comparisons between the proposed approaches.

[13] In particular, for units that correspond to discourse roots—17% of all the units—the argumentative parent is predicted correctly 95% of the times.

- **Syntactic features**: lemma, POS; dependency-tree parent; dependency-tree relation;
- **Discourse features**: discourse function and parent in the discourse tree.

BiLSTM. For the BiLSTM networks we used the implementation made by the Ubiquitous Knowledge Processing Lab of the Technische Universität Darmstadt [23].[14] We used one BiLSTM layer with 100 recurrent units with Adam optimizer and naive dropout probability of 0.25. For the parent attachment task (Pa) we included an additional BiLSTM layer of 100 recurrent units. We used a Softmax classifier as the last layer of the network, except in the case of the task B (boundary prediction), in which we used a CRF classifier. We used a batch size of 10 and trained the networks for 30 epochs for each task and each training-test split of our cross-validation setting. The tokens were encoded as the concatenation of 300 dimensional dependency-based word embeddings [10] and 1024-dimensional ELMo word embeddings [21]. In addition to the tokens, the BiLSTMs are fed with the same features used for the implementation of the baselines (sentence boundaries and position, lemma of the syntactic root and discourse function), which are encoded as 10-dimensional embeddings.

6 Results and Analysis

The experiments are evaluated with the ConNLL criteria for named-entity recognition. A true positive is considered when both the boundaries and class (type, function, parent) match. A post-processing filter is run in order to ensure that all the BIO-encoded identified units are well-formed. In the case of impossible sequences (e.g. an I tag without the preceding B), the labels are changed to the most frequent ones in the argumentative unit (the boundaries are considered to be correct). In the reported results the predicted boundaries are considered.

Table 2. F1-measures with and without discourse info.

Algorithm	Function (B+Fu)		Type (B+Ty)		Attachment (B+Pa)	
	RST	No RST	RST	No RST	RST	No RST
Baseline	57.04	46.52	56.03	43.84	47.06	31.26
BiLSTM-MT	**71.88**	62.22	**68.78**	68.03	45.70	45.05
BiLSTM-ST	71.38	68.50	67.81	66.18	**47.90**	43.61
CRF	62.51	53.33	65.77	61.62	44.96	39.81

Table 2 shows the F1-measure obtained in average for each task with and without discourse information, respectively. Explicitly incorporating discourse information significantly contributes to the identification of the argumentative

---

[14] https://github.com/UKPLab/elmo-bilstm-cnn-crf.

function. It has also a more moderate but positive effect in predicting the argumentative units' types and attachment. It can be observed that, even with the limited amount of training data available and without optimizing their hyperparameters, the neural models perform considerably better than more traditional sequence labelling algorithms such as CRF. In particular, for the prediction of the argumentative units' functions. No strong conclusions can be drawn from these results with respect the advantage of training the tasks separately or jointly in a multi-task framework. Diverging results can be attributed to the different difficulty levels of the tasks and the small number of training examples. As more annotated data becomes available, we will be in a better position to explore how these tasks relate to each other and their mutual effect in a joint-learning setting. More experiments with different values for the hyperparameters and different regularization strategies need to be conducted in order to explore the effects of the inductive biases introduced by means of training several tasks in parallel. In this sense, we believe that it could be productive to explore other multi-task learning architectures, that account for the differences of difficulties of the various tasks, such as the hierarchical architecture proposed in [25].

6.1 Error Analysis

In terms of the observed errors, we see the same patterns in all the experimental settings, with the numbers varying according to the respective performances of the systems. In particular, for the **Ty** task, the highest rate of errors are due to mis-classifying units of type *means* as either *proposal*, which accounts for 21% of all the errors (in average), or *result*, which accounts for 11% of the errors. The mis-classification in the other direction: units of type *proposal* or *result* being mis-classified as *means* is less frequent but still significant, as it accounts for 9% and 5% of all the errors, respectively. Of significance are also the errors generated by the mis-classification of units of type *assertion* as either *proposal* or *result*, giving origin to 11% and 9% of all the errors produced in average by all the systems. In the case of the identification of the argumentative function (**Fu**), the main source of errors are due to the mis-classification between the classes *support* and *detail*, which accounts for 59% of all the errors (with roughly the same number of errors in both directions). In the case of the parent attachment task (**Pa**), the two most frequent errors are due to missing one argumentative unit (for instance, attaching a unit to the adjacent unit instead of the following one in the text), which accounts for 30% of all the errors and in assigning the wrong direction to the relation, which accounts for 35% of all the errors.

7   Conclusions

In this work we addressed the problem of identifying argumentative components and relations in scientific texts, a domain that has been recognized as particularly challenging for argument mining. We presented work aimed at assessing the potential value of exploiting existing discourse-annotated corpora for the extraction of argumentative units and relations in texts. Our motivation lies in the fact that discourse analysis, in general, and in the context of the RST framework, in particular, is a mature research area, with a large research community that have contributed a considerable number of tools and resources–including corpora and parsers–which could prove valuable for the advancement of the relatively newer area of argument mining. In order to test our hypothesis, we proposed and pilot-tested an annotation schema that we used to enrich, with a new layer of argumentative structures, a subset of an existing corpus that had previously been annotated with discourse-level information. The resulting corpus was then used to train and evaluate neural and non-neural models. Based on the obtained results, we conclude that the explicit inclusion of discourse data contributes to improve the performance of the argument mining models.

The results of this preliminary study are auspicious and motivate us to expand it. In particular, we aim at extending our argumentative layer of annotations to the full SciDTB corpus in an iterative process of semi-automatic annotation and evaluation. We believe that this enriched corpus would become a valuable resource to advance the investigation of argument mining in scientific texts. In order to identify arguments in un-annotated abstracts, we will also analyze the results obtained by training our models with discourse annotations obtained automatically, by means of existing RST parsers. In a complementary line, we will explore the potential offered by jointly learning to predict argumentative and discourse annotations in a multi-task environment. The models thus obtained could then be used to identify argumentative structures when no discourse annotations are available.

Acknowledgments. This work is (partly) supported by the Spanish Government under the María de Maeztu Units of Excellence Programme (MDM-2015-0502).

References

1. Aharoni, E., et al.: Context-dependent evidence detection, 3 July 2018. US Patent App. 14/720,847
2. Biran, O., Rambow, O.: Identifying justifications in written dialogs by classifying text as argumentative. Int. J. Semant. Comput. **5**(04), 363–381 (2011)
3. Bornmann, L., Mutz, R.: Growth rates of modern science: A bibliometric analysis based on the number of publications and cited references. J. Assoc. Inf. Sci. Technol. **66**(11), 2215–2222 (2015)
4. Cabrio, E., Tonelli, S., Villata, S.: From discourse analysis to argumentation schemes and back: Relations and differences. In: Leite, J., Son, T.C., Torroni, P., van der Torre, L., Woltran, S. (eds.) CLIMA 2013. LNCS (LNAI), vol. 8143, pp. 1–17. Springer, Heidelberg (2013). https://doi.org/10.1007/978-3-642-40624-9_1
5. Caruana, R.: Multitask learning. Mach. Learn. **28**(1), 41–75 (1997)
6. Finkel, J.R., Grenager, T., Manning, C.: Incorporating non-local information into information extraction systems by gibbs sampling. In: Proceedings of the 43rd Annual Meeting of the Association for Computational Linguistics, pp. 363–370. Association for Computational Linguistics (2005)

7. Green, N.: Identifying argumentation schemes in genetics research articles. In: Proceedings of the 2nd Workshop on Argumentation Mining, pp. 12–21 (2015)
8. Hyland, K.: Hedging in Scientific Research Articles, vol. 54. John Benjamins Publishing, Amsterdam (1998)
9. Kirschner, C., Eckle-Kohler, J., Gurevych, I.: Linking the thoughts: Analysis of argumentation structures in scientific publications. In: Proceedings of the 2nd Workshop on Argumentation Mining, pp. 1–11 (2015)
10. Komninos, A., Manandhar, S.: Dependency based embeddings for sentence classification tasks. In: Proceedings of the 2016 Conference of the North American Chapter of the Association for Computational Linguistics: Human Language Technologies, pp. 1490–1500 (2016)
11. Lauscher, A., Glavaš, G., Eckert, K.: ArguminSci: A tool for analyzing argumentation and rhetorical aspects in scientific writing. In: Association for Computational Linguistics (2018)
12. Lauscher, A., Glavaš, G., Ponzetto, S.P.: An argument-annotated corpus of scientific publications. In: Proceedings of the 5th Workshop on Argument Mining, pp. 40–46 (2018)
13. Li, S., Wang, L., Cao, Z., Li, W.: Text-level discourse dependency parsing. In: Proceedings of the 52nd Annual Meeting of the Association for Computational Linguistics (Volume 1: Long Papers), vol. 1, pp. 25–35 (2014)
14. Liakata, M., Saha, S., Dobnik, S., Batchelor, C., Rebholz-Schuhmann, D.: Automatic recognition of conceptualization zones in scientific articles and two life science applications. Bioinformatics **28**(7), 991–1000 (2012)
15. Liakata, M., Teufel, S., Siddharthan, A., Batchelor, C.: Corpora for the conceptualisation and zoning of scientific papers (2010)
16. Lippi, M., Torroni, P.: Argumentation mining: State of the art and emerging trends. ACM Trans. Internet Technol. **16**(2), 10 (2016)
17. Mann, W.C., Matthiessen, C., Thompson, S.A.: Rhetorical Structure Theory and text analysis. In: Mann, W.C., Thompson, S.A. (eds.) Discourse Description: Diverse Linguistic Analyses of a Fund-Raising Text. Pragmatics & Beyond New Series 16, pp. 39–78 (1992)
18. Moens, M.F.: Argumentation mining: Where are we now, where do we want to be and how do we get there? In: Post-Proceedings of the 4th and 5th Workshops of the Forum for Information Retrieval Evaluation, Article no. 2 (2013). https://dl.acm.org/citation.cfm?doid=2701336.2701635
19. Morey, M., Muller, P., Asher, N.: How much progress have we made on RST discourse parsing? A replication study of recent results on the RST-DT. In: Conference on Empirical Methods on Natural Language Processing, pp. 1319–1324(2017)
20. Peldszus, A., Stede, M.: Rhetorical structure and argumentation structure in monologue text. In: Proceedings of the Third Workshop on Argument Mining, pp. 103–112 (2016)
21. Peters, M., et al.: Deep contextualized word representations. In: Proceedings of the 2018 Conference of the North American Chapter of the Association for Computational Linguistics: Human Language Technologies, Volume 1 (Long Papers), vol. 1, pp. 2227–2237 (2018)
22. Radev, D.R., Muthukrishnan, P., Qazvinian, V., Abu-Jbara, A.: The ACL anthology network corpus. Lang. Resour. Eval. **47**(4), 919–944 (2013)
23. Reimers, N., Gurevych, I.: Reporting score distributions makes a difference: Performance study of LSTM-networks for sequence tagging. In: Proceedings of the 2017 Conference on Empirical Methods in Natural Language Processing, pp. 338–348 (2017)

24. Ruder, S.: An overview of multi-task learning in deep neural networks. arXiv preprint arXiv:1706.05098 (2017)
25. Sanh, V., Wolf, T., Ruder, S.: A hierarchical multi-task approach for learning embeddings from semantic tasks. arXiv preprint arXiv:1811.06031 (2019)
26. Sonntag, J., Stede, M.: GraPAT: A tool for graph annotations. In: Proceedings of the 2014 The International Conference on Language Resources and Evaluation, pp. 4147–4151 (2014)
27. Stab, C., Kirschner, C., Eckle-Kohler, J., Gurevych, I.: Argumentation mining in persuasive essays and scientific articles from the discourse structure perspective. In: ArgNLP, pp. 21–25 (2014)
28. Stede, M., Afantenos, S.D., Peldszus, A., Asher, N., Perret, J.: Parallel discourse annotations on a corpus of short texts. In: Proceedings of the 2016 The International Conference on Language Resources and Evaluation (2016)
29. Strehl, A., Ghosh, J.: Cluster ensembles—a knowledge reuse framework for combining multiple partitions. J. Mach. Learn. Res. **3**, 583–617 (2002)
30. Teufel, S.: Argumentative Zoning: Information extraction from scientific text. Ph.D. thesis, University of Edinburgh (1999)
31. Teufel, S., Siddharthan, A., Batchelor, C.: Towards discipline-independent argumentative zoning: Evidence from chemistry and computational linguistics. In: Proceedings of the 2009 Conference on Empirical Methods in Natural Language Processing: Volume 3-Volume 3, pp. 1493–1502. Association for Computational Linguistics (2009)
32. Wing, B., Baldridge, J.: Hierarchical discriminative classification for text-based geolocation. In: Proceedings of the 2014 Conference on Empirical Methods in Natural Language Processing. Association for Computational Linguistics, Doha, Qatar, October 2014
33. Yang, A., Li, S.: SciDTB: Discourse dependency treebank for scientific abstracts. In: Proceedings of the 56th Annual Meeting of the Association for Computational Linguistics (Volume 2: Short Papers), vol. 2, pp. 444–449 (2018)

TAGS: Towards Automated Classification of Unstructured Clinical Nursing Notes

Tushaar Gangavarapu$^{(\boxtimes)}$ (iD), Aditya Jayasimha, Gokul S. Krishnan (iD),
and Sowmya Kamath S. (iD)

Healthcare Analytics and Language Engineering (HALE) Lab,
Department of Information Technology, National Institute of Technology Karnataka,
Surathkal, Mangaluru, India
tushaargvsg45@gmail.com, adityajayasimha@gmail.com, gsk1692@gmail.com,
sowmyakamath@nitk.edu.in

Abstract. Accurate risk management and disease prediction are vital
in intensive care units to channel prompt care to patients in critical conditions and aid medical personnel in effective decision making. Clinical
nursing notes document subjective assessments and crucial information
of a patient's state, which is mostly lost when transcribed into Electronic Medical Records (EMRs). The Clinical Decision Support Systems
(CDSSs) in the existing body of literature are heavily dependent on the
structured nature of EMRs. Moreover, works which aim at benchmarking
deep learning models are limited. In this paper, we aim at leveraging the
underutilized treasure-trove of patient-specific information present in the
unstructured clinical nursing notes towards the development of CDSSs.
We present a fuzzy token-based similarity approach to aggregate voluminous clinical documentations of a patient. To structure the free-text
in the unstructured notes, vector space and coherence-based topic modeling approaches that capture the syntactic and latent semantic information are presented. Furthermore, we utilize the predictive capabilities
of deep neural architectures for disease prediction as ICD-9 code group.
Experimental validation revealed that the proposed Term weighting of
nursing notes AGgregated using Similarity ($TAGS$) model outperformed
the state-of-the-art model by 5% in AUPRC and 1.55% in AUROC.

Keywords: Healthcare analytics · Disease group prediction ·
Natural Language Processing · Risk assessment systems · Deep learning

1 Introduction

Risk assessment and disease prediction in Intensive Care Units (ICUs) have had
a prominent impact on clinical care and management [13]. As per US healthcare

This work is funded by the Government of India's DST-SERB Early Career Research
Grant (ECR/2017/001056) to Sowmya Kamath S.
T. Gangavarapu and A. Jayasimha—contributed equally to this work.

© Springer Nature Switzerland AG 2019
E. Métais et al. (Eds.): NLDB 2019, LNCS 11608, pp. 195–207, 2019.
https://doi.org/10.1007/978-3-030-23281-8_16

reports, more than 30 million patients visit hospitals annually, and 83% of these hospitals have adopted the Electronic Medical Record (EMR) systems [6]. In the recent years, a rapid increase in the adoption of EMRs in the hospitals of developed countries is also observed, which has prompted significant research towards modeling the patient data for diverse clinical tasks like mortality, length of stay, and hospital readmission prediction using various machine and deep learning approaches [15]. Such works have further been employed towards determining diagnostic measures needed to design and implement effective healthcare policies [8]. Despite these trends in western countries, hospitals in developing countries are yet to gain momentum in the implementation of EMRs.

Caregivers in developing countries most often resort to a human evaluation of available clinical notes for decision making and cause-effect inference [10]. Clinicians and nurses document subjective assessments and crucial information about a patient's state, which is often lost when transcribed into EMRs [4]. Clinical nursing notes remain largely unexplored for mining and modeling the rich and valuable patient-specific information. It is challenging to utilize unstructured clinical nursing notes to predict the clinical outcomes and events primarily due to their sparsity, rawness, complex linguistic and temporal structure, high-dimensionality, rich medical jargons, and abundant abbreviations [7]. How effectively the rich information embedded in unstructured clinical text is extracted and consolidated, determines the efficacy of their usage [14]. Due to the diverse manifold nature of prevalent disease symptoms, there is often a need for assigning multiple labels to a patient record in the database [1]. Risk assessment as ICD-9[1] code group prediction using clinician's notes can help in recognition of the onset and severity of the disease. Such assessments, when preceded by a timely response and effective communication by interdisciplinary care team members have been reported to result in a reduction in the hospital mortality rate [3].

Most state-of-the-art works [8,12] present machine learning models built on structured EMR data to facilitate various clinical prediction tasks. The few works that adopt deep learning models [5,13] neglect the rich patient-specific information present in the clinical nursing notes. In this paper, we utilize term weighting, word embedding (Doc2Vec), and coherence-based topic modeling (Latent Dirichlet Allocation (LDA)) approaches to structure the clinical nursing notes for capturing both the syntactic and semantic relationships between the textual features of the nursing notes, to aid in the accurate prediction of the ICD-9 code group. Deriving optimal data representations and eliminating redundant data from the nursing notes is achieved using a fuzzy similarity based data cleansing approach. Furthermore, we report the results of our exhaustive experimentation with three deep architectures including Multi-Layer Perceptron (MLP), Long Short Term Memory (LSTM), and Convolutional Neural Network (CNN).

The rest of this paper is structured as follows: Sect. 2 discusses relevant work in the area of our work. Section 3 presents our detailed methodology for deriving optimal data representations. Data modeling and deep architectures used in

---

[1] International Statistical Classification of Diseases and Related Health Problems.

ICD-9 code group prediction are discussed in Sect. 4. The experiments, evaluation, and results are discussed in detail in Sect. 5. Finally, we conclude with a summary and future research possibilities in Sect. 6.

2   Related Work

The prediction of prominent clinical outcomes and benchmarking the performance of the proposed machine and deep learning models is greatly facilitated by the availability of sizeable public patient datasets such as MIMIC-III [9]. In 2016, Pirracchio [12] presented a super learner algorithm which was an ensemble of various machine learning models. For the task of ICU mortality prediction, the super learner algorithm outperformed various severity scores including SAPS-II, SOFA, and APACHE-II. The preponderance of machine learning approaches over traditional prognostic scoring systems was emphasized. However, the obtained results were not benchmarked against the latest machine and deep learning models. The clinical task of mortality prediction was presented as a case study by Johnson *et al.* [8], who highlighted the challenges in replicating the results reported by 28 related and recent prominent publications on publicly available MIMIC-III database. They used an extracted set of features from the database and compared the reported performance against gradient boosting and logistic regression models. In order to take into consideration the significant heterogeneity in the studies and ensure a fairer comparison between approximate approaches, Johnson *et al.* [8] emphasized the need to improve the way of reporting the performance of clinical prediction tasks.

Harutyunyan *et al.* [5] used multitask recurrent neural networks to develop a comprehensive deep learning model and benchmarked their outcomes on four disparate clinical prediction tasks on the MIMIC-III database. Their work showed encouraging results in clinical prediction tasks with the use of deep learning approaches. However, their obtained performance was only benchmarked against a standard logistic regression model and LSTM model, and did not benchmark against machine learning models (including super learner) or severity scoring systems. An extensive set of benchmarking results on various clinical tasks such as the prediction of the ICD-9 code group, length of stay, and several versions of in-hospital mortality was presented by Purushotham *et al.* [13] on MIMIC-III, against various severity scoring systems and machine learning models. More recently, Krishnan and Kamath [10] benchmarked their performance for the ICU mortality prediction task. They used Word2Vec embeddings of the electrocardiogram reports in the MIMIC-III database and an unsupervised data cleaning approach using K-means clustering, followed by an extreme learning machine classifier for the prediction task.

Our work advances the efforts of these previous state-of-the-art approaches by exploring the potential use and availability of unstructured clinical notes, an under-tapped resource of rich patient-specific information. The EMR coding process often decimates the treasure-trove of information present in the clinician's notes. Our work addresses this issue by designing a clinical processing and representation generation methodology based on clinical concept extraction and topic modeling using deep learning models. Furthermore, we present an exhaustive comparative study to evaluate the performance of the proposed fuzzy similarity based data cleansing approach across a variety of deep learning models in the clinical task of multi-label ICD-9 code group prediction.

3   Materials and Methods

In this section, we present a brief description of the Natural Language Processing (NLP) pipeline depicted in Fig. 1. We also discuss the preprocessing steps employed to derive optimal data representations for the multi-label classification task of ICD-9 code group prediction.

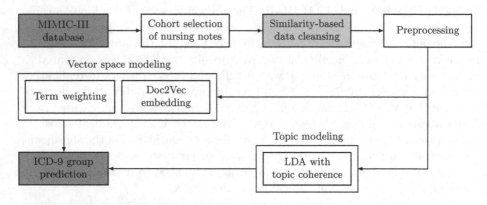

Fig. 1. NLP pipeline used to predict the ICD-9 code group.

3.1   Dataset and Cohort Selection

The Massachusetts Institute of Technology Lab for Computational Physiology developed MIMIC-III, a freely accessible large healthcare database. It comprises comprehensive and diverse de-identified healthcare data of more than 40,000 intensive care patients. Also provided are 223,556 nursing notes corresponding to 7,704 distinct ICU patients (*diagnoses_icd* table) extracted from 2,083,180 note events (*noteevents* table). For the preparation of our datasets, two criteria were employed in selecting the MIMIC-III subjects. First, the age at the time of a subject's admission to the ICU was used to identify subjects with age less than 15 (*patients* and *admissions* tables), and their records were removed. Second, only the first admission of each MIMIC-III subject to the hospital was considered, and all later admissions were discarded. Such selection was made to assure the prediction with the earliest detected symptoms aiding in faster risk assessment. Both steps were carried out in accordance with the existing literature [8,10,13]. Overall, the dataset elicited from the selected tables of the database encompassed nursing notes corresponding to 7,638 subjects with a median age of 66 years (Quartile Q_1–Q_3: 52–78 years).

3.2   Data Cleaning and Aggregation

It was observed that the data extracted from the MIMIC-III database had erroneous entries due to various factors including outliers, noise, duplicate or incorrect records, and missing values. First, we identified and removed the erroneous entries in these nursing notes by using the *iserror* attribute of the *noteevents* table set to 1. Second, the MIMIC-III subjects with duplicate records were identified and such records were deduplicated. After handling these erroneous entries, the resulting data corresponded to 6,532 patients.

A significant challenge in modeling the voluminous nursing notes and facilitating multi-label ICD-9 code group classification is the aggregation of multiple such notes of a specific MIMIC-III subject. Such notes may have many similar terms which can affect the vector representations significantly. Monge-Elkan (ME) token-based fuzzy similarity scoring approach is coalesced with Jaro internal scoring scheme to enable decision-making while handling multiple near-duplicate nursing notes of a subject. ME similarity handles alternate names, clinical abbreviations, and medical jargons. Jaro similarity, as an internal scoring strategy effectively handles spelling errors and produces a normalized similarity score between 0 and 1. Given two nursing notes η_p and η_q with $|\eta_p|$ and $|\eta_q|$ tokens ($\mathcal{C}_i^{(p)}$s and $\mathcal{C}_j^{(q)}$s) respectively, their ME similarity score with Jaro internal scoring is computed using,

$$\text{ME}_{\text{Jaro}}(\eta_p, \eta_q) = \frac{1}{|\eta_p|} \sum_{i=1}^{|\eta_p|} \max \left\{ \text{Jaro}(\mathcal{C}_i^{(p)}, \mathcal{C}_j^{(q)}) \right\}_{j=1}^{|\eta_q|} \tag{1}$$

where, the Jaro similarity score of two given tokens $\mathcal{C}_m$ of length $|\mathcal{C}_m|$ and $\mathcal{C}_n$ of length $|\mathcal{C}_n|$ with c matching characters and t transpositions, is computed using,

$$\text{Jaro}(\mathcal{C}_m, \mathcal{C}_n) = \begin{cases} 0, & \text{if } c = 0 \\ \frac{1}{3} \left(\frac{c}{|\mathcal{C}_m|} + \frac{c}{|\mathcal{C}_n|} + \frac{2c-t}{2c} \right), & \text{otherwise} \end{cases} \tag{2}$$

A pair of nursing notes are merged only if the ME score of that pair is lower than a preset threshold. Only the first record is retained and the second is purged when the ME score of a pair of nursing notes is greater than the threshold. Note that only the nursing notes and not the ICD-9 code groups are merged or purged based on similarity. To facilitate multi-label classification, we merge the corresponding ICD-9 codes across multiple nursing notes of a patient. We refer to the resultant nursing note for a patient obtained as a result of merging as the *aggregate nursing note* of that patient. In this study, the fuzzy-similarity threshold (θ) was empirically determined to be 0.825.

3.3   Data Preprocessing

The next step in the NLP pipeline is preprocessing the nursing notes to achieve data normalization. Preprocessing includes tokenization, stop-word removal, and stemming/lemmatization. Firstly, multiple spaces, punctuation marks, and special characters are removed. During tokenization, we split the text in each clinical nursing note into numerous smaller words (tokens). Using the NLTK English stopword corpus, stopwords among the generated tokens are removed. Furthermore, any references to images (file names such as '*MRI_Scan.jpg*') are removed, and character case folding is performed. Token removal based on its length based was not performed to mitigate any loss of vital medical information (such as '*MRI*' in '*MRI Scan*'). Finally, suffix stripping was facilitated by stemming, followed by lemmatization which aimed at converting the stripped words to their base forms. The words appearing in less than ten nursing notes were removed to lower the computational complexity and avoid overfitting.

3.4   Feature Modeling of Clinical Concepts

Let $\mathbb{S}$ denote the set of all aggregate nursing notes. Each aggregate nursing note η_i in $\mathbb{S}$ constitutes a variable length of tokens from a sizeable vocabulary $\mathbb{V}$, thus making $\mathbb{S}$ very complex. Therefore, a transformation of unstructured clinical text into an easier-to-use machine processable form is critically important. The performance and efficacy of the utilized deep architectures are heavily reliant on the optimal vector representations of the underlying corpus. We used vector space modeling and coherence-based topic modeling for feature modeling to enable an optimal representation of the patient cohort.

Vector Space Modeling. Vector space modeling of clinical concepts aims at representing each nursing note as a point in a multidimensional vector space of d dimensions (usually, $d \ll |\mathbb{V}|$). The term weighting scheme is a transformation of the Bag of Words (BoW) that assigns weight to tokens in an unsupervised manner. This scheme captures both the importance (occurrence frequency) and rarity (specificity) of a token in the given vocabulary. The weight ($W_i^{(p)}$) assigned to a term $w_i^{(p)}$ (of total $|w^{(p)}|$ terms) in a nursing note η_p (of total N nursing notes) occurring $f_i^{(p)}$ times is given by,

$$W_i^{(p)} = \begin{cases} \left(1 + log_2 f_i^{(p)}\right)\left(log_2 \frac{N}{|w^{(p)}|}\right), & \text{if } f_i^{(p)} > 0 \\ 0, & \text{otherwise} \end{cases} \tag{3}$$

The term weights of all the tokens in an aggregate nursing note are computed to obtain a vector ($\{W_i^{(p)}\}_{i=1}^{|\mathbb{V}|}$) in the machine processable form.

Although the term weighting scheme effectively captures the syntactic relation between the textual features, it often suffers from the curse of high dimensionality and sparsity. Furthermore, it does not capture the intuition that semantically similar nursing notes have similar representations (e.g., '*bone*' and '*fracture*'). Doc2Vec or Paragraph Vectors (PVs) overcome these shortcomings by efficiently learning the term representations in a data-driven manner. Doc2Vec numerically represents variable length documents as low dimensional, fixed length document embeddings. It is a simple neural network with one shallow hidden layer that learns the distributed representations and provides content-related measurement. It captures both semantic and syntactic textual features obtained from the nursing notes text corpus. The implementations in the Python Scikit-learn and Gensim packages were used on the transcribed clinical words, to extract the features modeled using vector space models. In this study, we used the PV distributed memory variant of Doc2Vec with a dimension size of 500 (trained for 25 epochs) due to its ability to preserve the word order in the nursing notes.

Coherence-Based Topic Modeling. Topic modeling aims at finding a set of topics (collection of terms) from a collection of documents (nursing notes) that best represents the documents in the corpus. LDA, a popular cluster analysis approach is a generative topic model based on the Bayesian framework of a three-layer structure including documents, topics, and terms. A soft probabilistic and flat clustering of terms into topics and documents into topics is facilitated by LDA. It posits that each nursing note and each term, belong, with a certain probability, to a set of topics. This topic modeling approach can capture the context of occurrence which is essential for accurate predictability by the underlying deep architectures.

Similar to other clustering methods, the challenge is to determine the correct number of LDA clusters. To cope with this issue, the Topic Coherence (TC) between topics is used to derive the optimal number of topics. TC is a way to evaluate the topic models with a much higher guarantee on human interpretability. In our work, we adopt coherence-based LDA as it accounts for the semantic similarity between the higher scoring terms. The implementation available in Python Gensim package was used implement LDA with TC. A normalized pointwise mutual information score was used as a confirmation measure due to its high correlation with human interpretability [2]. The number of topics for LDA models was set to 100 and the LDA matrix was built on a BoW representation of the nursing notes. Furthermore, the number of topics was determined to be 100, by comparing the coherence scores of several LDA models obtained by varying the number of topics from 2 to 500 (increments of 100).

4   ICD-9 Code Group Prediction

ICD-9 codes are a taxonomy of diagnostic codes used by medical personnel including doctors and public health agencies to classify diseases and a wide variety of symptoms, infections, disorders, causes of injury etc. Researchers have stressed the need to differentiate between full-code predictions and category-level (group) predictions due to the high granularity in the diagnostic code hierarchy [11]. Each code group[2] includes a set of similar diseases, and almost every health condition can be classified into a unique code group. In this research, we focus on

---

[2] The code ranges used for mapping can be found at http://tdrdata.com/ipd/ipd_SearchForICD9CodesAndDescriptions.aspx.

ICD-9 code group predictions as a multi-label classification problem, with each patient's nursing note mapped to more than one code group. All the ICD-9 codes for a given admission are mapped into 19 distinct diagnostic classes. Note that there are no records in the MIMIC-III database within the ICD-9 code range of 760–779. Furthermore, this study classifies all the Ref and V codes into the same code group.

4.1 Deep Neural Architectures

We used three deep neural architectures including MLP, LSTM, and CNN to make the ICD-9 code group predictions. We used the implementations available in the Python Keras package with Tensorflow backend for this purpose. All the presented deep models were trained to minimize a binary cross entropy loss function using an Adam optimizer, with a batch size of 128, for eight epochs.

Multi-layer Perceptron. MLP is a feed-forward neural network with an input layer, one or more hidden layers, and an output layer. The first layer takes the clinical terms in an aggregate nursing note as the input and uses the output of each layer as the input to the following layer. Each node of the hidden or output layer l is associated with a bias $(b^{(l)})$ and each node to node connection (from layer l to $l+1$) has a weight $(W^{(l,l+1)})$. A node in a layer l with an input $s^{(l)}$ is activated in the layer $l+1$ using an activation function $\mathbf{g}$ as $\mathbf{g}(W^{(l,l+1)} \cdot s^{(l)} + b^{(l+1)})$. Furthermore, MLP uses the backpropagation algorithm to calculate the gradient of the loss function, allowing it to learn an optimal set of weights and biases needed to minimize a suitable loss function. The ability of MLPs to solve problems stochastically enables them to approximate solutions even for extremely complex problems. In this research, we utilize an MLP network with one hidden layer of 75 nodes with a Rectified Linear Unit (ReLU) activation function and an output layer of 19 nodes with a sigmoid activation function.

Long Short Term Memory. LSTMs are a special type of Recurrent Neural Networks (RNNs) that effectively capture the long-term dependencies. LSTMs overcome the problem of vanishing gradients, typically observed in traditional RNNs. Capturing the context and long-term dependencies in the raw clinical text would be crucial in the accurate prediction of the ICD-9 code groups. A recurrent neuron in RNNs has a simple activation structure, similar to that in MLP. In LSTM networks, however, the recurrent neuron, termed as the LSTM memory cell is equipped with a much more complex structure. More specifically, given a nursing note η_i at a time step t, with an embedding of $s_t^{(i)}$, the output (h_t) and the state (c_t) of an LSTM memory cell can be given by,

$$c_t = f \odot c_{t-1} + i \odot g; \ h_t = o \odot \tanh(c_t) \tag{4}$$

where, $\odot$ represents element-wise multiplication, $\tanh(\cdot)$ is the hyperbolic tangent function, and i, f, o, and g are the values at the input gate, forget gate, output gate, and cell state respectively and are computed as ($\sigma(\cdot)$ denotes the sigmoid function),

$$
\begin{pmatrix} i \\ f \\ o \\ g \end{pmatrix} = \begin{pmatrix} \sigma \\ \sigma \\ \sigma \\ \tanh \end{pmatrix} W^{(l,l+1)} \begin{pmatrix} h_{t-1} \\ s_t^{(i)} \end{pmatrix}
\tag{5}
$$

Note that $W^{(l,l+1)}$ varies between the layers but is shared through time. In this study, the dimensions of the embedding and LSTM hidden state are 289 (17 time steps with 17 features each) and 300. The multi-label prediction is achieved by a sigmoid activation of the final LSTM output.

Convolutional Neural Network. CNN is a deep feed-forward neural network architecture which uses a variation of the MLP aimed at minimal processing. Let an aggregate nursing note η_i be modeled to produce an embedding of $s_{1:n}^{(i)}$, where $t_{k:k+l}$ refers to the concatenation of the terms $t_k, t_{k+1}, \ldots, t_{k+l}$. The computation of a new feature involving a convolution operation using a filter $f \in \mathbb{R}^{wn}$ on a window of w terms and bias b is given as $\mathbf{g}(f \cdot s_{k:k+w-1}^{(i)} + b)$. The filter f is applied to every possible window of terms in the embedding $\{s_{1:w}^{(i)}, s_{2:w+1}^{(i)}, \ldots, s_{n-w+1:n}^{(i)}\}$ to produce a feature map. The same process can be extended to multiple filters (with varying window sizes) to obtain multiple feature maps. The features from the penultimate layer are passed to a fully connected layer using an activation function. CNNs drastically reduce the number of hyper-parameters (weights and biases) to be learned by the network, thus reducing the training overhead. In this research, we employed one fully connected layer of 289 nodes with ReLU activation function, one CNN layer with 3×3 convolution window size and 19 feature map size. The multi-label prediction is achieved by a sigmoid activation of the final convolved output.

5   Experimental Results and Discussion

The experiments were performed using a server running Ubuntu OS with 56 cores of Intel Xeon processors, 128 GB RAM, 3 TB hard drive, and two NVIDIA Tesla M40 GPUs. To validate the proposed data modeling and prediction approaches, we performed an exhaustive benchmarking over the nursing notes data obtained from the MIMIC-III database. The primary challenge is the multi-label classification, where a set of ICD-9 code groups are predicted for each nursing note and a pairwise comparison of actual and predicted ICD-9 code groups for nursing note is performed. Seven standard evaluation metrics were used to evaluate the performance of each of the three deep learning models with respect to each of the four presented data modeling approaches. The evaluation metrics include Accuracy (ACC), F1 score, MCC score, Label Ranking Loss (LRL), Coverage Error (CE), Area Under the Precision-Recall Curve (AUPRC), and Area Under the

Table 1. Experimental results for ICD-9 code group prediction using MLP, LSTM, and CNN models

Data model	Classifier	Performance scores				
		ACC	F1	MCC	LRL	CE
TAGS (6,532 × 14,650)	MLP	**0.8130 ± 0.0005**	**0.6803 ± 0.0024**	**0.5704 ± 0.0020**	0.4199 ± 0.0024	18.5048 ± 0.0544
	LSTM	0.7946 ± 0.0011	0.6661 ± 0.0028	0.5365 ± 0.0027	0.4293 ± 0.0048	18.2477 ± 0.1010
	CNN	0.8049 ± 0.0007	0.6785 ± 0.0032	0.5594 ± 0.0022	**0.4124 ± 0.0047**	**18.1300 ± 0.1088**
Doc2Vec (6,532 × 500)	MLP	0.7903 ± 0.0019	0.6559 ± 0.0019	0.5212 ± 0.0032	0.4426 ± 0.0021	**18.6485 ± 0.0539**
	LSTM	**0.8005 ± 0.0017**	**0.6655 ± 0.0022**	**0.5386 ± 0.0032**	**0.4388 ± 0.0019**	18.6709 ± 0.0796
	CNN	0.7737 ± 0.0012	0.6381 ± 0.0028	0.4879 ± 0.0034	0.4599 ± 0.0033	18.6664 ± 0.0490
LDA (6,532 × 100)	MLP	0.7905 ± 0.0017	0.6397 ± 0.0027	0.5221 ± 0.0031	0.4610 ± 0.0030	18.8997 ± 0.0534
	LSTM	0.7842 ± 0.0013	0.6329 ± 0.0027	0.5078 ± 0.0014	0.4697 ± 0.0044	18.9252 ± 0.0607
	CNN	**0.8034 ± 0.0016**	**0.6643 ± 0.0013**	**0.5542 ± 0.0022**	**0.4361 ± 0.0018**	**18.6243 ± 0.0679**

ROC Curve (AUROC). Five-fold cross-validation was used to evaluate the predictability of the proposed models. Furthermore, the mean and standard errors (of the mean) of all the performance scores are presented. Table 1 shows the performance of all data modeling approaches and all deep prediction models processed using the proposed fuzzy token-based similarity approach ($\theta = 0.825$). We observe that the proposed Term weighting of nursing notes AGgregated using Similarity ($TAGS$) model, modeled with MLP outperforms more complex vector space and topic models.

Table 2. AUPRC and AUROC performance of the proposed ICD-9 code group prediction models

Data model	Classifier	Performance scores	
		AUPRC	AUROC
$TAGS$ (6,532 × 14,650)	MLP	**0.6291 ± 0.0027**	0.7738 ± 0.0013
	LSTM	0.5990 ± 0.0014	0.7646 ± 0.0025
	CNN	0.6153 ± 0.0031	**0.7817 ± 0.0023**
Doc2Vec (6,532 × 500)	MLP	0.5914 ± 0.0016	0.7562 ± 0.0013
	LSTM	**0.6076 ± 0.0033**	**0.7600 ± 0.0010**
	CNN	0.5686 ± 0.0030	0.7433 ± 0.0019
LDA (6,532 × 100)	MLP	0.5965 ± 0.0016	0.7497 ± 0.0017
	LSTM	0.5865 ± 0.0012	0.7431 ± 0.0017
	CNN	**0.6181 ± 0.0011**	**0.7649 ± 0.0011**

AUPRC measures the number of true positives from positive predictions and is more relevant since the data extracted from the MIMIC-III database is highly imbalanced. Most previous works including the state-of-the-art model [13] are heavily reliant on the structured nature of the EMRs modeled in the form of feature sets to aid the prediction of clinical outcomes. Table 2 presents the AUPRC and AUROC performance of the proposed ICD Code group prediction models. From Fig. 2, it can be noted that the proposed $TAGS$ model consistently outperforms the existing state-of-the-art model by 5% in AUPRC and 1.55% in AUROC. The previous works do not benchmark metrics other than AUROC and AUPRC presented in this study. We argue that the presented metrics aid in the measurement of various aspects of the predictive model's performance including precision and recall which are vital in critical clinical tasks. The richness and abundance of information captured by the unstructured nursing notes are often lost in the structured EMRs coding process [4]. From the results, it can be noted that the $TAGS$ model captures the discriminative features of the clinical nursing notes, eliminates redundancy, and purges anomalous data effectively aiding the deep learning classifier to learn and generalize, and such modeling results in the improvement of the clinical decision-making process.

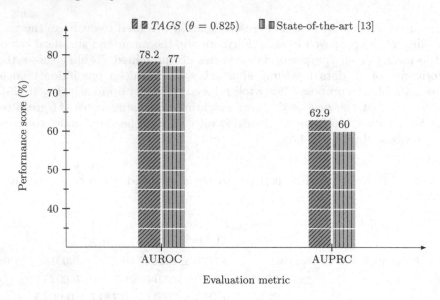

Fig. 2. Comparison of the *TAGS* model with the state-of-the-art model.

6   Concluding Remarks

Clinical nursing notes hold a treasure trove of patient-specific information. The voluminous and heterogeneous nature of the unstructured nursing notes with complex linguistic structure makes it hard to model them. In this paper, we presented a fuzzy similarity based matching approach to eliminate redundant and anomalous data resulting in reduced cognitive burden and enhancement in the clinical decision-making process. Vector space modeling and Coherence topic modeling approaches were built on the aggregated data to capture the syntactic and latent semantic information in the nursing notes and effectively leverage it for disease prediction. It was observed that the proposed *TAGS* model achieved superior performance when benchmarked against the structure EMR based state-of-the-art model by 5% in AUPRC and 1.55% in AUROC. Furthermore, its performance was benchmarked using seven evaluation metrics which are vital in the assessment of the predictive capability of the proposed models, especially in clinical tasks. Our model built on unstructured text eliminates the dependency on EMRs which is extremely useful in countries with low EMR adoption rates. As part of future work, we aim at validating the *TAGS* model on real-time clinical records. We also intend to improve the predictive capabilities of our models, focusing on building time-aware prediction architectures in real-time.

References

1. Baumel, T., Nassour-Kassis, J., Cohen, R., Elhadad, M., Elhadad, N.: Multi-label classification of patient notes a case study on ICD code assignment. arXiv preprint arXiv:1709.09587 (2017)

2. Bouma, G.: Normalized (pointwise) mutual information in collocation extraction. Proc. GSCL, 31–40 (2009)
3. Collins, S.A., Cato, K., Albers, D., Scott, K., et al.: Relationship between nursing documentation and patients' mortality. Am. J. Crit. Care **22**(4), 306–313 (2013)
4. Dubois, S., Romano, N., Kale, D.C., Shah, N., Jung, K.: Learning effective representations from clinical notes. arXiv preprint arXiv:1705.07025 (2017)
5. Harutyunyan, H., Khachatrian, H., Kale, D.C., Galstyan, A.: Multitask learning and benchmarking with clinical time series data. arXiv preprint arXiv:1703.07771 (2017)
6. Henry, J., Pylypchuk, Y., Searcy, T., Patel, V.: Adoption of electronic health record systems among us non-federal acute care hospitals: 2008-2015. ONC Data Brief **35**, 1–9 (2016)
7. Jo, Y., Lee, L., Palaskar, S.: Combining LSTM and latent topic modeling for mortality prediction. arXiv preprint arXiv:1709.02842 (2017)
8. Johnson, A.E., Pollard, T.J., Mark, R.G.: Reproducibility in critical care: a mortality prediction case study. In: Machine Learning for Healthcare Conference, pp. 361–376 (2017)
9. Johnson, A.E., et al.: MIMIC-III, a freely accessible critical care database. Sci. Data **3**, 160035 (2016)
10. Krishnan, G.S., Sowmya Kamath, S.: A supervised learning approach for ICU mortality prediction based on unstructured electrocardiogram text reports. In: Silberztein, M., Atigui, F., Kornyshova, E., Métais, E., Meziane, F. (eds.) NLDB 2018. LNCS, vol. 10859, pp. 126–134. Springer, Cham (2018). https://doi.org/10.1007/978-3-319-91947-8_13
11. Larkey, L.S., Croft, W.B.: Automatic assignment of ICD9 codes to discharge summaries. Technical report, University of Massachusetts at Amherst, Amherst, MA (1995)
12. Pirracchio, R.: Mortality prediction in the ICU based on MIMIC-II results from the super ICU learner algorithm (SICULA) project. Secondary Analysis of Electronic Health Records, pp. 295–313. Springer, Cham (2016). https://doi.org/10.1007/978-3-319-43742-2_20
13. Purushotham, S., Meng, C., Che, Z., Liu, Y.: Benchmarking deep learning models on large healthcare datasets. J. Biomed. Inform. **83**, 112–134 (2018)
14. Wang, Y., et al.: MedSTS: a resource for clinical semantic textual similarity. Lang. Resour. Eval., 1–16 (2018)
15. Waudby-Smith, I.E., Tran, N., Dubin, J.A., Lee, J.: Sentiment in nursing notes as an indicator of out-of-hospital mortality in intensive care patients. PLoS ONE **13**(6), e0198687 (2018)

Estimating the Believability of Uncertain Data Inputs in Applications for Alzheimer's Disease Patients

Fatma Ghorbel[1,2]([✉]), Fayçal Hamdi[1], and Elisabeth Métais[1]

[1] CEDRIC Laboratory, Conservatoire National des Arts et Métiers (CNAM),
Paris, France
fatmaghorbel6@gmail.com,
{faycal.hamdi,metais}@cnam.fr
[2] MIRACL Laboratory, University of Sfax, Sfax, Tunisia

Abstract. Data believability estimation is a crucial issue in many application domains. This is particularly true when handling uncertain input data given by Alzheimer's disease patients. In this paper, we propose an approach, called DBE_ALZ, to estimate quantitatively the believability of uncertain input data in the context of applications for Alzheimer's disease patients. In this context, data may be given by Alzheimer's disease patients or their caregivers. The believability of an input data is estimated based on its reasonableness compared to common-sense standard and personalized rules and the reliability of its authors. This estimation is based on Bayesian networks and Mamdani fuzzy inference systems. We illustrate the usefulness of our approach in the context of the Captain Memo memory prosthesis. Finally, we discuss the encouraging results de-rived from the evaluation of our approach.

Keywords: Applications for Alzheimer's disease patients ·
Uncertain data inputs · Data believability estimation · Probability theory ·
Fuzzy set theory · Patterns

1 Introduction and Motivation

In the context of the VIVA[1] project ("*Vivre à Paris avec Alzheimer en 2030 grâce aux nouvelles technologies*"), we are proposing a memory prosthesis, called Captain Memo [1], to help Alzheimer's patients to palliate mnesic problems. In this prosthesis, personal data of the patient are structured as a semantic knowledge base using an ontology called, PersonLink [2]. This multicultural and multilingual OWL 2 ontology enables storing, modeling and reasoning about interpersonal relationships (e.g. husband, half-brothers) and people description (e.g., name and lived events).

Captain Memo is conceived to be used by patients having earliest symptoms of Alzheimer. As this disease progresses, Captain Memo will use the stored data, entered by patients, to afterwards supply a set of services that help them in some of their

---

[1] http://viva.cnam.fr/.

© Springer Nature Switzerland AG 2019
E. Métais et al. (Eds.): NLDB 2019, LNCS 11608, pp. 208–219, 2019.
https://doi.org/10.1007/978-3-030-23281-8_17

daily activities. However, these particular users, living in a form of uncertainty, may introduce inconsistent data. For example, an Alzheimer's patient who has, in reality, only one son named Paul, could declare that he also has a daughter named Juliette. Captain memo should verify this information before introducing it in the knowledge base, as it will lead the system to infer wrong conclusions or, worst, to undecidable reasoning. For the previous example, the Captain Memo service that generates genealogic trees, will show that Juliette is the daughter of the patient and the sister of Paul. The wrong fraternity link was automatically inferred from the wrong entered input.

To deal with this issue, a process estimating data believability should be added to our prosthesis. This process, which assigns to each input a believability value, will allow Captain Memo to only accept input having a high believability degree. In the literature, only a few approaches address the assessment of data believability. In [6] and [7], the authors proposed a general context-independent estimation of believability. However, in the context of Alzheimer's disease, it is necessary to consider the particular profile of patients to correctly estimate the believability of data they entered. Indeed, the fuzziness and imprecision of input data related to the profile of each patient should be taken into account. Some of the existing applications intended to this category of users, deal with this problem by only considering inputs of the patients' caregivers (they assume that they should be more accurate than the patients' inputs). The issue is that, also these inputs have to be assessed as, for instance, caregivers may not necessarily know all information related to the patient's life.

In this paper, we propose an approach, called DBE_ALZ, to estimate quantitatively the believability of uncertain input data in the context of applications for Alzheimer's disease patients. Data may be given by Alzheimer's disease patients or their caregivers. Based on generated believability degrees, a set of decisions can be made (e.g., inferring or not based on given data). Our approach is based on probability and fuzzy set theories to reason about uncertain and imprecise knowledge.

The rest of the present paper is organized as follows. Section 2 is devoted to present the data believability research field. Section 3 details the proposed DBE_ALZ approach. Section 4 presents and discusses experimental results. Finally, Sect. 5 summarizes the main contributions of our work and gives some future directions.

2 Data Believability: Definition and Estimation

In [3, 4] and [5] data believability is defined by authors as "*the extent to which data are accepted or regarded as true, real and credible*". We have adapted this definition to our case by adding the relationship with the context of use. Thus, we defined believability as "*the extent to which data are, **in a specific context**, accepted or regarded as true, real and credible*".

The estimation of data believability is based on the definition of a set of dimensions. To each dimension, a set of metrics are defined. Dimensions are used to evaluate qualitatively the believability and the metrics are used to evaluate it quantitatively.

Only a few approaches have been proposed to estimate believability of the data e.g., [6] and [7]. Lee et al. [6] proposed three dimensions to evaluate the data's believability: *(i)* the believability of source which is defined as the data originating from a trustworthy source *(ii)* the believability according to internal common-sense standard which is the extent to which a data value is possible, consistent over sources (different sources agree on the data value) and consistent over time (data value is consistent with past data values), and *(iii)* the believability based on temporality of data which is the extent to which a data value is credible based on proximity of transaction time to valid times and derived from data values with overlapping valid times. These dimensions remain quite general and enable only qualitative estimation. Besides, they do not provide a formal definition of their metric. Prat and Madnick [7] proposed an approach to estimate quantitatively the data's believability based on provenance metadata i.e., the origin and subsequent processing history of data. This approach uses the believability assessment dimensions proposed by [6]. It is 3 aspects: *(i)* definition of metrics for assessing the believability of data sources *(ii)* definition of metrics for assessing the believability of data resulting from one process run, and *(iii)* assessment of believability as a whole based on all the sources and processing history of data.

Compared to existing approaches, our work provides dimensions and metrics that are suitable for a context of use 'Alzheimer's patients'. To assess data believability, profiles of both patients and caregivers are taken into account. Parameters such as fuzziness introduced by patients or caregiver knowledge are considered to better ad-just the calculation of the data believability degree.

3   DBE_ALZ Approach: Estimating the Believability of Uncertain Data Inputs in Applications for Alzheimer's Disease Patients

Our DBE_ALZ approach provides a quantitative estimation of the believability of uncertain data inputs in applications for Alzheimer's patients. In this context, data may be given by Alzheimer's patients or their caregivers. It generates, for each input, a degree of credibility C (C $\in$ [0, 1], 0 and 1 represent, respectively, completely unbelievable and completely believable). It is based on probabilities and fuzzy set theories and is composed of three modules: "data reasonableness estimation", "author reliability estimation" and "data believability estimation". Figure 1 summarizes this approach.

3.1   Data Reasonableness Estimation

This module estimates the reasonableness of the uncertain input data. It generates a score R (R $\in$ [0, 1], 0 and 1 represent, respectively, completely unreasonable and completely reasonable). It is based on the probability theory to deal with uncertainty and the Bayesian Network model. It is composed of three sub-modules: "input data pattern matching", "verification rule pattern generation" and "verification rule fulfillment".

Input Data Pattern Matching. This sub-module takes as input the given uncertain data IK. It returns the associated pattern PIK thanks to a matching process with the pre-established input data patterns base PI = {PI1, PI2... PIN}. For instance, if we have the

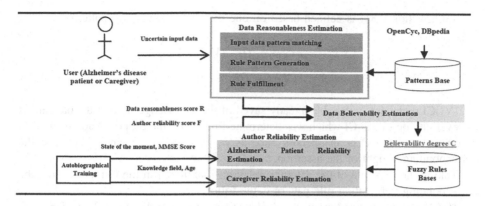

Fig. 1. Our DBE_ALZ approach.

following data: "Philippe is the father of Pierre". The corresponding pattern is "Per-son X father Person Y".

Verification Rule Pattern Generation. This sub-module takes as input the input data pattern PIK and returns the associated verification rule pattern PRK.

$R_P = \{R_{P1}, R_{P2}... R_{PN}\}$ is a verification rule patterns base. For each input data pattern IPK, a rule pattern PRK is associated. A verification rule pattern is defined as the following:

$$IF\ P_{IK}\ THEN\ A_{K/1}...A_{K/N}$$

The resulting part consists of a conjunction and/or disjunction of one or more assertions AK/I that ought to be fulfilled to confirm that IK is reasonable. We classify them into common sense and personalized assertions. Common sense ones estimate the data reasonableness regarding common sense standards. For instance, the father's age is always higher than the son's age. We use OpenCyc[2] ontology that proposes millions of pieces of knowledge that compose human common sense. Personalized assertions estimate the data reasonableness based on user's background. They depend on a number of parameters specified by an expert e.g., culture. For instance, in the United States, there is a high probability that the first name of a son is the same as that of the father (or at least includes the father first name). However, most likely it is not the same in many other cultures.

We associate a weight WK/I to each personalized assertion AK/I to estimate its validity according to the associated parameters. These weights are determined by interrogating, via a set of SPARQL queries, Linked Open Data datasets e.g., DBpedia[3] and Wikidata[4].

---

[2] http://www.cyc.com/opencyc/.

[3] http://dbpedia.org/.

[4] https://www.wikidata.org/.

An assertion AK/I is defined based on one of the following two patterns:

$$P_{A1} = \left\{ (V_1)_{C1}, \ OP, \ (V_2)_{C2,} W_{AI} \right\}$$
$$P_{A2} = \{ (V_1) \ C_1, \ OP, \ C; \ W_{AI} \}$$

(VI)CI represents a variable already saved in the knowledge base of the patient. Its believability degree is CI which was determined based on the proposed approach. C is a constant value. OP is an operator such as =, $\neq$, <, >, $\leq$, $\geq$ and $\in$. WAI is the corresponding weight of the assertion.

Taking the example of the paternity relation, mentioned in the last subsection, the corresponding verification rule pattern is the following:

IF ("Person X father Person Y") THEN ({Age (X) C1 > Age (Y) C2; WA1 = 1} $\bigwedge$ {Gender(X) C3 = "Man"; WA2 = 1} $\bigwedge$ {Last name(X) C4 = Last name(Y) C5; WA3 = 0.99} $\bigwedge$ {Nationality(X) C6 = Nationality(Y) C7; WA4 = 0.99} $\bigwedge$ {First name (X) C8 <> First name(Y) C9 WA5 = 0.99}).

Verification Rule Fulfillment. This sub-module takes as input the verification rule pattern PRK and returns the associated data reasonableness score R.

The verification rule pattern presents two sources of uncertainty. First, each personalized assertion is uncertain. Its certainty is equal to its associated weight. Second, each assertion refers to at least one variable VI that represents a data stored in the knowledge base of the patient. Its certainty is equal to the believability degree CI. We use Bayesian networks to reason about uncertain facts.

For each assertion AK/I, a Bayesian network pattern BNK/I is associated to determinate a score representing the extent to which the assertion is fulfilled. It is determined based on the pattern of the assertion, as shown in Table 1.

V_VI, V_WK/I and V_AK/I are probabilistic variables. They represent, respectively, a data stored in the knowledge base of the patient, the assertion's weight and the probability of fulfilling the assertion.

For each verification rule pattern PRK, we associate a Bayesian network pattern BNK. It is formed by the N Bayesian networks BNK/I and a probabilistic variable R. This variable depends on the probabilistic variables V_AK/I, as shown in Fig. 2.

Finally, we instantiate the Bayesian network pattern BNK based on the input data IK and the data saved in the knowledge base of the patient.

Taking the mentioned example of the paternity relation, the corresponding Bayesian Network pattern is shown in Fig. 3.

3.2   Author Reliability Estimation

This module estimates the reliability of the data's author (Alzheimer's disease patient or caregiver). It generates a score F (F $\in$ [0, 1], 0 and 1 represent, respectively, completely unreliable and completely reliable). It is composed of two sub-modules: "Alzheimer's patient reliability estimation" and "caregiver reliability estimation". The first sub-module is activated when the data input is given by the patient. The second one when it is given by a person from the patient's surroundings.

Table 1. Bayesian network patterns associated to assertion patterns.

Assertion pattern	Associated Bayesian network pattern
$P_{A1} = \{(V_1)_{C1}, \ OP, (V_2)_{C2} \ ; W_{AI}\}$	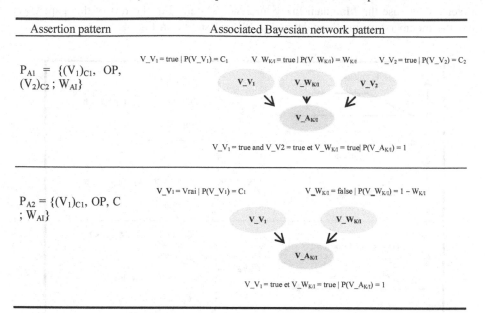
$P_{A2} = \{(V_1)_{C1}, \ OP, C \ ; W_{AI}\}$	

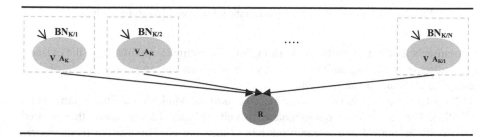

Fig. 2. Bayesian network pattern associated to verification rule pattern.

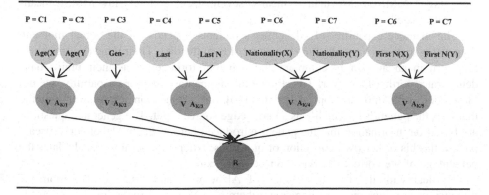

Fig. 3. An example of a Bayesian network pattern.

These two sub-modules are based on the fuzzy set theory to deal with imprecision. Precisely, we use the Mamdani fuzzy inference system. For the rest of this paper, we use the membership functions defined in [8] and shown in Fig. 4.

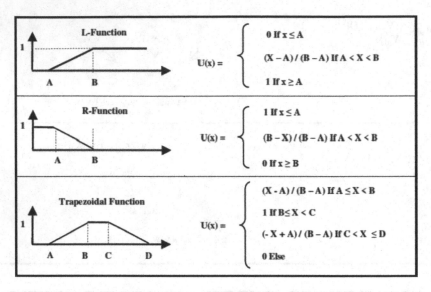

Fig. 4. L-Function, R-Function and Trapezoidal membership functions [8].

Alzheimer's Patient Reliability Estimation. To estimate qualitatively the Alzheimer's disease patient reliability score, two dimensions are proposed: "stage of Alzheimer" and "state of the moment".

To determinate the stage of Alzheimer, we used the Mini Mental State Examination (MMSE). It is a 30-point questionnaire, which includes 12 questions, that is used extensively in clinical and research settings to measure cognitive impairment. Scores range from 0 to 30, with higher scores indicating better performance. According to [9], MMSE better than 20 means "mild stage", MMSE between 10 and 19 means "moderate stage", MMSE below than 10 means "severe stage" and very low MMSE means "terminal stage".

The memory capability of Alzheimer's patients depends on the actual moment. Sometimes, patients have stunning moments of total lucidity. Their minds have seemed to return in an amazingly complete and coherent form, even as their brains have deteriorated further than ever. To estimate the momentary memory capability, we use our earlier work Autobiographical Training [10]. It is a "Question and Answer" training that uses the patient's private life as a knowledge source input. The generated questions are based on information that the patient introduced before. Each Alzheimer's disease patient has his or her own collection of questions. After each session, it calculates the percentage of the correct answers ("Successful Score").

To deal with the fact that "stage of Alzheimer" and "state of the moment" dimensions are imprecise, we implement them as a Mamdani fuzzy inference system.

This system takes as input two fuzzy variables related to the two dimensions and returns the patient reliability score F. The variable "Alzheimer_Stage" related to the "stage of Alzheimer" dimension has four linguistic labels {TerminalStage, SevereStage, ModerateStage and MildStage}. TerminalStage has a L-Function membership function which has as parameters A = 2 and B = 5. SevereStage has a trapezoidal member-ship function which has as parameters A = 2, B = 5, C = 8 and D = 12. ModerateStage has a trapezoidal membership function which has as parameters A = 8, B = 12, C = 18 and D = 22. MildStage has R-Function membership function which has as parameters A = 18 and B = 22. The variable "Momentary_State_Memory" related to the "state of the moment" dimension is based on the "Successful Score" returned by the "Question and Answer" training. It has the linguistic labels: {Confused, Average and Well-Remembered}. Confused has a L-Function membership which has as parameters A = 20 and B = 40. Average has a trapezoidal function membership which has as parameters A = 20, B = 40, C = 60 and D = 80. Well-Remembered has an R-Function membership function which has as parameters A = 60 and B = 80. The generated score is based on pre-established fuzzy rules base.

Caregiver Reliability Estimation. To estimate qualitatively the caregiver reliability score, two dimensions are proposed: "knowledge field" and "age".

The caregivers do not necessarily know all information related to the patient's life. For instance, a friend of the patient knows all their friends in common. However, he or she does not necessarily know, for example, all his or her family members and his or her colleague. To estimate the "knowledge field" dimension, we also use our earlier work Autobiographical Training. The caregiver response to a set of questions related to a specific field. After responding all questions, it calculates the percentage of the correct answers ("Successful Score"). Based on this score, we estimate the extent to which the caregiver is reliable on this specific field.

The reliability of the caregivers depends on their age. We consider that all users under the age of six and over ninety are unreliable. Users between the ages of thirty and sixty are very reliable. These values are validated by a neurologist doctor.

As the mentioned dimensions are imprecise, we infer the caregiver reliability score using a Mamdani fuzzy inference system. This system takes as input two fuzzy variables related to the two dimensions and returns the author reliability score F of the caregiver. For instance, the variable "Knowledge_Field" related to the "knowledge field" dimension has the linguistic labels {WeakKF, GoodKF and VeryGoodKF}. WeakKF has a L-Function membership function which has as parameters A = 20 and B = 40. GoodKF has a trapezoidal membership function which has as parameters A = 20, B = 40, C = 60 and D = 80. VeryGoodKF has a R-Function membership function which has as parameters A = 80 and B = 100. A fuzzy rules base is pre-established.

3.3   Data Believability Estimation

This module takes as input the inputs of the two other modules. It returns the believability degree of the input data. The author reliability score F and the data reasonableness score R are imprecise. Thus, we implement this module as a Mamdani fuzzy inference system that takes into account two fuzzy variables representing these scores. For instance, the variable "Reasonableness_Score" related to the reasonableness score has the linguistic labels {WeakRS, GoodRS and VeryGoodRS}. WeakRS has a L-Function membership function which has as parameters A = 0, 2 and B = 0, 4. GoodRS has a trapezoidal membership function which has as parameters A = 0,2, B = 0,4, C = 0,6 and D = 0,8. VeryGoodRS has a R-Function membership function which has as parameters A = 0,8 and B = 1. A fuzzy rules base is pre-established.

4   Experimentation

A Java-based prototype is implemented based on the DBE_ALZ approach. It uses jFuzzyLogic [11] (for implementing industry standards related to fuzzy logic) and JavaBayes [12] (for implementing Bayesian networks).

4.1   Application to the Captain Memo Memory Prosthesis

We integrate the DBE_ALZ prototype in Captain Memo to estimate the believability of the data given by Alzheimer's disease patients or their caregivers.

To facilitate the matching process between an input data and the corresponding pattern, we provide choices to select from dropdown lists; as shown in Fig. 5.

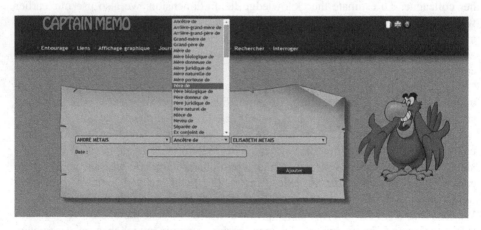

Fig. 5. Family links entry based on dropdown lists.

Based on the believability degrees generated by the DBE_ALZ approach, a set of corrective actions are proposed to guarantee the quality of the services offered by the Captain Memo. *(i)* Only data having a believability degree greater than 0,8 are saved in the knowledge base of the patient. *(ii)* Our approach is useful in case of contradictory input data. We rely only on the data having the higher believability degree.

4.2   Evaluation

This evaluation is done in the context of Captain Memo. A total of 12 Alzheimer's disease patients P = {P1 ... P12} and their associated caregivers C = {C1 ... C12} were recruited to participate in this evaluation study. All caregivers are first-degree relatives. {P1 ... P8} were early stage Alzheimer's patients. The others were moderate stage Alzheimer's disease patients. They were aged between 65 years old 73 years old (median = 68 years). Their MMSE scores ranged from 15 to 29 at the baseline. Most Alzheimer's patients were living in a nursing home in Sfax/Tunisia. We asked each patient's legal sponsor for the consent letter.

Two scenarios are proposed:

- "DBE_ALZ @ 2 weeks" scenario: We integrate the DBE_ALZ prototype in Captain Memo. All data entered by the Alzheimer patient P_i and having a credibility degree higher than 0,8 are saved in a $KB_{i/s1}$ (knowledge base that corresponds to the data entered by the patient P_i based on the first scenario). $KB_{i/s1}$ are generated after 2 weeks of using Captain Memo. Each caregiver C_i is requested to identify only the true facts given by the patient. The last ones formed the gold standard knowledge base $KB_{i/GS1}$.

- "DBE_ALZ @ 14 weeks" scenario: We integrate the DBE_ALZ prototype in Captain Memo. All data entered by the Alzheimer patient P_i and having a credibility degree higher than 0,8 are saved in a knowledge base $KB_{i/s2}$. $KB_{i/s2}$ are generated after 14 weeks of using Captain Memo. Each caregiver C_i is requested to identify only the true facts given by the patient. The last ones formed the gold standard knowledge base $KB_{i/GS2}$.

We compare the generated $KB_{i/s1}$ and $KB_{i/s2}$ knowledge bases against the golden standard ones. We use the Precision evaluation metric. $P_{i@2}$ ($|KB_{i/s1} \cap KB_{i/GS1}|/|KB_{i/s1}|$) and $P_{i@14}$ ($|KB_{i/s2} \cap KB_{i/GS2}|/|KB_{i/s2}|$) represent, respectively, the Precision associated to the patient P_i according to the first and second scenarios. Table 2 shows the results.

Table 2. Evaluation's results.

$P_{1@2}$	$P_{2@2}$	$P_{3@2}$	$P_{4@2}$	$P_{5@2}$	$P_{6@2}$	$P_{7@2}$	$P_{8@2}$	$P_{9@2}$	$P_{10@2}$	$P_{11@2}$	$P_{12@2}$	Mean
0,72	0,75	0,71	0,85	0,92	0,9	0,69	0,76	0,77	0,77	0,82	0,89	0,78
$P_{1@14}$	$P_{2@14}$	$P_{3@14}$	$P_{4@14}$	$P_{5@14}$	$P_{6@14}$	$P_{7@14}$	$P_{8@14}$	$P_{9@14}$	$P_{10@14}$	$P_{11@14}$	$P_{12@14}$	Mean
0,86	0,78	0,73	0,87	0,92	0,9	0,75	0,89	0,79	0,85	0,96	0,95	0,83

The overall means of the precision associated to "DBE_ALZ @ 14 weeks" scenario is better than the overall means of the precision associated to "DBE_ALZ @ 2 weeks" scenario. This value is ameliorated as the data reasonableness scores were improved. Indeed, the knowledge bases of the patients store more data from one navigation session to another. As a result, more fuzzy rules are activated to determinate these scores.

5 Conclusion and Future Work

In this paper, we presented the DBE_ALZ approach that estimates quantitatively the believability of uncertain and imprecise input data in applications for Alzheimer's disease patients. We used Bayesian networks and Mamdani fuzzy inference systems to take into account, when we calculate the input believability degree, the uncertainty and the imprecision introduced by Alzheimer's disease patients or their caregivers.

At the beginning, we elaborated a state of the art focusing on data believability estimation. This study showed that existing works do not consider, in the definition of data believability dimensions and metrics, the context of use. To the best of our knowledge, there is no approach that proposes dimensions or metrics to estimate the believability of input data in applications for Alzheimer's disease patients. In our approach we proposed two dimensions to estimate the believability of such data. The first one represents the reasonableness of the data. Compared to related work, this dimension is measured not only based on common-sense standard, but also on a set of personalized rules. The second dimension estimated the reliability of the Alzheimer's patients or their caregivers. Two metrics were used to estimate the reliability of the patients ("stage of Alzheimer" and "state of the moment") and two other metrics were used to estimate the reliability of caregivers ("age" and "knowledge field"). Finally, we proposed a prototype that implements our approach. This prototype was integrated in Captain Memo prosthesis. The evaluation of our approach, carried out with 12 Alzheimer's patients and their caregivers, showed promising results compared to the gold standard, especially after several uses of the prosthesis.

Future work will be devoted to evaluate the efficiency of the DBE_ALZ approach in estimating the believability of input data given by caregivers.

References

1. Métais, E., et al.: Memory prosthesis. In: Non-Pharmacological Therapies in Dementia (2015)
2. Herradi, N., Hamdi, F., Métais, E., Ghorbel, F., Soukane, A.: PersonLink: an ontology representing family relationships for the CAPTAIN MEMO memory prosthesis. In: Jeusfeld, Manfred A., Karlapalem, K. (eds.) ER 2015. LNCS, vol. 9382, pp. 3–13. Springer, Cham (2015). https://doi.org/10.1007/978-3-319-25747-1_1
3. Wang, R., Strong, D.: Beyond accuracy: what data quality means to data consumers. J. Manag. Inf. Syst. 12(4), 5–33 (1996)
4. Pipino, L., Lee, Y., Wang, R.: Data quality assessment. Commun. ACM 45, 211–218 (2002)
5. Hong, T.: Contributing factors to the use of health-related websites. J. Health Commun. 11(2), 149–165 (2006)
6. Lee, Y.W., Pipino, L.L., Fund, J.F., Wang, R.Y.: Journey to Data Quality. The MIT Press, Cambridge (2006)
7. Prat, N., Madnick, S.: Measuring data believability: a provenance approach. In: Hawaii International Conference on System Sciences, p. 393, Los Alamitos, CA, USA (2008)
8. Zadeh, L.: The concept of a linguistic variable and its application to approximate reasoning. Int. J. Inf. Sci. 9(1), 301–357 (1975)

9. Folstein, M.F., Folstein, S.E., McHugh, P.R.: Mini mental state: a practical method for grading the cognitive state of patients for clinician. J. Psychiatry Res. **12**, 189–198 (1975)

10. Ghorbel, F., Ellouze, N., Métais, E., Hamdi, F., Gargouri, F.: MEMO_Calendring: a smart reminder for Alzheimer's disease patients. In: International Conference on Smart, Monitored and Controlled Cities, p. 6, Sfax, Tunisie (2017)

11. Cingolani, P., Alcala-Fdez, J.: jFuzzyLogic: a robust and flexible fuzzy-logic inference system language implementation. In: International Conference on Fuzzy Systems, pp. 1–8 (2012)

12. Cozman, F.: The JavaBayes system. ISBA Bull. **7**(4), 16–21 (2001)

Deep Genetic Algorithm-Based Voice Pathology Diagnostic System

Rania M. Ghoniem[1,2(✉)]

[1] Department of Computer, Mansoura University, Mansoura, Egypt
Prof_rania@mans.edu.eg
[2] Department of Information Technology, College of Computer
and Information Sciences, Princess Nourah Bint Abdulrahman University,
Riyadh, Kingdom of Saudi Arabia
RMGhoniem@pnu.edu.sa

Abstract. Automatic voice pathology diagnosis is a widely investigated area by the research community. Recently, in the literature, most of the proposed solutions are based on robust feature descriptors, which are combined with machine learning algorithms. Despite of their success, it is practically difficult to design handcrafted features which are optimal for specific classification tasks. Nowadays, deep learning approaches, particularly deep Convolutional Neural Networks (CNNs), have significant breakthroughs in the recognition tasks. In this study, the deep CNN, which was mainly explored in image recognition purposes, is used for the purpose of speech recognition. An approach is proposed for voice pathology recognition using both deep CNN and Genetic Algorithm (GA). The CNN weights are initialized using the solutions produced by GA, which minimizes the classification error and increases the ability to discriminate the voice pathology. Moreover, three popular deep CNN architectures, which have been investigated in the literature for image recognition, are adapted for voice pathology diagnosis, namely: AlexNet, VGG16, and ResNet34. For comparison purposes, performance of the hybrid CNN-GA algorithm is compared to the performance of the conventional CNN, and to some other approaches based on hybridization of deep CNN and meta-heuristic methods. Experimental results reveal that the improvement in voice pathology classification accuracy for proposed method in comparison to the basic CNN was 5.4% and when compared with other meta-heuristic based algorithms was up to 4.27%. The proposed approach also outperforms the state of the art works on the same dataset with overall accuracy of 99.37%.

Keywords: Voice pathology recognition · Deep learning ·
Convolutional neural networks · Genetic algorithm · AlexNet ·
VGG16 · ResNet34

1 Introduction

Due to its noninvasive nature, the automatic classification of vocal fold pathology is being considered as an essential screening tool for supporting the clinicians. Several contributions have been introduced to classify the voice pathology through voice

© Springer Nature Switzerland AG 2019
E. Métais et al. (Eds.): NLDB 2019, LNCS 11608, pp. 220–233, 2019.
https://doi.org/10.1007/978-3-030-23281-8_18

analysis [1–3]. The main objective was to develop robust feature descriptors which can efficiently discriminate between normal and pathological voice. These features are separated into two classes, one depends on speaker or speech recognition methods, and the other utilizes voice quality measurements. Features of speaker or speech recognition methods involve Mel-frequency Cepstral Coefficients (MFCC), Relative Spectra Perceptual Linear Prediction (RASTA-PLP), and Linear Prediction Cepstral Coefficients (LPCC) [4, 5]. While the features of voice quality measurements involve Shimmer, Jitter, Harmonic-to-Noise ratio, Glottal to Noise Excitation rate, and Cepstral Peak Prominence [6].

Furthermore, the previously published work on voice pathology diagnosis mainly focuses on developing robust feature descriptors combined with machine learning algorithms. However, the methods based upon feature extraction have several drawbacks such as high-dimensionality of feature space. In order to overcome the limitations and improve the performance of the voice pathology recognition systems, the representation learning importance has to be emphasized instead of feature engineering. Deep learning [7–11] is a kind of representation learning techniques which learns hierarchical feature representation from a matrix of data. It has been proved to be efficient in identification of patterns existing in datasets.

Among the deep learning models, CNNs are the most widely applied methods in recognition tasks [12, 13] because of their high ability in accurately classifying patterns. However, the deep learning approaches, in particular, the deep CNN has been rarely conducted for automatic speech recognition in general [10, 14]. Moreover, too little work has been done on using the deep learning in voice pathology recognition [15]. In this study, the deep CNN, which has been used mainly for the image recognition task, is used for speech recognition. In addition, the paper overcomes the CNN drawbacks, such as stuck in the optima, by proposing a genetic algorithm for optimization of the CNN weights.

2 Literature Review

In [16], new weighted spectrum features of speech based upon the Jacobi–Fourier Moments were presented to recognize larynx pathologies. Disorders classification was implemented using a multi-class Fuzzy Support Vector Machine (SVM) model, in which the fuzzy memberships are calculated using Partition Index Maximization algorithm, and the kernel function parameters were optimized using Particle Swarm Optimization (PSO). In [17], the authors introduced a method for voice pathology classification based on an Interlaced Derivative Pattern (IDP), in which the directional information is used for detecting pathologies because of its encoding capability along the axes of time, frequency, and time-frequency. Furthermore, the SVM was utilized as a classification method.

The Long-Short-Term Memory was used in [18] as a classifier for voice pathology. The parameters Shimmer, Relative Jitter, and autocorrelation was utilized as input of the LSTM. Furthermore, a model based upon losses coupling was presented for voice pathology detection in [19], where the regression losses of two audios are firstly coupled by the model to learn jointly a transformation matrix of each audio. Secondly, the model uses ϵ-dragging technique for relaxing the zero-one regression targets. Thirdly, the correlation structure between classes is exploited by imposing low-rank constraint.

From the literature, the majority of works concentrate on the handcrafted features extracted from speech combined with the machine learning algorithms. This paper advances the state of the art on voice pathology diagnosis by:

- Introducing an approach for voice pathology diagnosis using hybridization of deep CNN and GA. This leads to initialization of the CNN weights using the solutions produced by GA that minimizes the classification error and increases the ability to discriminate the voice pathology,
- Adapting the three common deep CNN architectures, namely: AlexNet, ResNet34, and VGG16, which have been used frequently in the literature for image recognition to serve the purpose of speech recognition, and
- Extensively evaluating, the results of the proposed CNN-GA algorithm by comparing it to those of the ordinary CNN, and to some other approaches based on hybridization of deep CNN and meta-heuristic methods, i.e. the CNN-SA (CNN with Simulated Annealing Algorithm), the CNN-PSO (CNN with Particle Swarm Optimization), and CNN-BA approach (CNN and Bat Algorithm).

3 Proposed Methodology

The steps of the proposed voice pathology diagnostic system include: extraction of spectrogram segments, configuration of CNN classifier, optimization of CNN using the GA through the proposed hybrid CNN-GA.

3.1 Database

This study has been conducted using the database of Massachusetts Eye and Ear Infirmary (MEEI) [20] that is used in literature for voice pathology diagnosis [15, 16]. It comprises continuous speech signals that having a significant duration of 12 s for each of which. The database information is demonstrated in Table 1.

Table 1. MEEI Database information.

Classes	Number of females	Number of males	Total
Healthy	22	14	36
Vocal fold edemas or nodules	49	10	59
Unilateral vocal fold paralysis	30	29	59

3.2 Extraction of Spectrogram Segments

The spectrogram of speech signal is computed from the Fourier transform through multiplication of signal $s(z)$ by a sliding window $w(z)$. The sliding window location adds a time dimension and accordingly obtaining a time-varying frequency analysis.

$$S(i, k) = \sum_z s(z) \, w(z - i) \, e^{-j2\pi zk/Z} \qquad (1)$$

where $S(m, k)$ is the STFT of the speech signal $s(z)$, $w(z)$ is a windowing of size z, centered at time location i and Z is the discrete frequencies number. In this regard, the speech signals are blocked into short frames of 224 samples, with a 50 % predefined overlapping value. Each speech frame of a signal s is windowed using a *hamming* window. Then spectrogram matrix resulting from Eq. 1 is divided into segments of 224×224 samples in order to be manipulated by the CNN.

3.3 Configuration of CNN Classifier

CNN [7, 12, 13, 15] is a deep, feed-forward network that has been used efficiently in image analysis applications. As depicted in Figs. 1, 2, and 3, CNN architecture includes convolutional, pooling (sub-sampling), and fully connected layers. The outcome of each convolution is a feature map that is subsequently down-sampled through a pooling layer. Furthermore, the max-pooling is frequently used to retain the maximum neighborhood value in a feature map. This downscales the matrix of input sample and reduces the number of weights in every layer. The final layer of a CNN is fully connected that operates like those of a multilayer perceptrons with a preceding decision layer to predict the class that the input matrix is belonging to. Training of CNN is implemented using back-propagation and stochastic gradient. Typically, a desirable CNN architecture is found by testing various common network structures. In this study, three models of deep CNN architectures that were used in the literature for image recognition purposes are adapted for automatic speech recognition. The models are AlexNet, VGG16 and ResNet34 [7]. The *AlexNet* is an eight-layer CNN comprising five convolutional layers as well as three fully connected layers. It can be used as a feature extractor, through replacement of the output layer with a decision layer appropriating the given classification problem. The *ResNet34* was used to train the CNNs by reformulation of the network layers as learning residual functions in reference to the layer input, rather than learning unreferenced functions. The VGG16 comprises 13 convolutional layers, 5 pooling layers, and three fully connected layers.

To adapt these models to current dataset, all of the fully connected layers of each model have been replaced with a fully connected layer of 1024 neurons, and a *softmax* layer, and then fine-tuned it using current pathological speech dataset. The final decision layer of the three network architectures was altered to have three outputs, one output for each voice pathology class. Figures 1, 2, and 3 demonstrate the adapted AlexNet, ResNet34, and VGG16 convolutional neural network architecture for voice pathology diagnosis, respectively.

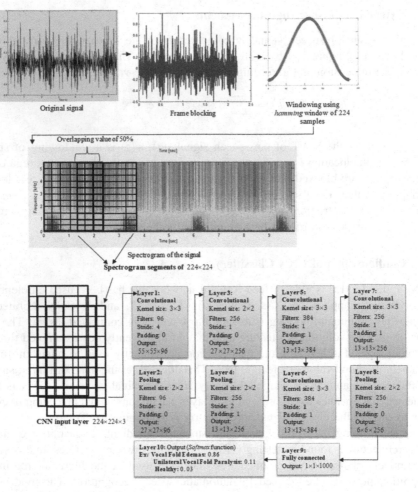

Fig. 1. Adapted AlexNet convolutional network architecture for voice pathology diagnosis

3.4 The Proposed Hybrid CNN-GA Algorithm for Voice Pathology Classification

Training the CNN using gradient descent algorithm may produce solutions that are stuck in the local optima. Furthermore, the performance of any trained CNN relies on its initial weights. Therefore, the GA is used in this work to search for the optimal weight set in between different initial weight sets. The GAs [21] are a random search algorithms that are inspired by natural genetic mechanism and biological natural selection. In the proposed algorithm presented as Algorithm 1, the spectrogram

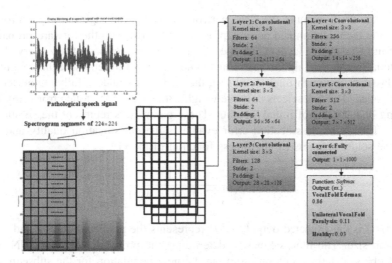

Fig. 2. Adapted ResNet34 convolutional neural network architecture for speech recognition

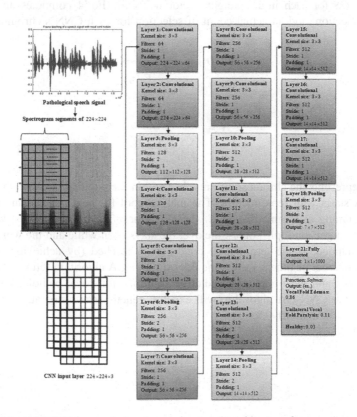

Fig. 3. Adapted VGG16 convolutional neural network architecture for speech recognition

segments are firstly input to the CNN network, which go through the steps of forward-propagation (convolution and pooling processes together with the forward-propagation in a last fully connected layer), thereafter, the probabilities of outputs for every class are computed. The genetic algorithm chromosomes are used to represent the weights of a CNN classifier, in which filter weights in the deep CNN and its fully connected layer are real coded. The GA selects only the weights that reduce the total error using Eq. 2, by trying distinct sets of initial weights. These parameters represent the solutions for the GA. The CNN classification error evaluates the classification quality and determines the solution fitness by the fitness function of the GA:

$$Fitness = \frac{1}{\frac{1}{S}\sum [A_k(t) - \hat{A}_K(t)]^2} \tag{2}$$

where $A_k(t)$ is the expected output, $\hat{A}_K(t)$ represents the predicted output, and S is the population size. The fitness function determines the probability of the CNN weight vector to be selected as a chromosome in the new population for the subsequent GA generation as shown at Eqs. 3 and 4. Equation 3 is used to compute the probability of selection POS for each initial weight vector w. While Eq. 4 computes an expected count of selection, and an actual count of selection for each CNN's chromosome.

$$POS(n) = \frac{F_x(w)}{\sum_{n=0}^{k} F_x(n)} \tag{3}$$

$$N = \frac{F_x(w)}{\sum_{n=0}^{k} F_x(n)/k} \tag{4}$$

Where k represents the chromosomes number within the population. A new generation of the GA starts with reproduction. The mating pool for next generation is selected through spinning a weighted roulette wheel, 6 times. Therefore, the best weight set representation gets multiple copies, the average stays even while and the worst dies off and are excluded. If the termination condition is reached (by achieving maximum generation number or fitness value), the hybrid CNN-GA algorithm returns the optimal CNN weights and the optimized predictive solutions for voice pathology diagnosis. Otherwise, the CNN-GA algorithm will execute genetic search operations (selection, crossover, and mutation).

4 Experimental Results

4.1 Performance Measures and Cross Validation

The performance validation metrics used in this work are sensitivity, accuracy, and specificity [22]. The sensitivity (SEN) defines number of true positives (pathological voice samples classified correctly) divided by number of positive cases. The specificity (SPE) defines number of true negatives (the signal carries disorder but is classified correctly) divided by number of negative cases. The classification accuracy (ACC) defines the proportion of correct results (true positives as well as true negatives) in population. Where TP is true positive, FN is false negative, TN is true negative, and FB is false positive.

$$SEN = \frac{TP}{TP + FN} \tag{5}$$

$$SPE = \frac{TN}{TN + FP} \tag{6}$$

$$ACC = \frac{TP + TN}{TP + FN + FP + FN} \tag{7}$$

The k-fold cross-validation was utilized for evaluating the classification quality of the CNN-GA algorithm, where k is equal to 10. The database was randomly divided into 10 sub-samples that are equal in size. Each single sub-sample is taken as a validation data set for performance testing, while the $k - 1$ sub-samples are obtained as a training data set. The k outcomes of the folds are then averaged to give a single estimation.

Algorithm 1: Proposed hybrid CNN-GA for voice pathology recognition

---

Input: Voice pathology dataset i.e. $CNN_{Training}$, $CNN_{Testing}$

The GA parameter set: $T \leftarrow$ GA iteration termination condition, $S \leftarrow$ Initial population size

The CNN parameter set: $Err_{Max} \rightarrow$ the maximum error of CNN, $Learning_{rate} \rightarrow$ the

CNN learning rate, $Iteration_{Max} \rightarrow$ the CNN maximum number of iterations,

$A_k(t) \leftarrow$ the CNN expected output, $\hat{A}_k(t) \leftarrow$ the CNN predicted output

Output: W^*, CNN weights that satisfy the best solution for voice pathology diagnosis.

1. Determine the topology architecture of CNN i.e. AlexNet, VGG16, and ResNet34.
2. Set the network layers: CNN_{input}, CNN_{conv}, $CNN_{pooling}$, CNN_{dense}, and CNN_{output}.

3. Initialize the CNN parameters: Err_{Max}, $Iteration_{Max}$, $Learning_{rate}$.
4. Extract spectrogram segments of 224×224 form each speech signal of the training set.
5. Train the CNN with the spectrogram segments then compute the CNN initial weights.
6. Initialize the parameters of genetic chromosomes.
7. Encode each initial weigh vector of CNN within a chromosome to get a pool of chromosomes
   $W = \{w_1, w_2, ..., w_k\}$.

8. Compute the fitness value of each CNN initial weight vector $w_i \in W$ using Eq. 2.

9. **While** $(t \leq T)$

10.   **For each** $I \leftarrow 1$ to S

11.   Select the best CNN'S chromosomes $w_x, w_y \in W (x \neq y)$ using the normalized fitness $F(n$

12. $\tilde{F}(w) = \dfrac{F(w)}{\displaystyle\sum_{i=1}^{k} F(w)}$.

13.   Cross over w_x and w_y using: $C(w_x, w_y \mid \alpha) \Rightarrow w'_x, w'_y$.

14.   Perform mutation on children w'_x, w'_y using: $M(w'_x \mid \beta) \Rightarrow \tilde{w}_x$ and
   $M(w'_y \mid \beta) \Rightarrow \tilde{w}_y$.

15.   Compute the fitness of children $F(\tilde{w}_x)$ and $F(\tilde{w}_y)$.

16.   **End For**
17. **End**

18. Obtain the optimal CNN's chromosome which satisfies $w^* = \arg\max_{w} F(w), w \in W$.

19. Train the CNN with the optimal weights.
20. Return the optimal solutions for voice pathology diagnosis.

4.2 Results and Discussion

In this section, the performance of the proposed hybrid CNN-GA approach is evaluated. The performance of the CNN-GA algorithm was tested using the three models of CNN on the MEEI database. From Table 2, AlexNet and VGG16 perform similarly, in terms of ACC, SEN, and SPE. The best CNN-GA results were obtained using the *ResNet34* architecture. This is due to the high representational ability of the residual networks. These results were supported in [7], where the three models were used for image recognition purposes. Furthermore, the solution results obtained from the algorithm using *ResNet34* architecture (Table 2) were also compared to those of Table 3 obtained using the basic CNN, the CNN-SA approach (CNN with simulated annealing algorithm [23]), the CNN-PSO (CNN with particle swarm optimization [24]), and the CNN-BA approach (hybrid of CNN and bat algorithm [25]). The Parameter setting of GA, SA, PSO, PA algorithms using the MEEI database is illustrated in Table 4.

Table 2. Performance evaluation of CNN-GA using the AlexNet, ResNet34, and VGG16

Epoch	AlexNet			ResResNet34			VGG16		
	ACC (%)	SEN (%)	SPE (%)	ACC (%)	SEN (%)	SPE (%)	ACC (%)	SEN (%)	SPE (%)
1	97.52	96.44	97.77	99.74	99.14	99.75	98.21	96.64	98.35
2	97.42	96.71	97.57	99.8	99.19	99.78	98.05	96.69	98.25
3	96.43	97.66	97.22	98.72	98.34	98.76	97.55	97.58	97.77
4	97.81	96.24	97.95	99.52	98.78	99.59	97.82	97.11	97.97
5	97.05	97.24	97.35	98.65	97.91	98.73	96.83	98.06	97.62
6	97.65	96.29	97.85	99.25	98.1	99.33	97.92	96.84	98.17
7	97.15	97.18	97.37	98.78	98.4	98.83	97.55	97.48	97.83
8	96.92	97.32	97.24	99.68	99.14	99.72	97.45	97.64	97.75
9	97.23	97.01	97.48	99.82	99.01	99.88	97.32	97.72	97.64
10	97.15	97.08	97.43	99.7	99.15	99.8	97.63	97.41	97.88

From Tables 2 and 3, it is obvious the superiority of the proposed method in all epochs. The ACC, SEN, and SPE results obtained from the hybrid CNN-GA algorithm are higher in comparison to those from the basic CNN, the CNN-SA algorithm, the CNN-PSO algorithm, and the CNN-BA algorithm, which refer that the optimal quantitative evaluation outcomes have been obtained, through using the proposed approach. The improvement of ACC of voice pathology classification in comparison to the basic CNN was 5.4% and when compared with other meta-heuristic based algorithms was up to 4.27%. Furthermore, it is obvious that the CNN-GA is more powerful than the basic CNN, the CNN-SA algorithm, the CNN-PSO algorithm, and the CNN-BA algorithm in term of classification accuracy, sensitivity, and specificity. Table 5 presents the comparison of the proposed voice pathology diagnosis approach to other

230   R. M. Ghoniem

Table 3. Performance evaluation using basic CNN algorithm, CNN-SA, CNN-PSO, CNN-BA

Epoch	CNN			CNN-SA			CNN-PSO			CNN-BA		
	ACC (%)	SEN (%)	SPE (%)	ACC (%)	SEN (%)	SPE (%)	ACC (%)	SEN (%)	SPE (%)	ACC (%)	SEN (%)	SPE (%)
1	93.88	94.09	94.16	94.13	93.97	94.25	94.18	94.39	94.52	94.97	94.39	95.96
2	92.01	92.7	92.81	93.95	93.82	94.03	95.12	94.17	95.32	95.38	94.17	95.48
3	93.98	94.1	94.3	92.88	92.91	93.09	95.33	94.31	95.49	95.66	94.31	95.76
4	92.25	93.07	93.13	91.32	90.56	91.41	93.99	93.96	94.07	94.12	93.96	94.32
5	92.46	92.03	92.71	94.98	94.77	95.2	94.97	94.42	95.13	95.11	94.42	95.35
6	94.96	95.1	95.29	94.24	94.21	94.37	94.83	94.75	95.05	94.55	94.75	95.23
7	92.34	92.4	29.41	95.05	94.83	95.16	93.92	94.78	94.57	94.17	94.78	95.38
8	91.54	91.78	91.99	94.29	93.9	94.38	94.23	94.97	94.71	94.28	94.97	95.42
9	93.57	93,91	94.1	95.13	95.31	95.65	94.96	95.18	95.22	95.17	95.18	95.36
10	92.28	92.3	92.7	92.42	92.96	93.11	95.94	95.91	96.16	96.21	95.91	96.52

Table 4. Parameter setting of GA, SA, PSO, and PA algorithms using the MEEI database

GA	SA	PSO	BA
Population size: 60	Number of neighborhood: 10	Particles number: 10	Number of bats: 10
Maximum generations: 60	Number of iterations: 10	Iterations number: 60	Number of iterations: 10
Crossover probability: 0.3		Acceleration $(c1)$: 2	
Mutation rate: 0.15		Acceleration $(c2)$: 2	
Reproduction rate: 0.18		Maximal inertia weight: 0.7	Constants $f_{min} = 0, f_{max} = 2$, and $\alpha = \gamma = 0.9$
weighted Roulette Wheel		Minimal inertia weight: 0.1	

previous work on voice pathology diagnosis. The database, feature extraction and classification algorithms that are used for diagnosis, and accuracy of classification are presented.

As seen in Table 5, the proposed deep genetic voice pathology diagnosis approach acts better than other works on the same dataset with overall accuracy of 99.37%. This result is due to the robustness of deep learning as a representation learning technique that can learns hierarchical feature representation from speech data. Moreover, the GA as a meta-heuristic method has optimized the performance of the deep CNN by selecting the optimal weights that reduce the classification error.

Table 5. Comparison of proposed approach to previous work on voice pathology diagnosis

Reference	Year	Database	Approach of voice pathology diagnosis	Accuracy (%)
[26]	2014	MEEI	Wavelet packet transform, multi-class linear discriminant analysis, and multi-layer neural network	97.33%
[27]	2016	MEEI	Features extracted from vocal tract area, principal component analysis, and support vector machine	99.22%
[16]	2017	MEEI	Weighted spectrum features based upon Jacobi–Fourier Moments, and a fuzzy support vector machine classifier optimized using PSO	97.6%
[28]	2017	MEEI	Multi-dimensional Voice Program (MDVP) parameters was investigated for detecting and classifying voice pathologies	88.21%
[15]	2018	MEEI	MFCC for feature extraction from 3-second samples, Deep Neural Network, SVM, and Gaussian Mixture Model for voice pathology classification.	99.14%
Proposed		MEEI	Spectrogram segments and hybrid CNN-GA method	99.37%

5 Conclusions

This paper proposes a classification approach for voice pathology recognition using deep CNN and Genetic Algorithm. The proposed approach is used to determine the optimal initial weights to train the CNN. A comparison between the hybrid CNN-GA approach and other approaches, involving the basic CNN, the CNN-SA algorithm, as well as the CNN-PSO algorithm, is shown in current experiments. Results were implemented using the MEEI database so as to evaluate the performance of proposed approach in voice pathology classification. The results taken from the proposed CNN-GA approach reveal performance that outperforms other existing methods. The improvement of voice pathology classification accuracy for proposed method in comparison to the basic CNN was 5.4% and when compared with other meta-heuristic based algorithms was up to 4.27%. Accordingly, the CNN-GA algorithm is powerful than the basic CNN algorithm, the CNN-PSO algorithm, the CNN-SA algorithm, and the CNN-BA algorithm in terms of average classification accuracy, sensitivity, and specificity evaluation measures. The best CNN-GA results were obtained using the ResNet34 architecture. The proposed approach also acts better than the state of the art works on the same dataset with overall accuracy of 99.37%. In conclusion, the proposed approach is highly efficient from the viewpoint of solution quality. It can serve as a robust technique in voice pathology diagnosis.

References

1. Al-Nasheri, A., Muhammad, G., Alsulaiman, M., Ali, Z.: Investigation of voice pathology detection and classification on different frequency regions using correlation functions. J. Voice **31**, 3–15 (2017)
2. Kohler, M., Mendoza, L.A.F., Lazo, J.G., Vellasco, M., Cataldo, E.: Classification of Voice Pathologies Using Glottal Signal Parameters. Anais do 10. Congresso Brasileiro de Inteligência Computacional (2016)
3. Ali, Z., Elamvazuthi, I., Alsulaiman, M., Muhammad, G.: Automatic voice pathology detection with running speech by using estimation of auditory spectrum and cepstral coefficients based on the all-pole model. J. Voice **30**, 757-e7 (2016)
4. Hossain, M.S., Muhammad, G.: Cloud-assisted speech and face recognition framework for health monitoring. Mob. Networks Appl. **20**, 391–399 (2015)
5. Cordeiro, H., Meneses, C., Fonseca, J.: Continuous speech classification systems for voice pathologies identification. In: Camarinha-Matos, L.M., Baldissera, T.A., Di Orio, G., Marques, F. (eds.) DoCEIS 2015. IAICT, vol. 450, pp. 217–224. Springer, Cham (2015). https://doi.org/10.1007/978-3-319-16766-4_23
6. Kay Elemetrics, Multi-Dimensional Voice Program (MDVP) [Computer Program] (2012)
7. Fu, Y., Aldrich, C.: Flotation froth image recognition with convolutional neural networks. Miner. Eng. **132**, 183–190 (2019)
8. Traore, B.B., Kamsu-Foguem, B., Tangara, F.: Deep convolution neural network for image recognition. Ecol. Inf. **48**, 257–268 (2018)
9. Fang, L., Jin, Y., Huang, L., Guo, S., Zhao, G., Chen, X.: Iterative fusion convolutional neural networks for classification of optical coherence tomography images. J. Vis. Commun. Image Represent. **59**, 327–333 (2019)
10. Fayek, H.M., Lech, M., Cavedon, L.: Evaluating deep learning architectures for speech emotion recognition. Neural Networks **92**, 60–68 (2017)
11. Tu, Y.-H., et al.: An iterative mask estimation approach to deep learning based multi-channel speech recognition. Speech Commun. **106**, 31–43 (2019)
12. Angrick, M., Herff, C., Johnson, G., Shih, J., Krusienski, D., Schultz, T.: Interpretation of convolutional neural networks for speech spectrogram regression from intracranial recordings. Neurocomputing **342**, 145–151 (2019)
13. Hossain, M.S., Muhammad, G.: Emotion recognition using deep learning approach from audio–visual emotional big data. Inf. Fusion. **49**, 69–78 (2019)
14. Palaz, D., Magimai-Doss, M., Collobert, R.: End-to-end acoustic modeling using convolutional neural networks for HMM-based automatic speech recognition. Speech Commun. **108**, 15–32 (2019)
15. Fang, S.-H., et al.: Detection of pathological voice using cepstrum vectors: a deep learning approach. J. Voice (2018)
16. Ghoniem, R.M., Shaalan, K.: FCSR - fuzzy continuous speech recognition approach for identifying laryngeal pathologies using new weighted spectrum features. In: Proceedings of the International Conference on Advanced Intelligent Systems and Informatics 2017 Advances in Intelligent Systems and Computing, pp. 384–395 (2017)
17. Muhammad, G., et al.: Voice pathology detection using interlaced derivative pattern on glottal source excitation. Biomed. Signal Process. Control **31**, 156–164 (2017)
18. Guedes, V., Junior, A., Fernandes, J., Teixeira, F., Teixeira, J.P.: Long short term memory on chronic laryngitis classification. Procedia Comput. Sci. **138**, 250–257 (2018)
19. Wu, K., Zhang, D., Lu, G., Guo, Z.: Joint learning for voice based disease detection. Pattern Recogn. **87**, 130–139 (2019)

20. Eye, M., Infirmary, E.: Voice Disorders Database, (Version 1.03 Cd-Rom). Vol (Kay Elemetrics Corp., Lincoln Park N, ed.). Kay Elemetrics Corp., Lincoln Park (1994)
21. Song, R., Zhang, X., Zhou, C., Liu, J., He, J.: Predicting TEC in China based on the neural networks optimized by genetic algorithm. Adv. Space Res. **62**, 745–759 (2018)
22. Ghoniem, R., Refky, B., Soliman, A., Tawfik, A.: IPES: an image processing-enabled expert system for the detection of breast malignant tumors. J. Biomed. Eng. Med. Imaging **3**, 13–32 (2016)
23. Rere, L.R., Fanany, M.I., Arymurthy, A.M.: Simulated annealing algorithm for deep learning. Procedia Comput. Sci. **72**, 137–144 (2015)
24. Silva, G.L.F.D., Valente, T.L.A., Silva, A.C., Paiva, A.C.D., Gattass, M.: Convolutional neural network-based PSO for lung nodule false positive reduction on CT images. Comput. Meth. Programs Biomed. **162**, 109–118 (2018)
25. Yang, X.-S.: A new metaheuristic bat-inspired algorithm. In: Nature Inspired Cooperative Strategies for Optimization (NICSO 2010) Studies in Computational Intelligence, pp. 65–74 (2010)
26. Akbari, A., Arjmandi, M.K.: An efficient voice pathology classification scheme based on applying multi-layer linear discriminant analysis to wavelet packet-based features. Biomed. Signal Process. Control **10**, 209–223 (2014)
27. Muhammad, G., et al.: Automatic voice pathology detection and classification using vocal tract area irregularity. Biocybernetics Biomed. Eng. **36**, 309–317 (2016)
28. Al-Nasheri, A., et al.: An investigation of multidimensional voice program parameters in three different databases for voice pathology detection and classification. J. Voice **31**, 113-e9 (2017)

An Arabic-Multilingual Database
with a Lexicographic Search Engine

Mustafa Jarrar[(✉)] ⓘ and Hamzeh Amayreh ⓘ

Birzeit University, Birzeit, Palestine
mjarrar@birzeit.edu, hamayreh@staff.birzeit.edu

Abstract. We present a lexicographic search engine built on top of the largest Arabic multilingual database, allowing people to search and retrieve translations, synonyms, definitions, and more. The database currently contains about 150 Arabic multilingual lexicons that we have been digitizing, restructuring, and normalizing over 9 years. It comprises most types of lexical resources, such as modern and classical lexicons, thesauri, glossaries, lexicographic datasets, and (bi/)tri-lingual dictionaries. This is in addition to the Arabic Ontology – an Arabic WordNet with ontologically cleaned content, which is being used to reference and interlink lexical concepts. The search engine was developed with the state-of-the-art design features and according to the W3C's recommendation and best practices for publishing data on the web, as well as the W3C's Lemon RDF model. The search engine is publicly available at (https://ontology.birzeit.edu).

Keywords: Arabic · Multilingual lexicons · Online dictionary ·
Language resources · Lexical semantics · Lexicographic search ·
W3C lemon · RDF · NLP

1 Introduction and Motivation

The increasing demands to use and reuse dictionaries (of all types) in modern applications have shifted the field of lexicography to be a multidisciplinary domain, engaging ontology engineering [18, 19, 22], computational linguistics [1, 17, 25], and knowledge management [12, 16, 23, 24]. Dictionaries are no more limited to hard copies and are not only used by humans; they are becoming important for IT applications that require natural language processing [6–8, 20], in addition to the need to access them electronically. In response to these demands, there have been several efforts to digitize, represent, and publish them online. As will be discussed later, the ISO37 has released more than 50 standards in the past 15 years related to terminology and lexical resources, in addition to several W3C recommendations e.g., SKOS, Lemon [22], and the Linguistic Linked Open Data Cloud [7].

Although there are many lexicons available on the internet for most languages, especially English, few Arabic lexicons are available in digital forms [1, 5, 25]. This lack of such digital resources has limited the progress in Arabic NLP research [21], and has also led many people to use statistical machine translation tools (e.g., Google Translate) in place of dictionaries [4].

© Springer Nature Switzerland AG 2019
E. Métais et al. (Eds.): NLDB 2019, LNCS 11608, pp. 234–246, 2019.
https://doi.org/10.1007/978-3-030-23281-8_19

In this paper, we present the digitization of 150 Arabic multilingual lexicons and a lexicographic search engine built on top of them, covering many domains such as, natural sciences, technology, engineering, health, economy, art, humanities, philosophy, and more. The digitization process was carried out over 9 years, as most lexicons had to be manually typed, then restructured and normalized. The copyright owners of all lexicons were contacted individually for a permission to digitize and use their lexicons. To the best of our knowledge, our database is currently the largest Arabic lexicographic database, compromising about 2.4 million multilingual lexical entries and about 1.1 million lexical concepts. The search engine was designed according to W3C's recommendations, especially the Lemon model which is important for referencing and linguistic data linking purposes. The ranking strategy used in the search engine is a combination metric of lexicon-renown and concept-citation.

The rest of this paper proceeds as follows: In Sect. 2, we overview related work. We elaborate on the construction of the lexicographic database in Sect. 3. Section 4 presents the search engine, its architecture, URLs design, ranking strategies, and usability. Finally, we conclude and discuss future work in Sect. 5.

2 Related Work

We first present recent standards for representing and publishing linguistic data, then we overview related digital lexicographic resources and repositories.

In response to the increasing demands to use and reuse linguistic resources in modern applications, there have been many efforts to standardize the way linguistic resources are structured, represented, and published on the web. The ISO37 produced over 50 standards in the recent years in this direction. For example, the ISO24613 is a lexical markup framework (LMF) to represent lexicons in a machine-readable format; the ISO860 is concerned with the harmonization of concepts, concept systems, definitions and terms; the ISO16642 supports the development, use, and exchange of terminological data between different IT applications. W3C has also developed several recommendations related to linguistic data sources. For example, SKOS provides a way to represent thesauri, classification schemes, subject headings, and taxonomies within the framework of the Semantic Web. The W3C's Lemon RDF model [22] aims at enabling lexicons to be used by ontologies and NLP applications. It can be used to describe the properties of lexical entries and their syntactic behavior, encouraging reuse of existing linguistic data. The importance of Lemon is that, it was developed based on the W3C recommendations for Open Linked Data [2]. The Linguistic Linked Open Data Cloud (LLOD) [7] was initiated as a collaborative effort to interlink the lexical entries of different linguistic resources using Lemon.

A new ambitious project, called PanLex, aims at building the world's largest lexical database [3], with 2500 dictionaries for 5700 languages. Its objective is to be a bilingual translation-oriented database, offering about 1.3 billion translation pairs. Compared with our work, PanLex offers only bilingual translations, rather than a lexicographic database with definitions, synonyms, and other lexicographic features, and it does not support a large number of Arabic lexicons.

There have also been other related initiatives aiming to integrate wordnets with other resources. BabelNet [24] is a multilingual encyclopedic dictionary covering 284 languages, as an integration of many wordnets, Wiktionary, Wikipedia, GeoNames, and more. BabelNet is a semantic network connecting concepts and named entities. Similarly, ConceptNet 5.5 [23] is an open multilingual knowledge graph connecting words and phrases with labeled edges. It links the Open Mind Common Sense with Multi Wordnet, Wiktionary, Wikipedia, and OpenCyc. Compared with our work, both, BabelNet and ConceptNet, aim at building encyclopedic knowledge graphs, and their linguistic information is limited to wordnets rather than targeting a large number of dictionaries as we try to do. Additionally, their support of Arabic is limited to Arabic Wikipedia and Arabic WordNet that is quite small (only 11 k synsets).

The number of available structured Arabic lexicons in digital format is indeed limited [1, 21, 25]. Earlier attempts to represent Arabic morphological lexicons in ISO LMF standard can be found in [25], and in [6] to represent Dutch bilingual lexicons, including Arabic. A lexicon called Al-Madar [1] was developed and represented using the ISO LMF standard. Similarly, Al-Qamus Almuhit was digitized and represented in the ISO LMF and later in the Lemon model [20]. A preliminary progress in digitizing several Hadith lexicons is reported in [21]. Nevertheless, none of the lexicons above is accessible online.

There are several online portals offering lexicographic search (e.g., almaany.com, lisaan.net, almougem.com, albaheth.info, ejtaal.net, alburaq.net), each comprises only a small number of lexicons. More importantly, most of the content in these portals is partially structured (i.e. available in flat text), as they allow people to search for a word, and the paragraphs that include this word as a headword will be retrieved.

It is worth noting that, modern Arabic lexicons are mostly the production of two authoritative institutions, the ALECSO that produced about 50 lexicons, and the Arabic Academy in Cairo that produced about 20 lexicons. The majority of these lexicons were digitized and included in our database. Additionally, SAMA[1] and Sarf[2] are two Arabic morphological databases that were designed for morphological analysis only. Both are being used to map between, and enrich, the lexical entries in our database.

3   Constructing the Lexicographic Database

This section presents our database, which contains about 150 lexicons that we have been digitizing from scratch. Although we were able to obtain some lexicons in digital flat text format, we had to type most lexicons manually. First, we tried to use OCR tools, but we failed due to their low quality. We also failed in crowdsourcing the digitization process among 300 students as most of them were uncareful. Afterwards, we contracted some careful students to type lexicons in MS Word format, and then gave the output to two experts to manually compare with original copies. The output was then converted into a preliminary ad hoc DB table. A lexicon that uses explicit and

---

[1] Developed by LDC, accessible at: https://catalog.ldc.upenn.edu/LDC2010L01.

[2] An open source project, accessible at: https://sourceforge.net/projects/sarf/.

steady markers (e.g. tab, comma, semicolon) to separate between different features was parsed and converted automatically; otherwise, such markers were manually added before parsing.

Lexicon Restructuring: the content of each lexicon was restructured separately, which was a semi-automated task (fully presented in [5]). Before overviewing this task, we first present a classification of our lexicons, and describe their internal structure and type of content:

- **Glossary**: a domain-specific lexicon, where each lexical entry is defined in a few lines. Advanced glossaries provide also synonyms, multilingual translation(s), and sometimes references to related lexical entries, e.g. similar, equivalent, or related.
- **Thesaurus**: sets of synonymous lexical entries. Each set might be lexicalized in one or more languages. A set might be also labeled with a part-of-speech tag.
- **Dictionary**: a list of lexical entries, each with some bi/trilingual translations.
- **Linguistic Lexicon**: a set of lexical entries, each with its linguistic features and sense(s). A lexical entry may have several meanings, which some lexicons designate into separate senses, while others combine them in one description. Lexicons may also provide linguistic features for each entry, e.g. root, POS, and inflections.
- **Semantic-variations lexicon**: a set of pairs of semantically close lexical entries and the differences between their meanings, (e.g. like $\sim$ love, pain $\sim$ ache).

To structure the content of such types of lexicons, we developed two general templates (see Fig. 1), where each lexicon was parsed and mapped to them, as the following:

1. Each lexical entry in every language, whether provided as a headword, a synonym, or a translation, was extracted and given a unique ID. The features of the lexical entry, (e.g. POS, lexical forms, inflections), were all extracted and stored in the Lexical Entry template, in an RDF-like format. Deciding whether two Arabic lexical entries of the same letters are the same is challenging, as they might be partially or non-diacritized [11].
2. Each meaning of every headword in every lexicon is considered a lexical concept, and is given an ID. This is straightforward in case of glossaries as each headword typically has only one meaning. Each set of synonymous entries in a thesaurus, and similarly each group of translations in a dictionary, is mapped into a lexical concept and is given an ID. However, in case of linguistic lexicons, the different senses of a lexical entry were each extracted, and mapped into a separate lexical concept. In case of references to other lexical concepts, (e.g. indicating semantic relations like related and similar), these relations were also extracted and stored in the Lexical Concept template. In this way, the Lexical Concept template was filled in, providing the concept ID, and if available its set of synonyms, definitions, examples, and relations.

Cleaning and Normalization: The content of these general templates was then cleaned and normalized before storing them in a relational database. As lexicons are typically designed to be printed and used as hard copies, new challenges are faced

Lexical Entry	
Lexical Entry ID	
Lexical Entry	

Feature/derivation	Value/LexicalEntryID

Lexical Concept	
Lexical concept ID	
Synonyms	*<lexicalEntry₁>* | *<lexicalEntry₂>* | ... | *<lexicalEntryₙ>*
Definition	
Example	

Relation Name	LexicalConceptID

Fig. 1. Lexical concept and lexical entry templates.

when converting them into a machine processable format. In what follows, we summarize some of these challenges – see [5] for more issues and details.

- *Challenges induced by ordering:* to maintain a proper alphabetical ordering in hard copies, many lexicons tend to re-arrange words in the lexical entry, such as: "accelerator (linear...)", "affinity (chemical)", "drawing (final)", "earth (the)", and "crush (to)". Detecting such cases and deciding whether to move the text between parentheses to the beginning or to keep the order intact is difficult. This is because parenthesis might be also used for other purposes, as in (e.g. "tube (pipe)", "academy (of art)"), which indicate synonymy and context, respectively. There are no markers that would help detecting and normalizing such lexical entries.

- *Subterm synonymy*: most lexicons use commas or other symbols to separate between synonyms (e.g. "benzene, benzol", "tie, bind"). Though it is easy to split them, we found many cases where the comma is used differently, e.g. to indicate a more specific meaning, as in "calomel electrode, calomel", "kelvin's scale, kelvin's absolute scale", and "liquid drier, drier". That is, if a term is synonymous with another term, and one is part of the other, it is likely to be a mistake or to indicate another more specific meaning. Such cases need to be manually reviewed and decided upon.

- *Long multiword lexical entries:* there are cases where a lexical entry is composed of many words, such as "buildings or other structures recurrent taxes on land". Such cases of long and "poor" lexical entries need to be manually reviewed. i.e. by considering it a definition, or excluding it.

- *Special characters*: The use of special characters in a lexical entry (e.g. quotations, punctuation marks, and brackets) is allowed if they were used intentionally as part of the lexical entry. Nevertheless, they are often introduced in lexicons as annotations. Therefore, they have to be filtered out and individually reviewed.

- *Character set*: Same characters and symbols have different encodings across different languages (e.g., the dash, quotations, punctuations, and whitespaces), which is not a problem in case of printed lexicons, but they are obstacles when digitizing lexicons. This issue is trickier in Arabic as there are also different versions of character sets and there are characters in Arabic which have the same orthography but with different encodings that need to be changed to use the same encoding version.

Addressing such challenges in a fully automatic manner is difficult. Therefore, we have developed a parsing framework, presented in [5], that first detect and filter out each individual issue (e.g., whether a lexical entry includes parenthesis, commas, subterm synonymy, long multiword entries, character set issue, etc.). The parsers then assign a category to each of these issues to indicate its nature. The output of the parsers includes also a suggested treatment, depending on the nature of the issue. Each category was then given to a linguist to review and confirm the suggested treatments. After normalizing lexical entries and features, the data was stored in a MySQL database and indexed for searching purposes. Table 1 illustrates some statistics about our database.

Table 1. Statistics of the Lexicographic Database – being extended.

Category	Lexical Concepts	Lexical entries	Synsets	Translations pairs	Glosses	Semantic relations
Total (Milions)	1.1 M	2.4 M	1.8 M	1.5 M	0.7 M	0.5 M
Sub Counts		1,100 K Arabic 1,100 K English 200 K French 3 K Others 1,300 K Single-word 1,000 K Multi-word	800 K Arabic 800 K English 200 K French 50 K Others	1,000 K English-Arabic 300 K English-French 200 K French-Arabic	400 K Arabic 300 K English 1 K Others	170 K Sub-super links 29 K Part-of links 260 K Has-Domain links 30 K Other links

Copyrights: We contacted all lexicons' owners individually to get an official permission to digitize and include their lexicons in the search engine, a process that took several years, and although some refused, most of them accepted. Their main motivation was that the search engine displays the copyright symbol and the lexicon's name below each retrieved result, keeping their rights reserved. Additionally, when the lexicon's name is clicked (see Fig. 3), it shows the author(s), publisher, and links to their websites and to bookstores to purchase their lexicon.

Referencing lexical concepts in the Arabic Ontology
The Arabic Ontology is part of the database and is accessible in the search engine[3]. It can be seen and used as a formal Arabic wordnet built on the basis of a carefully designed ontology [9, 15]. It consists currently of about 1.3 K concepts that are also mapped to WordNet, BFO, and DOLCE, in addition to 11 K concepts that are being validated and mapped. The Arabic ontology is currently being used to reference lexical concepts in all lexicons; such that, each lexical concept is mapped (e.g., equal, or subtype) to a concept in the ontology. In this way, lexical concepts across all lexicons would be semantically linked; and since the ontology is mapped with other resources, it implies that lexical concepts would also be mapped to these resources. Presenting these mappings is beyond the scope of this article.

---

[3] http://ontology.birzeit.edu/concept/293198.

Data Indexing: To prepare our database for efficient search, we built two indexes, depicted in Fig. 2. The *Lexical Concept Index* aggregates relevant information for each concept in one record. It includes the concept ID, Arabic and English synsets, gloss (i.e. definition), semantic relations, and other features that need to be retrieved by the search engine. The computed rank for each lexical concept (as we will discuss later) is also pre-calculated and stored in this *Index*. The *Term-Concepts Inverted Index* is an inverted index that links between lexical entries and their lexical concepts. This inverted index was built by first collecting all lexical entries from all synsets, which can be single or multiple words (we call it *Term*). Second, by linking each of these terms with its posts (i.e. lexical concepts). This index was implemented using MySQL's full-text index, especially that it supports the generation of concordances. For search effectiveness, each of the terms in the inverted index was normalized and stemmed. Currently, *Lexical Concept* contains about 2.2 million records, whereas the inverted index contains about 1.1 million records, each having 25 postings on average.

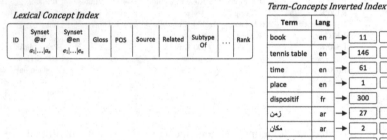

Fig. 2. Main Indexes.

4 Building a Lexicographic Search Engine

Figure 3 illustrates a screenshot of the search engine. It allows people to search for translations, synonyms and definitions from the 150 lexicons and the Arabic Ontology, and filter the results. The engine is designed based on a set of RESTful web services (Fig. 4), which query the database and return the results in JSON format that is then rendered at our front-end, and can be also used by third-party applications.

4.1 URLs Design

The URLs in the search engine are designed according to the W3C's Best Practices for Publishing Linked Data [2], including the Cool URIs, simplicity, stability, and linking best practices, as described in the following URL schemes. This allows one to also explore the whole database like exploring a graph:

- *Term*: Each term (i.e., affix, word, or multiword expression) is given a URL: `http://{domain}/term/{term}`, which retrieves the set of all lexical concepts, in all lexicons, that are lexicalized using this term, i.e. that have this term as

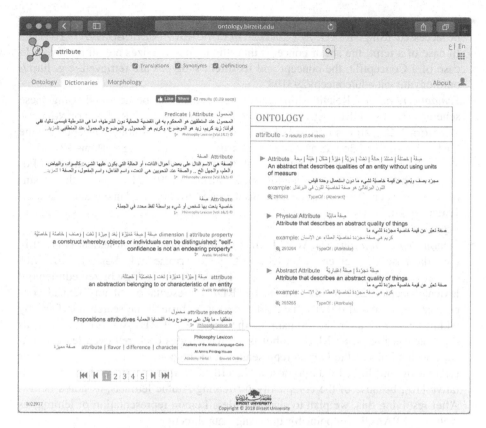

Fig. 3. Screenshot of the search engine's frontend.

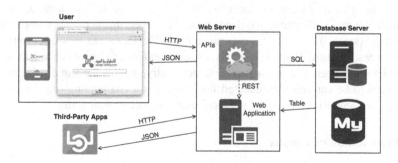

Fig. 4. Search Engine Architecture.

a separate lexical entry or among a synset. In order to keep the URLs Cool, simple and stable, URL parameters (e.g., filters and page number) are passed internally without treating them as part of the URL, e.g. http://ontology.birzeit.edu/term/virus

• *Lexical Concept:* Each lexical concept in all lexicons is given a URL based on its unique *LexicalConceptID*: http://{domain}/lexicalconcept/ {lexicalConceptID}

- *Ontology Concept:* Each concept in the Arabic Ontology has a ConceptID and can be accessed using: `http://{domain}/concept/{ConceptID | Term}`. In case of a term, the set of concepts that this term lexicalizes are all retrieved. In case of a ConceptID, the concept and its direct subtypes are retrieved, e.g. http://ontology.birzeit.edu/concept/293198
- *Semantic relations:* Relationships between concepts can be accessed using these schemes: (i) the URL: `http://{domain}/concept/{RelationName}/{ConceptID}` allows retrieval of relationships among ontology concepts. (ii) the URL: `http://{domain}/lexicalconcept/{RelationName}/{lexicalConceptID}` allows retrieval of relations between lexical concepts. For example, http://ontology.birzeit.edu/concept/instances/293121 retrieves the instances of the concept `293121`. The relations that are currently used in our database are: `{subtypes, type, instances, parts, related, similar, equivalent}`.
- *Lemon Representation:* The W3C Lemon representation of each lexical concept in the database is given a URL: `http://{domain}/lemon/lexicalconcept/{lexicalConceptID}`, e.g. http://ontology.birzeit.edu/lemon/lexicalconcept/1520098340. That is, the RDF representation of any lexical concept can be accessed directly (i.e., not necessarily through the search interface) by adding `/lemon` after the domain in the concept's URL. Additionally, and as illustrated in Fig. 5, an RDF symbol is shown besides each retrieved lexical concept, which links to the Lemon representation of the concept. This is important for referencing and linked data purposes. Nevertheless, this support of Lemon is tentative [10], because of the complexity of treating Arabic lemmas, as noted below. After resolving this, we plan to also provide a Lemon representation of lemmas as well as a SPARQL endpoint for querying data directly.

Remark on Lemma URLs: Each lemma is given a unique LemmaID and a URL: `http://{domain}/lemma/{LemmaID}`, which retrieves the lemma, its morphological features, inflections, and derivations. However, this is partially implemented at this stage (see [10]), as lexical entries in Arabic lexicons are less often lemmas – unlike the case in most English lexicons where a lexical entry is often a lemma (i.e., canonical form). Therefore, each Arabic lexical entry, within the same or across lexicons, needs to be carefully lemmatized first, which is a challenging ongoing task. At this stage, we tentatively consider a lexical entry as a canonical form.

4.2 Presentation of Results

The search engine supports the retrieval of three types of results, each presented in a separate tab, namely Ontology, Dictionaries, and Morphology:

- **Ontology tab**: results in this tab are ontology concepts retrieved only from the Arabic ontology. The tab also allows expanding and exploring the ontology tree.
- **Dictionaries tab**: results in this tab are lexical concepts retrieved from the lexicons. As discussed earlier, a lexical concept can be, for example, a row in a thesaurus (set of synonymous terms), a term(s) and its definition as found in glossaries, or a set of multilingual translations as found in bi/trilingual dictionaries. Figure 5 illustrates a

التسوية levelling | grading

تحريك التربة أثناء إعداد الأرض للري للوصول إلى سطح مستو أو سطح ذي انحدار منتظم.

Hydrology Lexicon ©

```
...
@prefix aot: <http://ontology.birzeit.edu/term/>.
@prefix aoc: <http://ontology.birzeit.edu/lexicalconcept/>.
@prefix aor: <http://ontology.birzeit.edu/lexicon/>.

<aoc:1623> a ontolex:LexicalConcept;
ontolex:isEvokedBy <aot:Lex-grading>;
ontolex:isEvokedBy <aot:Lex-levelling>;
ontolex:isEvokedBy <aot:Lex-تسوية>;
skos:definition "...سطح أو مستو سطح إلى للوصول للري الأرض إعداد أثناء التربة تحريك"@ar;
skos:inScheme <aor:Hydrology_Lexicon_1>.

<aot:lex-grading> a ontolex:LexicalEntry, ontolex:Word;
ontolex:canonicalForm [ontolex:writtenRep "grading"@en];
skos:inScheme <aor:Hydrology_Lexicon_1>.
<aot:lex-levelling> a ontolex:LexicalEntry, ontolex:Word;
ontolex:canonicalForm [ontolex:writtenRep "levelling"@en];
skos:inScheme <aor:Hydrology_Lexicon_1>.
<aot:lex-تسوية> a ontolex:LexicalEntry, ontolex:Word;
ontolex:canonicalForm [ontolex:writtenRep "تسوية"@ar];
skos:inScheme <aor:Hydrology_Lexicon_1>.
```

Fig. 5. Example of a lexical concept and its Lemon representation.

lexical concept retrieved from the Hydrology Lexicon. The first line represents the set of synonymous terms in Arabic and English, separated by the symbol "|". The gloss is presented in the second line. The RDF symbol in the third line refers to the Lemon RDF representation of this concept.

- **Morphology tab:** results in this tab are linguistic features, lemma(s), inflections, and derivations of the searched term. This tab is not fully functional yet because our linguistic data is not fully integrated since most dictionaries are not lemmatized.

Additionally, we plan to introduce a forth Dialect tab to allow users to also view the dialectal features [13, 14]. In this way, the four tabs would, to more or less, reflect the different language levels (ontology, meaning, syntax, and dialect).

4.3 Ranking of Search Results

Ranking lexical concepts based on their relevancy was a challenging task. People use lexicographic search for different purposes [4], e.g. searching for translations, definitions, synonyms and/or others. In what follows, we present three ranking strategies: *citation, lexicon's renown*, and a *hybrid* approach which we have adopted.

The **citation strategy (R$_{cit}$)** ranks each lexical concept based on the frequency of its terms –by counting how many times each of its terms appears as a lexical entry in all lexicons, i.e. the concept's rank is the summation of its terms' frequencies.

$$R = \sum_{n=1}^{|A|} \sum_{m=1}^{k} F_{a_{nm}}$$

$$R_{cit} = \frac{R - R_{min}}{R_{max} - R_{min}}$$

Where, A: is the set of synonyms of a lexical concept, in all languages. k: is the number of lexicons. $F_{a_{nm}}$: is the number of times a_n appears as a lexical entry in lexicon m where $a_n \in A$. R_{cit}: is the citation rank of the concept normalized to be between [0-1].

Our assumption is that the more the concept's terms appear in lexicons, the more this concept is likely to be important. However, its disadvantage is that it decreases the rank of uncommon concepts that are likely to be searched for. Additionally, it scatters the results of the same lexicon across pages, which might confuse users.

The **lexicon renown ranking strategy ($\mathbf{R_{ren}}$)** does not assign a specific rank for each lexical concept. Rather, it assigns each lexicon a rank based on its renown, whether it is general or domain-specific and of high or low quality. Since these criteria are subjective, each lexicon was manually ranked, with respect to other lexicons, by a group of experts. The rank of a lexical concept, then, is given the rank of its lexicon. This allows the renowned results to appear first, but the disadvantage is that linguistic-oriented lexicons are always promoted first, while e.g. the user might be looking for specialized translations. To overcome this, we implemented three types of filters that can be used to show only *translations*, *synonyms*, and/or *definitions*.

The **hybrid ranking strategy ($\mathbf{R_{hyb}}$)** is a combination metric of both strategies above. It ranks the results based on the lexicon renown, and then uses the citation strategy to rank each lexicon's results, which can be obtained by the summation of both ranks:

$$R_{hyb} = R_{ren} + R_{cit}$$

4.4   Usability and Performance Evaluation

We summarize two experiments to evaluate the usability of the search engine, the full details can be found in [4]. First: a subjective **user satisfiability survey** of 12 questions was distributed and answered by 620 users. The answers to these 12 questions are categorized as: efficient (75%), effective (80%), learnable (83%), and good design (73%); and 90% responded that they will use the search engine again. Second: a more objective **controlled usability experiment** was also conducted in a lab, involving 12 users that we arranged into four groups. Each group was given eight tasks to answer. The tasks required users to find synonyms, translations, definitions, and semantic variations between terms. Two groups were asked to use Google Translate, and the other two to use ours. The accuracy of the groups' answers was evaluated by an expert, which was 73% using ours compared to 38% using Google Translate.

Performance: The search engine is currently deployed on a Linux virtual server with average resources (8-core CPU and 16 GB RAM). To estimate its response time (i.e. both backend and frontend processing and retrieval), an experiment was conducted on three user machines. Each was installed in a different location and connected to a different internet service provider. They were programmed to simultaneously send 1 million requests at the rate of 1000 requests/minute, and record the frontend-to-frontend response time for each request. Although the response time is impacted by the network traffic, the experiment showed that it ranged between (0.001 s) and (0.200 s) for all requests.

5   Conclusion and Future Work

We presented a large Arabic multilingual linguistic database, which contains about 150 lexicons of different types, and discussed the different phases carried out to structure, normalize and index this database. We introduced a lexicographic search engine with state-of-the-art design, respecting the W3C's recommendations and best practices.

We plan to continue digitizing more lexicons and adding more functionalities to the search engine, specially the support for French and other languages. Our priority is to lemmatize all lexical entries and then link them across all lexicons. This will enable the interlinking of our lexicographic database with the Linguistic Data Cloud.

Acknowledgments. The authors are thankful to Mohannad Saidi, Mohammad Dwaikat, and other students and former employees who helped us in the technical development and digitization phases. We would like to also thank John P. McCrae for helping us in representing our lexical data in the W3C lemon model. We are also thankful to all lexicon owners, especially the ALECSO team who provided us with many lexicons and supported us in the digitization process.

References

1. Khemakhem, A., Gargouri, B., Hamadou, A.B., Francopoulo, G.: ISO standard modeling of a large Arabic dictionary. Nat. Lang. Eng. **22**(6), 849–879 (2016)
2. Hyland, B., Atemezing, G., Villazón-Terrazas, B.: Best practices for publishing linked data. World Wide Web Consortium (2014)
3. Kamholz, D., Pool, J., Colowick, S.M.: PanLex: building a resource for panlingual lexical translation. In: LREC 2014 (2014)
4. Al-Hafi, D., Amayreh, H., Jarrar, M.: Usability Evaluating of a Lexicographic Search Engine. Technical Report. Birzeit University (2019)
5. Amayreh, H., Dwaikat, M., Jarrar, M.: Lexicons Digitization. Technical Report. Birzeit University (2019)
6. Maks, I., Tiberius, C., Veenendaal, R.V.: Standardising bilingual lexical resources according to the lexicon markup framework. In: LREC 2018 Proceedings (2008)
7. McCrae, J.P., Chiarcos, C., Bond, F., Cimiano, P., et al.: The Open Linguistics Working Group: Developing the Linguistic Linked Open Data Cloud. LREC (2016)
8. Helou, M.A., Palmonari, M., Jarrar, M.: Effectiveness of automatic translations for cross-lingual ontology mapping. J. Artif. Intell. Res. **55**(1), 165–208 (2016). AI Access Foundation
9. Jarrar, M.: The arabic ontology - an arabic wordnet with ontologically clean content. Appl. Ontol. J. (2019, Forthcoming). IOS Press
10. Jarrar, M., Amayreh, H., McCrae, J.: Progress on representing Arabic Lexicons in Lemon. In: The 2nd Conference on Language, Data and Knowledge (LDK 2019). Leipzig, Germany (2019)
11. Jarrar, M., Zaraket, F., Asia, R., Amayreh, H.: Diacritic-based matching of Arabic Words. ACM Trans. Asian Low-Resource Langu. Inf. Process. **18**(2), 10 (2018)
12. Jarrar, M., Ceusters, W.: Classifying processes and basic formal ontology. In: The 8th International Conference on Biomedical Ontology (ICBO), Newcastle, UK (2017)
13. Jarrar, M., Habash, N., Alrimawi, F., Akra, D., Zalmout, N.: Curras: an annotated corpus for the Palestinian Arabic Dialect. J. Lang. Resources Eval. **51**(3), 745–775 (2017)

14. Jarrar, M., Habash, N., Akra, D., Zalmout, N.: Building a corpus for Palestinian Arabic: a preliminary study. In: Workshop on Arabic Natural Language Processing (EMNLP 2014). Association for Computational Linguistics (ACL), Qatar, pp. 18–27 (2014)
15. Jarrar, M.: Building a formal Arabic ontology (Invited Paper). In: Proceedings of the Experts Meeting on Arabic Ontologies and Semantic Networks at ALECSO, Tunis (2011)
16. Jarrar, M., Meersman, R.: Ontology engineering – the DOGMA approach. In: Dillon, T.S., Chang, E., Meersman, R., Sycara, K. (eds.) Advances in Web Semantics I. LNCS, vol. 4891, pp. 7–34. Springer, Heidelberg (2008). https://doi.org/10.1007/978-3-540-89784-2_2
17. Jarrar, M., Keet, M., Dongilli, P.: Multilingual verbalization of ORM conceptual models and axiomatized ontologies. Technical report. Vrije Universiteit Brussel (2006)
18. Jarrar, M.: Position paper: towards the notion of gloss, and the adoption of linguistic resources in formal ontology engineering. In: The Web Conference (WWW 2006). ACM (2006)
19. Jarrar, M.: Towards methodological principles for ontology engineering. Ph.D. Thesis. Vrije Universiteit Brussel (2005)
20. Khalfi, M., Nahli, O., Zarghili, A.: Classical dictionary Al-Qamus in lemon. In: 4th IEEE International Colloquium on Information Science and Technology. IEEE (2016)
21. Soudani, N., Bounhas, I., Elayeb, B., Slimani, Y.: An LMF-based normalization approach of Arabic Islamic dictionaries for Arabic word sense disambiguation: application on hadith. J. Islamic Appl. Comput. Sci. **3**(2), 10–18 (2015)
22. Cimiano, P., McCrae, J.P., Buitelaar, P.: Lexicon Model for Ontologies. Final Community Group Report. World Wide Web Consortium (2016)
23. Speer, R., Chin, J., Havasi, C.: ConceptNet 5.5: an open multilingual graph of general knowledge. In: The 31st AAAI Conference on Artificial Intelligence (2016)
24. Navigli, R., Ponzetto, S.P.: BabelNet: The automatic construction, evaluation and application of a wide-coverage multilingual semantic network. AI 193 (2012)
25. Salmon-Alt, S., Akrout, A., Romary, L.: Proposals for a normalized representation of Standard Arabic full form lexica. In: The International Conference on Machine Intelligence (2005)

Bug Severity Prediction Using a Hierarchical One-vs.-Remainder Approach

Nonso Nnamoko(iD), Luis Adrián Cabrera-Diego(iD), Daniel Campbell(iD), and Yannis Korkontzelos(✉)(iD)

Department of Computer Science, Edge Hill University, Ormskirk, UK
{nnamokon,diegol,campbeld,Yannis.Korkontzelos}@edgehill.ac.uk

Abstract. Assigning severity level to reported bugs is a critical part of software maintenance to ensure an efficient resolution process. In many bug trackers, e.g. Bugzilla, this is a time consuming process, because bug reporters must manually assign one of seven severity levels to each bug. In addition, some bug types may be reported more often than others, leading to a disproportionate distribution of severity labels. Machine learning techniques can be used to predict the label of a newly reported bug automatically. However, learning from imbalanced data in a multi-class task remains one of the major difficulties for machine learning classifiers. In this paper, we propose a hierarchical classification approach that exploits class imbalance in the training data, to reduce classification bias. Specifically, we designed a classification tree that consists of multiple binary classifiers organised hierarchically, such that instances from the most dominant class are trained against the remaining classes but are not used for training the next level of the classification tree. We used *FastText* classifier to test and compare between the hierarchical and standard classification approaches. Based on 93,051 bug reports from 38 Eclipse open-source products, the hierarchical approach was shown to perform relatively well with 65% Micro F-Score and 45% Macro F-Score.

Keywords: Bug severity · Imbalanced data · Text mining ·
Machine learning · Multi-class classification · *FastText*

1 Introduction

A critical part of software maintenance is debugging, i.e. the identification and removal of concealed bugs. In a small software project involving a handful of developers, the task is relatively straightforward because bug reporting is likely to occur less frequently. However, debugging is a major concern in larger projects where the number of reported bugs is usually quite high, prompting tool support to aid the development team in tracking bugs, verifying their severity and managing their resolution. Bugzilla[1] has become a very popular tracking system for

---

[1] www.bugzilla.org.

© Springer Nature Switzerland AG 2019
E. Métais et al. (Eds.): NLDB 2019, LNCS 11608, pp. 247–260, 2019.
https://doi.org/10.1007/978-3-030-23281-8_20

this task, largely due to the spread of open source software development. Generally, a developer reports what they see as a defect in the software and manually assigns one of seven severity level i.e., *blocker, critical, major, normal, minor, enhancement* or *trivial* to indicate the degree to which the said bug impacts the software quality. Based on the assigned severity, a moderator decides how soon the bug needs to be fixed from a business perspective and in some cases, re-assign severity to reflect its impact on the software.

Despite the benefits of Bugzilla, the process of reporting and tracking a bug is still inefficient as it takes much time and human resources [16]. Since bugs typically contain textual descriptions, machine learning and text mining techniques are likely to provide automated severity classification. These techniques have been previously applied on such textual bug descriptions to automate severity classification [1,5-7,12,15-17]. However, the standard learning algorithm typically assumes that classes within the training data are roughly balanced and learning from imbalanced data often leads to bias in favour of the dominant classes [2]. As such, previous studies that utilised bug reports from Bugzilla have removed the often dominant *normal* and *enhancement* classes to conduct a 5-class prediction task [15,17]; or a binary classification task involving *blocker, critical, major* as "severe bugs" and *minor, trivial* as "non-severe bugs" [5,6,12,16].

In this paper, we extend the granularity of the classification task to include all seven categories available in Bugzilla. We tackle the imbalance issue through a hierarchical classification approach that exploits class imbalance in the training data, in order to reduce classification bias. Specifically, we designed a classification tree that consists of multiple binary classifiers organised hierarchically, such that instances from the most dominant class are trained against the remaining classes but are not included in training the next level of the classification tree. Detailed description of the approach is presented in Sect. 4. Using 93,051 bug reports downloaded from 38 Eclipse products within the Bugzilla tracking system, we tested the approach on *FastText*, a linear classifier based on a Neural Network [4]. We compared performance between our approach and standard training of *FastText*. We also trained a random guess classifier which forms the baseline on which our approach and indeed the standard approach was evaluated.

In Sect. 2 we provide details about related work and the necessary background for the techniques and tools used in our experiment. The experimental data and the proposed hierarchical approach are presented in Sects. 3 and 4. Section 5 presents our findings and discusses issues likely to threaten the validity of results. Section 7 summarises the study and points out future work.

2   Background and Related Work

Studies, that attempted to automate bug severity prediction, can be grouped into fine-grained (multi-class) and coarse-grained (binary class) ones.

As an early work, the *SEVERity ISsue* assessment algorithm, also known as *SEVERIS*, was proposed to assist test engineers in assigning severity levels to bug reports [7]. The approach is based on entropy and information gain, supported by a rule learner. To train the model, they used bug reports from six

NASA PITS projects in the PROMISE data repository[2] were used. *SEVERIS* used the Repeated Incremental Pruning to Produce Error Reduction (RIPPER) algorithm to perform fine grained prediction involving 5 bug severity levels. An average of 775 reports composed of 79,000 terms were examined with optimisation F-Scores ranging from 65% to 98% for individual severity levels.

In other related fine-grained bug severity prediction studies, Tian et al. [15] proposed an algorithm called *INSPect* (acronym for *Information Retrieval based Nearest Neighbour Severity Prediction Algorithm*). *INSPect* combines the *k*-Nearest Neighbour (*k*-NN) algorithm with the BM25-based document similarity algorithm [14], to predict bug severity. The authors used three different sets of data to replicate and compare their approach to *SEVERIS*. The results show that their approach produced better results than *SEVERIS*. Chaturvedi and Singh [1] also utilised the NASA PITS data to provide a comparative analysis with *SEVERIS*. They applied 6 different classification algorithms namely; Naïve Bayes (NB), *k*-NN, Naïve Bayes Multinominal, Support Vector Machine, J48 Decision tree and RIPPER. The accuracy results ranges from 29% to 97% which is lower than the top limit obtained by *SEVERIS*. Zhang et al. [17] extended the INSPect algorithm by adding topic modelling; a statistical approach used to discover abstract topics that occur in a collection of documents [11]. By using this approach, the authors found the topic(s) that each bug report belongs to. These topics were introduced to *INSPect* as additional features to produce better prediction with F-Score ranging from 13.96% to 80.25%.

From a coarse-grained bug severity perspective, Lamkanfi et al. [5] used a combination of text mining techniques and Naïve Bayes classifiers to classify bug reports into 'severe' and 'non-severe' classes. Their main contribution is that predictive performance is directly dependent on the training data size. By gradually increasing the training set during experiments, they concluded that an average of 500 reports per bug class is required to obtain a stable and reliable prediction. Indeed, their experiments with larger datasets resulted to stable and improved prediction of the severity levels with precision between 65%–83% and recall between 62%–84%. In a successive study, Lamkanfi et al. [6] presented a comparison of several classifiers using the same approach. This time, performance was measured with Area Under Curve (AUC) and results range from 51% to 93%. Gegick et al. [3] studied a binary bug severity prediction using SAS text miner with the aid of Singular Value Decomposition. Overall, the approach allowed to identify 77% 'severe' bugs that were labelled as 'non-severe' by bug reporters. Yang et al. [16] compared three feature selection algorithms to determine the best features for training a Naïve Bayes classifier. The results showed that the application of feature selection can improve the results. Starting from a baseline AUC of 74%, the authors were able to reach 77% on a given training dataset. Roy and Rossi [12] applied feature selection on bi-grams to train a Naïve Bayes classifier. The results showed that performance is data/project dependent as the addition of bi-grams worsened performance in some cases.

---

[2] promise.site.uottawa.ca/SERepository.

```
<Bug id="242xxx" severity="normal" status="CLOSED">
  <summary>Javadoc is not in the build zips</summary>
  <Comments>
    <Comment bugId="242xxx" commentId="130xxx" product="Acceleo">
      <text>The javadoc creation fails, probably needing some tweaking in the build scripts...</text>
    </Comment>
    <Comment bugId="242xxx" commentId="176xxx" product="Acceleo">
      <text>We've removed the javadoc from the documentation altogether : it was redundant with the SDK.</text>
    </Comment>
  </Comments>
</Bug>
```

Fig. 1. Sample bug report from Bugzilla (anonymised for presentation purposes)

There are many commonalities among the studies presented so far. For example, they mostly use text mining and machine learning techniques. In addition, they all took a project specific approach in which experiments and results are presented in single project context. Put simply, the authors took a naive approach during experiment by considering bug reports from different projects to be independent of each other. Therefore, results are analysed and presented on a project-by-project bases. One of the few differences between the studies is the level of granularity applied to the bug severity. While some applied fine-grained categories involving 5 severity categories [1,7,13,15,17], others simply condensed the 5 severity categories to binary classifcation [3,5,6,12,16]. It must be noted that most of these studies utilised data from Bugzilla which had 7 severity categories originally. Unfortunately, none of them utilised all 7 bug severity categories in their experiments. The *Enhancement* category is always disregarded as a non-bug because it is believed to refer to feature improvements. Additionally, the *Normal* category is often removed as it is the default option for reporting bug severity in Bugzilla and many researchers [1,5–7] suspect that bug reporters do not bother to assess and set correct severity for the bug being reported. Lamkanfi et al. [5,6] believes that the *normal* category represents a grey zone and Singh et al. [13] suggests they may confuse the classifier because they are often disproportionate in size compared to the other categories which creates a huge class imbalance. Thus, two limitations we address in our work are: (1) to extend the severity granularity by including all 7 categories in a cross-project experiment; and (2) to mitigate the class imbalance issue presented by the *normal* and *enhancement* severity categories through a hierarchical classification approach.

3 Dataset

Each Bugzilla bug report is assigned one of seven severity labels shown in Table 1. We downloaded 93,051 bug reports from Bugzilla[3]. The data was split into 80% training and 20% testing, taking the distribution of the severity classes into account, as shown in Table 1.

Figure 1 shows a sample bug report assigned the *"normal"* severity label. A brief description of the bug can be seen in the *summary* element, with further

---

[3] Bug reports were downloaded from 38 Eclipse related products.

Table 1. Size of the data downloaded from Bugzilla

Level	Description	Total	Training	Test
Normal	Default/average bugs	61,421	49,136	12,285
Enhancement	New feature or functionality change in existing feature	15,156	12,125	3,031
Major	Loss of function in an important area	7,594	6,075	1,519
Minor	Loss of function that affect few people or with easy workaround	3,504	2,803	701
Critical	Crashes, loss of data in a widely used and important component	2,655	2,124	531
Blocker	Blocks further development or testing	1,371	1,097	274
Trivial	Cosmetic problem e.g. misspelled words	1350	1,080	270

description about the bug in the *comment* elements. For the experiment presented in this paper, a concatenation of textual information from the *summary* and *comment* elements is used as features. This is to ensure that our bug severity predictor is applicable to threads from other bug tracking systems such as GitHub that does not support the wealth of elements available in Bugzilla.

4   Method

In this section, we present our classification approach that would, in theory reduce classification bias to any class. Our aggregate consists of multiple one vs. remainder classifiers (i.e. binary sub-classifiers) organised hierarchically, so that instances from the dominant class[4] are trained against all other classes, but are not included in the next classification level. This is similar to the *one vs. all* strategy which involves training a single classifier per class, with the instances of that class as *positive* and all other instances as *negatives*. The slight difference in our approach is that the binary classifiers are arranged in descending order based on class size and positive instances from the preceding binary classifier are disregarded in subsequent classification. For example, our data contains more instances of the *normal* class than of any other class. Therefore, the first binary classifier is trained on instances of the *normal* class against instances of the other classes. In the second binary classifier, instances from the *normal* class are eliminated from the training data, so that those from the next dominant class, i.e., *enhancement* are trained against the remaining classes, i.e., *major, minor, critical, blocker* and *trivial* grouped together. The process is repeated until the binary classifier between the *blocker* and *trivial* classes. Figure 2 depicts the one vs. remainder method.

---

[4] Dominance refers to size. A dominant class contains more instances than another.

4.1   Experiment Setup

The hierarchical approach described in Sect. 4, was trained using *FastText* [4], a simple neural network classifier that implements an improved version of Word2Vec [8,9]. We chose *FastText* because it performs comparably better than the more complex deep learning algorithms, in less training time and without using a GPU [18].

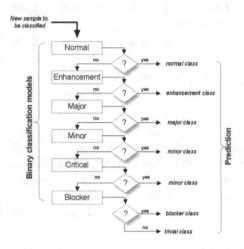

Fig. 2. A hierarchical one vs. remainder multi-label classification approach

The training data was lemmatised and tokenised using NLP4J[5] before being fed to the classifiers. In the proposed hierarchical approach, each binary *FastText* classifier has a set of parameters to be optimised. We used *Bayesian Optimisation* [10], which explores how parameters, i.e. learning rate (lr), vector dimension (dim), different numbers of word n-grams, where $n \in [1,3]$, minimum threshold of occurrence (min) and epoch, affect the results. As objective function either the median or the average was used, whichever achieves the lowest Macro F-score on a 10-fold cross validation basis. This function considers all classes as equally important, despite their disproportionate distribution in the data. The binary models were then organised hierarchically in order of descending class size for training and testing using the best parameters. Initially, we used all seven severity categories in the data. To investigate if the *normal* category in Bugzilla introduces noise, we subsequently performed an experiment without it.

For evaluation and comparison purposes, we performed a classification experiment based on random guess on the test data. The results were used as baseline to assess the hierarchical approach. 10 iterations were conducted and the average Macro and Micro F-Score recorded. We also compared the hierarchical approach against standard multi-class training using *FastText* i.e., non hierarchical approach. Again, the experiments was conducted in two phases - first with all the

7 severity categories and secondly without the *normal* category. We used the *Bayesian Optimisation* method to obtain the parameters that lead to optimum performance. Unlike the hierarchical approach that requires several binary models with different parameter sets, only one set of parameters is necessary for a standard multi-class training using *FastText*. All optimisation experiments were conducted in 35 optimisation iterations to obtain the best parameter sets shown in Table 2. For simplicity, we use suffixes *incl* and *excl* to differentiate between *FastText* trained with and without the *normal* instances respectively.

Table 2. Parameters used to train Hierarchical (H) and Standard (S) *FastText* models

	Classifiers	lr	dim	n-grams	min	epoch	Macro F-Score(%)
H	**Normal vs. Other**	0.20	100	3	10	5	66.87
	Enhancement vs. Other	0.11	200	3	30	20	88.03
	Major vs. Other	0.27	200	3	40	15	64.29
	Minor vs. Other	0.19	135	2	30	25	70.72
	Critical vs. Other	0.26	25	3	2	40	69.22
	Blocker vs. Other	0.19	50	2	4	35	87.29
S	$FastText_{incl}$	0.43	150	3	40	30	33.75
	$FastText_{excl}$	0.33	75	3	20	25	47.42

5 Result Analysis

In this section, we present the results obtained from *FastText* trained with both hierarchical and standard approaches. This includes results with and without the *normal* severity category in the training and test data. Aggregate measures derived from confusion matrix, such as *precision, recall, micro F-Score*, and *macro F-Score* were used for evaluation. For simplicity, the performance with and without the *normal* classifier/instances are presented separately in this section. We use the random guess classifier shown in Table 3 as a baseline.

Table 3. Results (%) of experimentation with the Random Guess classifier

	Normal	Enhancement	Major	Minor	Critical	Blocker	Trivial
Precision	65.80	16.23	8.16	3.66	2.81	0.12	0.13
Recall	49.80	25.05	12.42	6.09	3.07	0.12	0.14
F-Score	56.69	19.69	9.84	4.57	2.93	0.00	1.31
	Macro F-Score: 13.58			**Micro F-Score: 38.32**			

5.1 Experiments Including *normal*

Table 4 shows a combined confusion matrix of predictions using $FastText_{incl}$ in hierarchical and standard classification, marked in *blue* and *black*, respectively.

The hierarchical approach using $FastText_{incl}$, performed considerably better in classifying the *normal* and *enhancement* classes compared to the other classes. For example, the first classifier model in the hierarchical structure, *Normal vs. Other* predicted 10, 398 out of 12, 285 *normal* bugs correctly, but misclassified 3, 380 instances as *normal*. The misclassified instances represent 18.16% of the total test set; thus, reducing the number of test data instances for the subsequent classifier models. This dynamic continued at the next level of classification, i.e. *Enhancement vs. Other* in which a further 12.11% of the test set was misclassified. This had a knock on effect at subsequent nodes as shown in Table 5, where performance degrades massively further down the hierarchy.

Table 4. Combined confusion matrix of the $FastText_{incl}$ models showing both hierarchical (H) and standard (S) methods

	Normal		En/ment		Major		Minor		Critical		Blocker		Trivial	
	H	**S**	**H**	**S**	**H**	**S**	**H**	**S**	**H**	**S**	**H**	**S**	**H**	**S**
Normal	10398	9003	1003	1281	417	1234	161	240	132	299	71	102	103	126
Enhancement	847	789	2082	2121	47	65	26	29	16	11	5	4	8	12
Major	1239	929	61	76	157	397	12	18	33	75	12	19	5	5
Minor	517	461	63	73	34	57	56	80	5	10	10	2	16	18
Critical	404	301	13	16	51	128	3	2	46	70	14	14	0	0
Blocker	211	150	9	7	24	53	1	2	6	25	24	36	0	1
Trivial	162	139	27	30	9	11	24	27	4	3	3	2	41	58

Similar performance was obtained with the standard classification approach. Although *FastText* is designed to handle multi-class tasks intrinsically, the classifier seems to also struggle with the data as evident from Table 4. For example, $FastText_{incl}$ correctly predicted only 9, 003 out of 12, 285 *normal* instances using the standard approach. This is 1, 395 short of its correct predictions of the *normal* class with the hierarchical approach. However, the misclassified instances (2, 769) are lower than the hierarchical approach. That said, the misclassified instances represent 14.88% of the total test set. A further 7.97% instances was misclassified into the *enhancement* class.

In terms of F-Score, the standard $FastText_{incl}$ achieved a slightly better Macro F-Score (32.79%) than hierarchical (30.48%); but Micro F-Score is higher with the hierarchical approach (68.80%) compared to 63.26% obtained with the standard approach. Nonetheless, both classification approaches using *FastText* performed better than classification based on random guess which produced Macro F-Score of 13.58% and Micro F-Score of 38.32% (see Table 3).

A closer look at Table 4 revealed some interesting pattern that is worth discussing. Both approaches misclassified large proportions of the actual *normal* class as *enhancement* or *major* (see the "Normal" row in Table 4). Likewise, large proportions of both *enhancement* or *major* class instances are classified as *normal* (see the "Normal" columns in Table 4). This is probably an indication of noise in the data, particularly the *normal* class. Perhaps, bug reporters found it difficult to differentiate between *normal* vs. *enhancement* bug categories. This may explain the high misclassification of *normal* as *enhancement* classes but a deeper analysis is required to understand why instances of the *major* class was misclassified as normal in such proportion. In theory, *normal* and *major* bugs would have distinguishing terms within the dictionary which suggests that the noise may be because Bugzilla has the *normal* severity category as its default setting; and some bug reporters had not bothered to change it.

Table 5. Results (%) of *FastText$_{incl}$* [H: Hierarchical, S: Standard]

	Normal		En/ment		Major		Minor		Critical		Blocker		Trivial	
	H	S	H	S	H	S	H	S	H	S	H	S	H	S
P	75.47	76.48	63.92	58.85	21.24	20.41	19.79	20.10	19.01	14.20	17.27	20.11	23.70	26.36
R	84.64	73.28	68.69	69.98	10.34	26.14	7.99	11.41	8.66	13.18	8.76	13.14	15.19	21.48
F1	79.79	74.85	66.22	63.93	13.91	22.92	11.38	14.56	11.90	13.67	11.62	15.89	18.51	23.67
H	Macro F-Score: 30.48%						Micro F-Score: 68.80%							
S	Macro F-Score: 32.79%						Micro F-Score: 63.26%							

We suspect that performance could be higher if the noisy data was excluded. Particularly with the hierarchical approach, Table 5 may not reflect the individual predictive power of the classifiers at each node of the hierarchy during optimisation (see Table 2); where the *Enhancement vs. Other* classifier achieved the highest Macro-F-Score of 88.03%, followed by 87.29% for the *Blocker vs. Other* classifier. We expect similar performance at these nodes of the hierarchical classification approach during testing. To examine this, we performed a calculation based on elimination in which the performance at each node in the hierarchy was re-calculated by ignoring the test instances already seen/classified at preceding nodes. For example, to calculate the performance of the *Enhancement vs. Other* classifier, we ignored instances already seen/classified by the *Normal vs. Other* classifier i.e., the first row and column of the confusion matrix for the hierarchical approach shown in Table 4. We present the modified results in Table 6 which shows that both *Blocker vs. Other* and *Enhancement vs. Other* still maintained high performance relative to the other classifiers. It is worth noting that *Enhancement vs. Other* performed better than *Normal vs. Other* in the hierarchical structure, which suggests that the poor performance of the approach may be caused by the huge misclassification rate at the apex. Based on this observation, we excluded the *normal* class instances from the data and repeated the experiments with both hierarchical and standard approach.

Table 6. Modified results (%) of Hierarchical based experimentation with $FastText_{incl}$ (excluding misclassified instances)

	Normal	Enhancement	Major	Minor	Critical	Blocker	Trivial
Precision	75.47	92.37	57.09	66.67	82.14	88.89	100.00
Recall	84.64	95.33	71.69	64.37	76.67	100.00	93.18
F-Score	79.79	93.83	63.56	65.50	79.31	94.12	96.47

5.2   Experiments Excluding *normal*

Table 7 shows a combined confusion matrix of predictions using $FastText_{excl}$ in hierarchical and standard classification. Again, the predictions with hierarchical approach are marked with *blue* while the standard predictions are in *black*.

Table 7. Combined confusion matrix of $FastText_{excl}$ models showing both hierarchical (H) and standard (S) methods

	Enhancement		Major		Minor		Critical		Blocker		Trivial	
	H	**S**	**H**	**S**	**H**	**S**	**H**	**S**	**H**	**S**	**H**	**S**
Enhancement	2636	2652	143	213	126	90	29	43	20	11	45	22
Major	197	167	1001	1041	214	84	312	182	137	33	45	12
Minor	87	152	109	237	230	235	24	27	14	13	69	37
Critical	54	38	174	341	51	17	118	106	42	28	15	1
Blocker	28	24	68	139	33	9	42	44	59	57	9	1
Trivial	29	59	24	43	47	71	6	13	2	6	87	78

Table 8. Results (%) of $FastText_{excl}$ [H: Hierarchical, S: Standard]

	En/ment		Major		Minor		Critical		Blocker		Trivial	
	H	S	H	S	H	S	H	S	H	S	H	S
Precision	87.90	85.77	52.52	51.69	43.15	46.44	25.99	25.54	24.69	38.51	44.62	51.66
Recall	86.96	87.50	65.90	68.53	32.81	33.52	22.22	19.96	21.53	20.80	32.22	28.89
F-Score	87.43	86.62	58.45	58.93	37.28	38.94	23.96	22.41	23.00	27.01	37.42	37.05
H	Macro F-Score: 44.59%						Micro F-Score: 65.30%					
S	Macro F-Score: 45.16%						Micro F-Score: 65.90%					

Both hierarchical and standard $FastText_{excl}$ performed considerably better than $FastText_{incl}$ which was trained with the noisy *normal* class. For example, the hierarchical $FastText_{incl}$ misclassified 18.16% of the total test set at the apex but misclassification at the same node was reduced to 5.74% with $FastText_{excl}$. Similar performance was obtained with the standard classification approach using $FastText_{excl}$ which misclassified 6.96% of the total test data at the apex, compared to 14.88% misclassified by $FastText_{incl}$ at the same node. The evidence of improvement without the *normal* class can be seen in Table 8. Clearly, the Macro F-Score of both hierarchical and standard $FastText_{excl}$ are better than those recorded for $FastText_{incl}$ (Table 5). The Micro F-Score is higher with hierarchical $FastText_{incl}$ but this is likely because the test data is larger with the normal class included.

That said, there is very little distinction between the performance of hierarchical and standard $FastText_{excl}$. The standard $FastText_{excl}$ achieved a slightly better Macro F-Score (45.16%) than hierarchical (44.59%). Likewise, the Micro F-Score is slightly higher with the standard approach (65.90%) compared to 65.30% obtained with the hierarchical approach. Nonetheless, both classification approaches using $FastText_{excl}$ performed better than classification based on random guess which produced Macro F-Score of 13.58% and Micro F-Score of 38.32% (see Table 3).

Again, we believe that the results in Table 8 does not reflect the individual predictive power of $FastText_{excl}$ classifiers observed at each node of the hierarchical structure during optimisation (see Table 2). Although the *Enhancement vs. Other* Macro-F-Score of 87.90% is very close to 88.03% obtained during optimisation; the same cannot be said about the *Blocker vs. Other* classifier, that produced the second best Macro F-Score of 87.29% during optimisation but reduced to 24.69% during testing. This suggests that we are still loosing test instances due to misclassification at preceding nodes in the hierarchical structure.

To confirm this, we re-calculated performance for the hierarchical classifiers in Table 8 using the elimination process applied earlier; in which instances already seen/classified at preceding nodes are ignored. The modified results, in Table 9, shows that both *Blocker vs. Other* and *Enhancement vs. Other* maintained high performance relative to the other classifiers. Nonetheless, only 59 out of 274 *blocker* instances reached the *Blocker vs. Other* node of the hierarchy, which suggests that a large number of misclassification still occurs at preceding nodes.

Table 9. Modified results (%) of hierarchical-based experimentation with $FastText_{excl}$ (excluding misclassified instances)

	Enhancement	Major	Minor	Critical	Blocker	Trivial
Precision	87.90	58.57	68.25	67.43	86.76	97.52
Recall	86.97	72.75	63.71	71.08	96.72	90.63
F-Score	87.43	64.89	65.90	69.21	91.47	93.94

6   Discussion

In general, the proposed hierarchical approach performed similar to the standard classification approach, even with extensive misclassification observed down the hierarchy, which affects the less dominant classes heavily. For the misclassification observed in both hierarchical and standard $FastText_{incl}$, the default severity setting when reporting a bug on Bugzilla i.e., *normal* category, may be culpable.

Another reason for misclassification may be the class distribution of the data, especially given the level of granularity applied in our experiments. It is likely that one or more of the severity classes share similar vocabulary, thus are linearly inseperable. Perharps, performance would improve if the number of classes is reduced such that those with similar vocabulary are grouped together.

A further possible reason may lie within the separability of the severity classes available in the experimental data. Theoretically, classes such as *trivial* and *blocker* are easier to differentiate as they most likely use different vocabulary. However, instances from the *normal* and *enhancement* classes may share similar characteristics which may pose difficulties for a classifier. The hierarchical approach performed generally similar to the standard approach, even with the huge loss of test instances at each node. We believe that a hierarchical approach based on disparity rather than class size may improve our results. For instance, the *Blocker vs. Other* classifier which performs binary classification between *blocker* and *trivial* bugs, generally performed well. Therefore, arranging the classifiers hierarchically based on class disparity may well improve our results.

The cross-product approach of our experiment may also influence performance. All methods in the state-of-the-art were evaluated on individual products [1,5,7]. Their project-specific approach is unlikely to have an adverse effect on results because the presentation and annotation of the bug reports are likely to be consistent. In our case, we combined bug reports from 38 different projects, which may not follow similar bug presentation and/or annotation.

The hierarchical approach has been applied specifically to address the class imbalance in our dataset. Perhaps, further experimentation with data of similar characteristics from other domains is required to validate its generalisability.

7   Conclusion

We have investigated fine-grained bug severity classification using *FastText*. We acknowledge that several research works have been reported in this area, but none to our knowledge has gone to the level of granularity investigated in this paper. As our task is a multi-class one involving highly imbalanced classes in the experimental data, we explored a one vs. remainder approach based on hierarchical arrangement of binary classifiers. We investigated this method using *FastText* classifier, and compared the results to the same classifiers trained in a standard manner. We also used the performance of a random guess classifier as baseline. Our experiments show that the hierarchical method performed generally similar to the standard approach but better than a classifier based on random guess. However, the results might just be a signal to the difficulty of the task. For example, some severity categories such as major and critical might be hard to differentiate. Furthermore, it might be the case, that the annotation of the severity categories is not an easy task to do. There are indications from the experiment that the normal category may contain samples from the other severity categories. Thus, these aspects should be taken into consideration when interpreting the results presented in this work. Deeper experiments and analysis are required in order to draw a final conclusion on the results.

Acknowledgments. This research work is part of the CROSSMINER Project, which has received funding from the European Union's Horizon 2020 Research and Innovation Programme under grant agreement No. 732223.

References

1. Chaturvedi, K.K., Singh, V.B.: Determining Bug severity using machine learning techniques. In: 2012 CSI 6th International Conference on Software Engineering. CONSEG 2012, pp. 1–6. IEEE (2012). https://doi.org/10.1109/CONSEG.2012.6349519
2. Chawla, N.V., Bowyer, K.W., Hall, L.O., Kegelmeyer, W.P.: Smote: synthetic minority over-sampling technique. J. Artif. Intell. Res. **16**(1), 321–357 (2002). http://dl.acm.org/citation.cfm?id=1622407.1622416
3. Gegick, M., Rotella, P., Xie, T.: Identifying security bug reports via text mining: an industrial case study. In: Proceedings - International Conference on Software Engineering, pp. 11–20. IEEE (2010). https://doi.org/10.1109/MSR.2010.5463340
4. Joulin, A., Grave, E., Bojanowski, P., Mikolov, T.: Bag of tricks for efficient text classification. In: Proceedings of the 15th Conference of the European Chapter of the Association for Computational Linguistics, Valencia, Spain, vol. 2, pp. 427–431 (2017)
5. Lamkanfi, A., Demeyer, S., Giger, E., Goethals, B.: Predicting the severity of a reported bug. In: Proceedings - International Conference on Software Engineering, pp. 1–10. IEEE (2010). https://doi.org/10.1109/MSR.2010.5463284
6. Lamkanfi, A., Demeyer, S., Soetens, Q.D., Verdonckz, T.: Comparing mining algorithms for predicting the severity of a reported bug. In: Proceedings of the European Conference on Software Maintenance and Reengineering. CSMR, pp. 249–258. IEEE (2011). https://doi.org/10.1109/CSMR.2011.31
7. Menzies, T., Marcus, A.: Automated severity assessment of software defect reports. In: IEEE International Conference on Software Maintenance. ICSM, pp. 346–355 (2008). https://doi.org/10.1109/ICSM.2008.4658083
8. Mikolov, T., Chen, K., Corrado, G., Dean, J.: Efficient estimation of word representations in vector space. arXiv preprint arXiv:1301.3781 (2013)
9. Mikolov, T., Sutskever, I., Chen, K., Corrado, G.S., Dean, J.: Distributed representations of words and phrases and their compositionality. In: Advances in Neural Information Processing Systems, pp. 3111–3119 (2013)
10. Močkus, J., Tiešis, V., Žilinskas, A.: The application of Bayesian methods for seeking the extremum. In: Szegö, G.P., Dixon, L.C.W. (eds.) Towards Global Optimisation, vol. 2, pp. 117–128, North-Holland (1978)
11. Ramage, D., Hall, D., Nallapati, R., Manning, C.D.: Labeled LDA: a supervised topic model for credit attribution in multi-labeled corpora. In: EMNLP 2009 Proceedings of the 2009 Conference on Empirical Methods in Natural Language Processing, pp. 248–256 (2009). https://doi.org/10.3115/1699510.1699543

12. Roy, N.K.S., Rossi, B.: Towards an improvement of bug severity classification. Proceedings - 40th Euromicro Conference Series on Software Engineering and Advanced Applications. SEAA 2014, pp. 269–276 (2014). https://doi.org/10.1109/SEAA.2014.51

13. Singh, V.B., Misra, S., Sharma, M.: Bug severity assessment in cross project context and identifying training candidates. J. Inf. Knowl. Manag. **16**(01), 1750005 1–30 (2017). https://doi.org/10.1142/S0219649217500058, http://www.worldscientific.com/doi/abs/10.1142/S0219649217500058

14. Sun, C., Lo, D., Khoo, S.C., Jiang, J.: Towards more accurate retrieval of duplicate bug reports. In: Proceedings of the 2011 26th IEEE/ACM International Conference on Automated Software Engineering. ASE 2011, pp. 253–262 (2011). https://doi.org/10.1109/ASE.2011.6100061

15. Tian, Y., Lo, D., Sun, C.: Information retrieval based nearest neighbor classification for fine-grained bug severity prediction. In: Proceedings - Working Conference on Reverse Engineering. WCRE, pp. 215–224 (2012). https://doi.org/10.1109/WCRE.2012.31

16. Yang, C.Z., Hou, C.C., Kao, W.C., Chen, I.X.: An empirical study on improving severity prediction of defect reports using feature selection. In: Proceedings - Asia-Pacific Software Engineering Conference, APSEC. vol. 1, pp. 240–249. IEEE (2012). https://doi.org/10.1109/APSEC.2012.144

17. Zhang, T., Chen, J., Yang, G., Lee, B., Luo, X.: Towards more accurate severity prediction and fixer recommendation of software bugs. J. Syst. Software **117**, 166–184 (2016). https://doi.org/10.1016/j.jss.2016.02.034

18. Zolotov, V., Kung, D.: Analysis and optimization of fast text linear text classifier. arXiv preprint arXiv:1702.05531 (2017)

A Coherence Model for Sentence Ordering

Houda Oufaida[1]([✉]), Philippe Blache[2], and Omar Nouali[3]

[1] Ecole Nationale Supérieure d'Informatique ESI, Oued Smar, Algiers, Algeria
h_oufaida@esi.dz
[2] Aix Marseille Université, CNRS, LPL UMR 7309, 13604 Aix en Provence, France
blache@lpl-aix.fr
[3] Centre de Recherche sur l'Information Scientifique et Technique CERIST,
Ben Aknoun, Algiers, Algeria
onouali@cerist.dz

Abstract. Text generation applications such as machine translation and automatic summarization require an additional post-processing step to enhance readability and coherence of output texts. In this work, we identify a set of coherence features from different levels of discourse analysis. Features have either positive or negative input to the output coherence. We propose a new model that combines these features to produce more coherent summaries for our target application: extractive summarization. The model use a genetic algorithm to search for a better ordering of the extracted sentences to form output summaries. Experimentations on two datasets using an automatic coherence assessment measure show promising results.

Keywords: Coherence features · Coherence model ·
Sentence ordering · Automatic summarization · Genetic algorithm

1 Introduction

Coherence and cohesion are key elements for text comprehension [1]. Coherence involves logical flow of ideas around an overall intent. It reports a conceptual organization of discourse and can be observed at the semantic level of discourse analysis. Coherence is essential to text comprehension. Indeed, with a lack of coherence, the text loses quickly its informational value.

Dealing with text coherence remains a difficult issue for several NLP applications such as machine translation, text generation and automatic summarization. Most of automatic summarization systems rely on extractive techniques which extract complete sentences from source texts to form summaries. This ensures that the summary is grammaticality correct but in no case its coherence. Considering coherence of extractive summaries involves dealing with sentence informativness input against summary's flow. Several elements contribute

© Springer Nature Switzerland AG 2019
E. Métais et al. (Eds.): NLDB 2019, LNCS 11608, pp. 261–273, 2019.
https://doi.org/10.1007/978-3-030-23281-8_21

to text coherence such as discourse relations [2], sentences' connection by mean of common entities patterns [3] and thematic pregression [4].

In the automatic summarization task, it is fundamental to generate intelligible summaries. Extractive techniques succeed in selecting relevant information but mostly fail to ensure their coherence. Only few of these techniques considered coherence as an additional feature in the summary's extraction process. It is a difficult task that tackles with multi level discourse analysis: syntactic level in which connectors are used to improve text cohesion, semantic level in which textual segments are regrouped around common concepts and finally, global level in which sentences are presented in a logical flow of ideas.

In this paper we deal with coherence as an optimization problem. We identify a set of coherence features that have positive or negative impact on summaries coherence. The intuition is that positive input features such as original thematic ordering in the source text/texts and shared entities of adjacent sentences contribute to local and global coherence. These features should be maximized whereas negative input features such as redundancy should be minimized.

The rest of the paper is organized as follows: we first introduce a review of the very few works in the field. Second, we describe how our coherence model combines between coherence features to have better sentences ordering within system summaries. Details and discussion of our experiments are presented in which the coherence model is introduced as a post processing step. Finally, we conclude our work with interesting perspectives.

2   Related Work

Early approaches of automatic summarization use sentence compression techniques to improve summaries' coherence. The main idea is to reproduce human summarization process, namely: i-identify relevant sentences ii-compress and reformulate these sentences iii-reorder compressed sentences iv-add discourse elements to make a cohesive summary.

Probably the most referenced work is Rhetorical Structure Theory (RST) discourse analysis [2]. A set of discourse relation markers from an annotated corpus are used to define two elements for each relation: nucleus and satellite. The analysis generate a tree in which the nucleus parts of the top levels are the most relevant ones. [5] train an algorithm on collections of (texts, summaries) to discover compression rules using a noisy-channel framework. The assumption is that the compressed form is the source of a signal which was affected by some noise i.e. optional text. The model learns how to restore the compressed form and assesses the probability that it is grammaticality correct. More recently, [6] define the concept of textual energy of elementary discourse units. It reflects the degree of each segment's informativeness: the more the segment shares words with other segments the more it is informative. Less informative segments are eliminated and the remaining segments grammaticality is estimated by mean of a language model.

[4] study the thematic progression in the source texts and identify which thematic ordering is better for the output summaries. The authors define three

strategies for sentence ordering: (1) majority ordering which is a generalization of ordering by sentence position and reflects, for each couple of themes, how many sentences from the first theme precede sentences from the second one (2) chronological ordering in which themes are ordered according to the publication date and (3) Augmented ordering which add a cohesion element that regroups themes whose sentences appear in the same text blocks. Sentences in the output summary are assigned to themes and follow the thematic ordering. Augmented ordering seems to be the best alternative for news articles.

[3] define local coherence as a set of sentence transitions required for textual coherence. An entity-based representation of the source text is used to model coherent transitions. The intuition is that consecutive segments (sentences) about same entities are more coherent. The model estimates transition patterns probabilities from a collection of coherent texts.

More recently, [7] introduce a joint model that combines between coherence and sentence salience in the sentence extraction process. A discourse graph is first generated in which vertices correspond to sentences and positive edge weights correspond to coherent transitions between each couple of sentences i.e. the second sentence could be placed after the first sentence in a coherent text. It is based on syntactic information such as deverbal noun reference, event/entity continuation and RST discourse markers.

Defining deep learning architectures for various NLP tasks including coherence models was recently investigated. [8] train a three level neural network to model sentence composition to form coherent paragraphs. Here, positive examples are coherent sentence windows and negative examples are sentence windows in which a sentence was randomly replaced. Sentence vectors are induced from the sequence of its word embeddings using recurrent neural networks. The neural network is trained using pairs of original articles and articles with randomly permuted sentences, window size is three consecutive sentences. [9] propose to generalize the entity based coherence model initially proposed by [4] using a neuronal architecture. The model maps grammatical roles within entity grid to a continuous representation (a real valued vector learned by back propagation). Entity transition representations of a given sentence sequence are used by convolution, pooling and linear projection layers to finally compute a coherence score. The model is trained on a set of ordered coherent/less coherent document pairs and compared to several coherence models for two tasks: sentence ordering and summary coherence rating.

In the previous work, various features are used to improve output coherence. RST discourse analysis is certainly of value to define a global coherence model. However, it requires deep text analysis which is not available for most languages.

In this work, we have defined a set of coherence features. Each feature is supposed to help the model to give higher or lower coherence score according to a particular sentence ordering. The model combines between features and selects an ordering that maximises the coherence score. We assume that these features, once applied together, complement each other and lead to better coherence. We use genetic algorithm to select a coherent ordering. The advantage is that the model can be easily alimented by additional and language specific features. Features can be added to the fitness function by specifying its contribution to the output ordering. The next section describes, in detail, the proposed coherence model.

3   Coherence Model

In our coherence model, we propose to combine state-of-the-art features using a genetic algorithm. These features are domain independent and could be automatically extracted for a large number of languages.

3.1   Coherence Features

Positive Input Features positive input features are features who should be maximized in the output summary. They are assumed to help the model to produce more coherent summaries.

Sentence Position: sentence position feature is based on the assumption that sentence ordering in source text is coherent and a coherent summary should follow the original ordering. In multi-document summarization, this ordering is generalized using publication date in a way that the first sentence in the first document is given the label "1" and the last sentence in the most recent document is given the label "n", "n" being the number of sentences in all source documents.

Shared Entities: it is an important feature based on the assumption that sentences discussing same entities should appear in the same textual segment. [10] defines textual continuity as "*a linear progression of elements with strict recurrence*" which puts forward that coherent development of text should not introduce a sudden break.

Shared entities feature was introduced by [3], it requires part of speech tagging. In practice, noun phrases tag set depends on target language and the Part of Speech tagger used (NN, NNP, NNS, NNPS, etc. for English Penn Treebank tag set).

We use the number of shared noun phrases between each couple of adjacent sentences (S_1, S_2) in the candidate summary R as a positive input feature (1) and (2).

$$Shared\_Entities(S_1, S_2) = \frac{2 \times |Entities(S_1) \cap Entities(S_2)|}{|S_1| + |S_2|} \tag{1}$$

$$Entities\_Score(R) = \sum_{i=1..|R|-1} Shared\_Entities(S_i, S_{i+1}) \tag{2}$$

Thematic Ordering: Thematic progression is a key feature in information ordering and text comprehension. Presenting information in a logical progression is important especially in summaries with a limited size. Following [4], we want to make the thematic progression of summaries similar to source texts. We define a precedence matrix (PM) of topics. Each entry $PM[C_i, C_j]$ corresponds to the percentage of sentences from topic C_i that appear, in the source texts, before sentences from the second topic C_j. Figure 1 presents a precedence matrix of a document cluster extracted from the MultiLing 2015 dataset.

Topics (Clusters)	C_0	C_1	C_2	C_3	C_4	C_5
C_0	0.000	0.335	0.285	0.564	0.631	0.521
C_1	0.665	0.000	0.438	0.764	0.787	0.782
C_2	0.715	0.562	0.000	0.865	0.858	0.867
C_3	0.436	0.236	0.135	0.000	0.594	0.486
C_4	0.369	0.213	0.142	0.406	0.000	0.437
C_5	0.479	0.218	0.133	0.514	0.563	0.000

Fig. 1. Precedence matrix example

Different possible strategies for thematic ordering extraction could be considered. A first strategy is to order topics according to their precedence value. We define precedence value of a target topic as the sum of remaining topics precedence value to the target topic (sum per column) (3). Topic with minimum precedence will be the first topic to be mentioned in the summary.

$$Precedence\_Score(C_j) = \sum_{i=1..|C|} Precedence(C_i, C_j) \qquad (3)$$

Another strategy is to build thematic ordering gradually. The algorithm starts with couple of topics with a strong precedence score (C_2 and C_5 in the example). Then the algorithm search for another couple of topics that maximizes precedence scores for the just selected topics at the beginning/end of the previous ordering. Algorithm 1 repeats these steps until finding a complete ordering which includes all topics. We compare system summary ordering against source texts thematic ordering using the distance between the two ordering vectors (4). Summary vector is likely to be not complete, we complete the shortest vector by the value of the last item (last topic number). Formula (4) computes the thematic ordering score which is inversely proportional to the distance between summary ordering vector (Sum_Ord) and the source texts vector ($Source\_Ord$).

$$Thematic\_Ordering\_Score = \frac{1}{Distance(Sum\_Ord, Source\_Ord)} \qquad (4)$$

Negative Input Features

Redundancy: Bringing new information in each sentence is essential to the semantic coherence of any text. In the context of automatic summarization, it is critical to present new relevant information in each single sentence. We use a sentence relatedness measure proposed in [11] to compute information redundancy between each couple of sentences (S_1, S_2). Formula (5) uses word embeddings to find for each word in the first sentence (w_i) its matching word in second sentence (most related word using distance between embeddings vectors). It also do the same for words from the second sentence (w_j) and the final relatedness score will be the mean between all matching scores.

$$Relatedness(S_1, S_2) = \frac{\sum_i Match(w_i) + \sum_j Match(w_j)}{|S_1| + |S_2|} \tag{5}$$

```
1: Input:
       Precedence[,] : precedence matrix
2: Initialise:
       Ordering = {}
3: Ordering= (C_{Max_i}, C_{Max_j}) = Max{Precedence(C_i, C_j), ∀i, j < |C|}
4: do
5:     Max_i=Max{Precedence(*, C_j), ∀j < |C|}
6:     Max_j=Max{Precedence(C_i, *), ∀i < |C|}
7:     Ordering = Ordering ∪ {((C_{Max_i}, C_j))}
8:     Ordering = Ordering ∪ {((C_i, C_{Max_j}))}
9: while |Ordering| < |C|
10:    Return: Ordering R
```
Algorithm 1. Pseudo algorithm for thematic ordering extraction

We define a redundancy score for each system summary as the sum of all relatedness scores of included sentences (6). This feature is competing with the continuity defined by the shared entities feature. Indeed, if two sentences mention the commun entities, they are similar to a certain degree.

$$Redundancy\_Score = \sum_{i,j=1..|R||i \neq j} Relatedness(S_i, S_j) \tag{6}$$

3.2   Coherence Model

Our problem is ordering relevant sentences in the most possible coherent way. We have defined a set of positive/negative input features that improve/worsen summary's coherence. Obviously, evaluating a coherence score for each possible ordering is not feasible. Indeed, a summary of 250 words in English contains approximately 13 to 17 phrases (a sentence contains, in average, 15 to 20 words).

Each coherence feature defined in the previous section is an objective to be attended (maximize or minimize) in the output summary ordering. Figure 2 presents an overview of the coherence model steps.

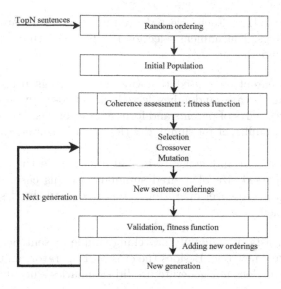

Genetic algorithm steps

Fig. 2. Coherence model

Model Parameters

Fitness Function: Each coherence feature is integrated to the fitness function according to its sense of contribution. For example, (Shared entities, +), (Thematic ordeing, +), (Sentence similarity, −1) is a fitness function. We define several possible combinations and evaluate coherence for each target fitness function.

Ordering Codification: Each candidate summary ordering is represented by a vector of sentences IDs. Vector size is equal to the number of sentences included in the system summary with respect to the summary's size.

Initial Population: Searching the best coherent ordering begins with a random ordering of selected sentences. Each solution is evaluated using the fitness function.

Coherence Assessment: Feature values are calculated for each ordering (chromosome) in the population. An ordering is better than another if it has higher feature values.

Selection: It consists of selecting best coherent orderings from the population to form the next generation. Each ordering which fits the best fitness function (coherence features) is more likely to be selected in the next generation. We use the tournament selection method since it tends to converge quickly towards satisfactory output [12]. Each selected ordering will be a parent of the next generation orderings. Tournament selection is repeated n times until having the complete set of parents.

Crossover: The parents are used to form new orderings using the crossover operator. Two parents are randomly selected and a two-point crossover operator is applied to merge parts of parents and form new orderings. We believe that two points crossover is sufficient for summaries (less then 20 sentences for a summary of 250 words).

Crossover operation may generate invalid orderings in the case of duplicate sentences or surpassed size of desired summary. In this case, invalid children are ignored and the crossover operation is repeated until the desired number of orderings is reached.

Mutation: It consists of randomly switching couple of sentences in the target ordering to create a new one. Besides the crossover operator, mutation assists in genetic diversity. It does not generate invalid summaries since it keeps the same sentences.

Final Output: The purpose of the development stage is to make sentence orderings more coherent across generations until reaching the maximum number of generations to be explored. Here, the ordering which fits, the most, fitness function is selected from the last generation as the final output.

4   Experimentation

The main goal of the experimentation is to assess the input of each coherence feature to enhance output coherence. We have implemented our solution under DEAP Package [13] which implements a set of evolutionary algorithms for optimisation problems: genetic algorithms, particle swarm optimization and differential evolution. We have opted for a dynamic fitness function which allows users to define couples of (feature, input sense) to be considered.

4.1   Coherence Assessment

It is a difficult task to assess text coherence from different levels; local and global coherence and in all its aspects: rhetorical organization, cohesion and readability. Using a coherence metric is a first quick option to assess coherence features input within our coherence model.

We use Dicomer metric [14] which is based on a model that captures statistical distribution of intra and inter-discourse relations. The model uses a matrix of discourse role transitions of terms from adjacent sentences. The nature of transition patterns and their probability are used to train an SVM classifier. The classifier learns how to rank original texts and texts in which sentence ordering is shuffled. Three collections of texts and summaries from TAC conferences are used to train the classifier.

4.2   Datasets

Since our target task is text summarization, we use two summarization datasets. The MultiLing 2015 dataset [15] is a collection of 15 document sets of news articles from the WikiNews website. Each document set contains 10 news texts about the same event such as 2005 London bombings or the 2004 tsunami. The task is to provide a single fluent summary of 250 words maximum.

The second dataset is DUC 2002 single document summarization dataset[1]. In our experiment, we use randomly selected 100 news articles and produce system summaries that not exceed 100 words. For each document, a human made summary is provided as a reference.

4.3   Summarization System

We use a multilingual summarizer [11] to generate extractive summaries. The summarizer first performs sentence clustering to identify main topics within source texts. Second, terms are ranked according to their relevance to each topic using minimum Redundancy and Maximum Relevance feature selection method [16] (mRMR). Finally, a score is assigned to each sentence according to the terms mRMR scores. The system summary keeps top relevant sentences up to the summary maximum size.

Top relevant sentences could be extracted from different source documents and paragraphs which necessarily affects summaries' coherence. Finding a better ordering of output sentences will improve summary's coherence

4.4   Genetic Algorithm Parameters

In addition to fitness function, there is a set of parameters that should be fixed such as crossover and mutation probabilities, population size and number of generations. For our experimentations, and in order to allow reasonable computation times, we have fixed population size at 300 individuals, the number of generations at 300, mutation probability at 0.01 and crossover probability at 0.001.

We deliberately decrease the crossover probability since crossover operator generated invalid individuals (summaries that contain duplicate sentences or exceed size limit).

4.5   Evaluation Protocol

As described in Table 1, we define eight configurations for output summary generation: Baseline, thematic ordering and genetic ordering.

Baseline the first configuration represents our baseline: ordering sentences following the original source text ordering. We assume that baseline ordering introduces gaps between sentences since sentences' sequence is broken.

---

[1] https://duc.nist.gov/duc2002/.

Topline we consider as a topline, Dicomer scores of reference summaries. Since reference summaries are human made, we assume that it is an upper bound for Dicomer coherence scores.

Rule this configuration combines between our baseline (original ordering) and thematic ordering (see pseudo algorithm 1). Sentences follow first thematic ordering and within each topic, sentences are ordered following their original ordering.

Coherence model ordering we define several configurations according to the number of positive/negative input features and the number of sentences to be considered as an input. Here, shared entities feature is combined with thematic ordering, sentence position in the fitness function. Sentence relevance and redundancy penalty features are considered when the model take as an input sentences that exceed the size limit (125% and 150% in our configurations). Then, the model selects a subset of sentences that optimize fitness function score with respect to summary size.

Table 1. Configurations of output summaries ordering

Baseline	SUMBA	[TopN, Position]
Topline	SUMMA	Model summary A MultiLing 2015
Topline	SUMMB	Model summary B MultiLing 2015
Topline	SUMMC	Model summary C MultiLing 2015
Topline	SUMMD	Model summary C DUC 2002
Rule	SUMTP	[Thematic, Position]
Genetic	SUMG1	[+Entity,+Thematic,+Position]
Genetic	SUMG2	[+Thematic]
Genetic	SUMG3	[+Entity]
Genetic	SUMG4	[+Entity,+Thematic]
Genetic	SUMG5	[125%, +Entity,+Thematic,+Position,+Relevance,- Redundancy]
Genetic	SUMG6	[150%, +Entity,+Thematic,+Position,+Relevance,- Redundancy]

4.6 Results and Discussion

Figures 3 and 4 report Dicomer coherence scores for each configuration. Topline (Human reference summaries) coherence scores reach an upper bound of 1.9 for MultiLing 2015 dataset and 1.87 for DUC 2002 dataset.

Baseline system summaries following original orderings (SUMBA) coherence scores is 1.41 for Multiling dataset and 1.29 for DUC 2002 Dataset. Thematic ordering combined with shared entities (SUMG2, SUMG4) present best coherence scores for system summaries for both DUC 2002 dataset with a value of 1.34 and Multiling dataset with a value of 1.59. It is the maximum coherence value of

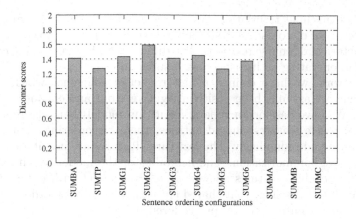

Fig. 3. MultiLing 2015 Dicomer coherence scores

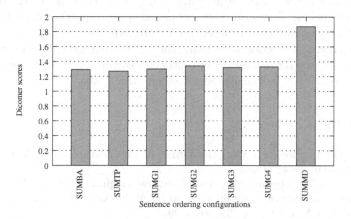

Fig. 4. DUC 2002 Dicomer coherence scores

system summaries using our coherence model. However, coherence model scores are average and range from 1.27 when five features are considered (SUMG5, SUMG6) to a value of 1.38 when shared entities are considered along with thematic ordering and sentence position feature for the Multiling dataset (SUMG1).

Baseline coherence scores are particularly high compared to other configuration results. When we examine output summaries of the TopN configuration, we find that TopN sentences are similar (contain most relevant terms) leading to some degree of topical coherence.

5   Conclusion

Dealing with text coherence is a challenging task in the NLP field. Taking into account coherence is critical to design efficient tools for text generation which is essential to a range of NLP tasks such as automatic summarization, dialog systems and machine translation. Modeling coherence involves syntactic and semantic levels of discourse analysis: entity-transition patterns, thematic ordering and rhetorical discourse relations. The difficulty with is in defining coherence features and operating all its aspects in a single model.

In this work, we have defined a first model of coherence which combines features that, we assume, have positive/negative input and enhance/affect text coherence. We have designed a genetic algorithm model that take into account a set of coherence features: shared entities, thematic ordering, sentence position, relevance and redundancy. The last three features are useful for the target task: extractive summarization. The flexibility of the model and its ability to easily include/exclude features allowed us to experiment with different feature combinations. Results show that shared entities and thematic ordering features significantly contribute to the output coherence.

Due to the nature of source texts (news texts which contains significant amount of date phrases), the results are strongly affected by the dissolution of temporal sequences. Temporal relations are also an important aspect of global coherence and should be considered for future experimentations [17]. Another possible interesting direction is to make the model task independent. Some features that we have defined, such as sentence position and relevance, are task-related and could not be considered for other NLP tasks.

References

1. Slakta, D.: L'ordre du texte (The Order of the Text). Etudes de Linguistique Appliquee **19**, 30–42 (1975)
2. Barzilay, R.: The rhetorical parsing of natural language texts. In: Proceedings of the Eighth Conference on European Chapter of the Association for Computational Linguistics. EACL 1997, pp. 96–103 (1997)
3. Barzilay, R., Lapata, M.: Modeling local coherence: an entity-based approach. In: Proceedings of the 43rd Annual Meeting of the Association for Computational Linguistics. ACL 2005, pp. 25–30 (2005)
4. Barzilay, R., Elhadad, N., McKeown, K.R.: Inferring strategies for sentence ordering in multidocument news summarization. J. Artif. Intell. Res. **1**(17), 35–55 (2002)
5. Knight, K., Marcu, D.: Summarization beyond sentence extraction: a probabilistic approach to sentence compression. J. Artif. Intell. **139**(1), 91–107 (2002)
6. Molina, A., Torres-Moreno, J.-M., SanJuan, E., da Cunha, I., Sierra Martínez, G.E.: Discursive sentence compression. In: Gelbukh, A. (ed.) CICLing 2013. LNCS, vol. 7817, pp. 394–407. Springer, Heidelberg (2013). https://doi.org/10.1007/978-3-642-37256-8_33
7. Christensen, J., Soderland, S., Etzioni, O.: Towards coherent multi-document summarization. In: Proceedings of the 2013 Conference of the North American Chapter of the Association for Computational Linguistics: Human Language Technologies. HLT-NAACL 2013, pp. 1163–1173 (2013)
8. Li, J., Hovy, E.: A model of coherence based on distributed sentence representation. In: Proceedings of the 2014 Conference on Empirical Methods in Natural Language Processing. EMNLP 2014, pp. 2039–2048 (2014)

9. Nguyen, D. T., Joty, S.: A model of coherence based on distributed sentence representation. In: Proceedings of the 55th Annual Meeting of the Association for Computational Linguistics. ACL 2017, pp. 1320–1330 (2017)
10. Charolles, M.: Introduction aux problèmes de la cohérence des textes: Approche théorique et étude des pratiques pédagogiques. Langue française 1(38), 7–41 (1978)
11. Oufaida, H., Blache, P., Nouali, O.: Using distributed word representations and mrmr discriminant analysis for multilingual text summarization. In: Biemann, C., Handschuh, S., Freitas, A., Meziane, F., Métais, E. (eds.) NLDB 2015. LNCS, vol. 9103, pp. 51–63. Springer, Cham (2015). https://doi.org/10.1007/978-3-319-19581-0_4
12. Razali, N. M., Geraghty, J.: Genetic algorithm performance with different selection strategies in solving TSP. In: Proceedings of the World Congress on Engineering, pp. 1–6 (2011)
13. Fortin, F., Rainville, D., Gardner, M., Parizeau, M., Gagné, C.: DEAP: evolutionary algorithms made easy. J. Mach. Learn. Res. 13(1), 2171–2175 (2012)
14. Lin, Z., Liu, C., Ng, H. T., Kan, M. Y.: Combining coherence models and machine translation evaluation metrics for summarization evaluation. In: Proceedings of the 50th Annual Meeting of the Association for Computational Linguistics: Long Papers. ACL 2012, vol. 1, pp. 1006–1014 (2012)
15. Giannakopoulos, G., et al.: MultiLing 2015: multilingual summarization of single and multi-documents, on-line fora, and call-center conversations. In: Proceedings of the 16th Annual Meeting of the Special Interest Group on Discourse and Dialogue, SIGDIAL 2015, pp. 270–274 (2015)
16. Peng, H., Long, F., Ding, C.: Feature selection based on mutual information: criteria of max-dependency, max-relevance, and min-redundancy. IEEE Trans. Pattern Anal. Mach. Intell. (8), 1226–1238 (2005)
17. Muller, P., Tannier, X.: Annotating and measuring temporal relations in texts. In: Proceedings of the 20th International Conference on Computational Linguistics. COLING 2004, p. 50 (2004)

Short Papers

Unified Parallel Intent and Slot Prediction
with Cross Fusion and Slot Masking

Anmol Bhasin[1]([✉]), Bharatram Natarajan[1]([✉]), Gaurav Mathur[1]([✉]),
Joo Hyuk Jeon[2]([✉]), and Jun-Seong Kim[2]([✉])

[1] Samsung R&D Institute - Bangalore, Bengaluru, India
{anmol.bhasin,bharatram.n,gaurav.m4}@samsung.com
[2] Samsung Electronics Co., Ltd., Suwon, South Korea
{joohyuk.jeon,js087.kim}@samsung.com

Abstract. In Automatic Speech Recognition applications, Natural Language Processing (NLP) has sub-tasks of predicting the Intent and Slots for the utterance spoken by the user. Researchers have done a lot of work in this field using Recurrent-Neural-Networks (RNN), Convolution Neural Network (CNN) and attentions based models. However, all of these use either separate independent models for both intent and slot or sequence-to-sequence type networks. They might not take full advantage of relation between intent and slot learning. We are proposing a unified parallel architecture where a CNN Network is used for Intent Prediction and Bidirectional LSTM is used for Slot Prediction. We used Cross Fusion technique to establish relation between Intent and Slot learnings. We also used masking for slot prediction along with cross fusion. Our models surpass existing state-of-the-art results for both Intent as well as Slot prediction on two open datasets.

Keywords: Unified intent and slot · Cross fusion · Intent and Slot Prediction

1 Introduction

Voice Assistants such as Samsung's Bixby, Amazon's Alexa, Apple's Siri etc. are becoming popular nowadays. Natural Language Understanding (NLU) system plays a key role for such digital assistants where it processes input in the form of text coming from Automatic Speech Recognition systems. NLU aims to assign predefined relevant categories by predicting domain, intent and required slots for the spoken utterance. To perform these subtasks different methods have been proposed [1–8].

Out of these sub-tasks, our research work focuses on Intent and Slot Predication. Intent classification aims to understand the user's intent while slot extraction identifies the semantic information for executing user's intended action. For example take the utterance "I want to fly from baltimore to dallas round trip" from ATIS dataset (Table 1). Here the user intention is to fly (intent) and he wants to fly from Baltimore (source-slot) to Dallas (destination-slot). Intent identification is a classification problem producing single output for the entire spoken utterance, whereas slot extraction is a sequence-labelling problem that maps each word of input X to a label Ys. Figure 1 is

© Springer Nature Switzerland AG 2019
E. Métais et al. (Eds.): NLDB 2019, LNCS 11608, pp. 277–285, 2019.
https://doi.org/10.1007/978-3-030-23281-8_22

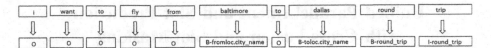

Fig. 1. Sample utterance from ATIS dataset representing semantic slots in IOB format

showing mapping of $X = (Xw\_1, \cdots, Xw\_T)$ to the corresponding slot label $Ys = (Ys\_1, \cdots, Ys\_T)$ in IOB format. A lot of work has been done in this area as discussed in the following sections.

1.1 Independent Intent Classification

In Intent classification, all the intents belong to one single domain, for example, in ATIS dataset all intents belong to 'Flight Reservation' domain. Schmidhuber et al. [9] proposed BiLSTM, which is widely used for this type of problem as discussed in some papers [10, 11]. Kim [6] used CNN network for classification problem and achieved state-of-the-art results. Kim used parallel convolution layers, which learnt sematic relations between the words and classified the text. We used similar parallel CNN network for our intent classification.

1.2 Independent Slot Prediction

Slot prediction is a sequence labeling problem, which predicts one label for each word. Since Slot label of a word depends on previous and next words, Bi-directional RNN based models outperformed in this task [12, 13]. Kurata et al. [14] proposed an encoder-labeler LSTM model that used encoder LSTM for slot filling. Shi et al. [15] combined semantic representation of RNN with the sequence level discriminative objective and proposed RSVM model.

1.3 Joint Intent Classification and Slot Prediction

In order to understand the relationship between intent prediction and slot filling, Zhang et al. [2] used GRU for slot filling and max-pooling layer for intent classification in a single model. Goo et al. [1] used slot gated modelling for learning the relationship between the two. Recently, Wang et al. [4] proposed a Bi-Model using BiLSTM for intent detection as well as for slot filling with two different cost functions considering cross impacts i.e. relation between intent and slot by passing one layer information between each other.

Inspired from the CNN work in text classification, we used CNN for intent detection and BiLSTM with CRF [16] (conditional random fields) for slot filling. We cross fused intent and slot to capture relationship between them. This aided in better understanding of the relationship between the intent and slots. We used sum of the loss from intent and slot models as the combined loss of cross-fused model during optimization. Use of masking for slots resulted in boosting of the accuracy.

2 Proposed Models

2.1 Model-1: Parallel Intent and Slot Prediction with Cross Fusion

For both Intent and Slot, we used same input sentence as shown in Fig. 2. We converted input sentence to sequence of word embedding using GloVe Embedding [17]. GloVe gives vector of dimension 300 (L1) for each word. We randomly initialized unseen words of size (L1). The max sentence length (L2) was chosen based on dataset being used (Table 1). In this way word embedding matrix of dimension L2 * L1 for input sentence was created.

For Intent classification word embedding matrix of a sentence was passed to 4 parallel CNN layers having different filter sizes f1 * L1, f2 * L1, f3 * L1 and f4 * L1. The filter height was set to f1 = 1, f2 = 2, f3 = 3 and f4 = 5, where height represents number of words to be convolved to capture unigram, bigram and higher n-grams. We took 128 such filters. The extracted convolution features then were concatenated and passed to dense layer.

For slot prediction on embedding matrix of dimension L2 * L1, we applied BiLSTM followed by single CRF layer and the output of CRF was passed to dense layer.

After the dense layer, we fused intent output with slot and slot information with intent. We tried three experiments with different fusion methods: average, concatenation and addition of the two inputs.

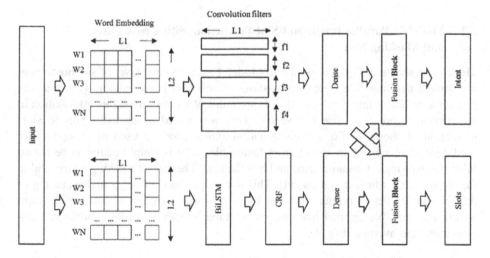

Fig. 2. Model-1: Intent and Slot Prediction with fusion layer

Cross Fusion Block

The learnings of the intent and slot were fused in this block. Due to dimensionality difference between intent learning I [batch_size * features] and slot learning S [batch_size * L2 * features], the fusion was done separately for intent classification and slot prediction. In case of intent classification, we reshaped slot learning and applied dense layer to maintain dimensional compatibility required for operation (op) as shown in Fig. 3a. In case of slot prediction, we broadcasted intent learning to maximum sentence length (L2) and fused with slot learning using operation (op) as shown in Fig. 3b. In both Fig. 3a and 3b 'op' means fusion operation either addition, average or concatenation, bs is the batch size and L2 is the word embedding size.

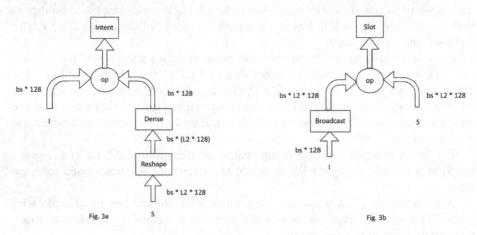

Fig. 3. The above figure explains fusion block.

2.2 Model-2: Parallel Intent and Slot Prediction with Cross Fusion and Masking Slot

Model-2 as shown in Fig. 3 is similar to Model-1 except in last layer of slot prediction, we applied masking. We generated masking information from training data in a way that for a particular intent, which all slots are valid. We represented this information in the form of word level one hot vector. This was added as an extra bias to slots prediction as shown in Eq. 1. This equation shows how we used masking for slot prediction, here x is the output of cross fusion block, W is weight matrix to be learnt, 'a' is the masking information used and b is the bias. The size of masking matrix ('a' in Eq. 1) was L2 * (distinct slot output). This technique aided in better understanding of the relationship between intent and slot by penalizing unnecessary slots for an intent during training. We applied masking on all three modes of fusions: addition, concatenation and average (Fig. 4).

$$Ys = \text{softmax}(W. \, x + a + b) \tag{1}$$

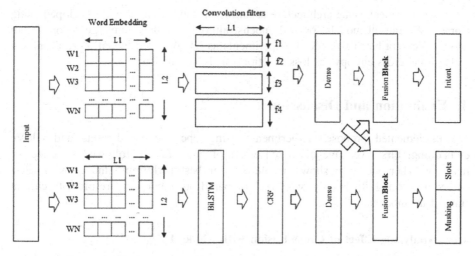

Fig. 4. Model-2: Intent and Slot Prediction with fusion layer and slot masking

3 Datasets and Experiment Detail

We evaluated our proposed architectures on two benchmarked datasets 'ATIS' and 'Snips' and all datasets are taken from the GitHub Source mentioned with Table 1.

Table 1. ATIS and Snips datasets used in experiment.

Dataset	Train data	Test data	Validation data	Vocabulary size	Slot	Intents	Max sentence length (L2)
ATIS[a]	4,978	893	500	722	120	21	50
Snips[b]	13,084	700	700	11,241	72	7	36

[a]https://github.com/yvchen/JointSLU/tree/master/data
[b]https://github.com/MiuLab/SlotGated-SLU/

ATIS Dataset: The ATIS (Airline Travel Information Systems) dataset consists of user-spoken utterances for flight reservation. It consists of 4,978 train utterances, 893 utterances as test data and 500 as validation data. The total number of intents to be predicted is 21 and total number of unique slots is 120. The maximum length of sentence present is 50.

Snips Dataset: Snips dataset is collected using Snips personal voice assistant. In Snips data, each intent is uniformly distributed. Train set consists of 13,084 utterances, validation and test have 700 utterances each. Number of unique labels present are 7 and 72 unique slots and maximum length sentence present is 36.

For experiment, same architecture as shown in Figs. 1 and 2 was developed using Keras[1]. We ran all models for 100 epochs, although models converge before 100th epoch. We kept batch size for all experiments as 64. We used 'adam' optimizers and 'categorical cross entropy' as loss function for the all models.

4   Evaluation and Discussion

We implemented proposed experiments using open source datasets and GloVe embedding. First, we have discussed effect of cross fusion, followed by effect of masking. Then we have shown results of our best performing model with latest embedding Bert [18]. Finally, we have compared our model best result with current state of art models.

4.1   Analyzing Effect of Cross Fusion with Model 1

To analyze effect of fusion, we are comparing results with and without cross fusion. Row 1 of Table 2 shows results without cross fusion where we used model same as Fig. 1 (Model 1) without cross fusion blocks. Table 2 also shows the accuracies achieved on ATIS and Snips datasets using model 1, with different fusion operation addition, average and concatenate. We found that cross fusion improves intent accuracy by 1.57% in ATIS and 1.14% in Snips and slots prediction by almost 0.3% in both datasets. Cross fusion helps to understand relationship between the slots and intent learnings by effective error propagation as compared to sequential type intent and slot models. Although we got similar results from all three fusion operations, 'addition' gave best results.

Table 2. Accuracy of proposed architecture Model-1 on two benchmark datasets with different cross fusion techniques and Model-1 without fusion.

Dataset model	ATIS		Snips	
	Intent	Slot (f1 score)	Intent	Slot (f1 score)
Model 1 (without fusion)	95.30	99.03	97	98.05
Model 1 (concatenate)	96.42	99.31	97.86	98.35
Model 1 (average)	96.87	99.3	98	98.39
Model 1 (addition)	96.87	99.32	98.14	98.38

4.2   Analyzing Effect of Masking with Model 2

In this experiment, we used masking with slots to boost its accuracy after cross fusion. As shown in Table 3, masking improved both Intent and Slot accuracies. Masking penalizes irrelevant slot and because of combined learning, it helped in improving intent accuracy.

---

[1] https://keras.io.

Table 3. Accuracy of proposed architecture Model-2 on two benchmark datasets with different cross fusion techniques.

Dataset model	ATIS		Snips	
	Intent	Slot (f1 score)	Intent	Slot (f1 score)
Model 2 (concatenate)	96.3	99.5	98	98.5
Model 2 (average)	97.42	99.53	98	98.39
Model 2 (addition)	97.42	99.54	98.14	98.44

4.3 Analyzing Effect of Embedding with Model 1 and Model 2

In all the experiments we used GloVe embedding of size L1 = 300. For analyzing embedding effect, we ran our best Model 1 and Model 2 using pre-trained Bert embedding model [Bert-Base, Uncased], keeping other hyper parameters exactly same. Bert gave word embedding considering contextual information of the sentence of size L1 = 768. Table 4 shows comparison of accuracy using Bert and GloVe with Model 1 and Model 2, using cross fusion operation as addition. Both embedding gave similar results but results with GloVe were marginally better.

Table 4. Accuracy of proposed architecture Model-1 and Model-2 using GloVe and Bert.

Dataset model	ATIS		Snips	
	Intent	Slot (f1 score)	Intent	Slot (f1 score)
Model 1 (Add and GloVe)	96.87	99.32	98.14	98.38
Model 1 (Add and Bert[a])	97.5	99.01	98	97.95
Model 2 (Add and GloVe)	97.42	99.54	98.14	98.44
Model 2 (Add and Bert[a])	97.38	99	98	97.88

[a]https://github.com/google-research/bert

4.4 Performance Comparison of State-of-the-Art Techniques

In this section, we compare our architecture best results (Model 1 and Model 2) with recent state-of-the-art models. From Table 4 we can observe that for Intent detection there is 0.25% improvement in ATIS and 1.14% in Snips datasets. In case of slots there is 1.78% improvement for ATIS and extensive improvement of 10.14% in Snips (Table 5).

Table 5. Comparison of Intent accuracy (%) and Slots (f1 score) with other state-of-art-models.

Dataset model	ATIS		Snips	
	Intent	Slot (f1 score)	Intent	Slot (f1 score)
Attention-based RNN [3], 2016	91.1	94.2	97.0	87.8
Bi-Directional RNN-LSTM [13], 2016	92.6	94.3	96.9	87.3
Slot-Gated (Full Attention) [1], 2018	93.6	94.8	97.0	88.8
Slot-Gated (Intent Attention) [1], 2018	94.1	95.2	96.8	88.3
Attention-Based CNN-BLSTM [5], 2018	97.17	97.76	-	-
Our Model 1 (best)	96.87	99.32	98.14	98.38
Our Model 2 (best)	97.42	99.54	98.14	98.44

5 Conclusions

Unified parallel Intent and Slot Prediction using cross fusion and slot masking is proposed. Both variants of model achieved state-of-the-art results. Cross fusion boosted the intent accuracy and masking improved slot prediction f1 score. The proposed architecture can be further extended to predict domain labels along with intent and slot. For cross fusion addition, concatenation and average is being used which can be further extended to more complex mathematical equations.

References

1. Goo, C.W., et al.: Slot-gated modeling for joint slot filling and intent prediction. In: Proceedings of Conference of the North American Chapter of the Association for Computational Linguistics: Human Language Technologies, vol. 2, pp. 753–757 (2018)
2. Zhang, X., Wang, H.: A joint model of intent determination and slot filling for spoken language understanding. In: IJCAI, pp. 2993–2999 (2016)
3. Liu, B., Lane, I.: Attention-based recurrent neural network models for joint intent detection and slot filling. arXiv:1609.01454 (2016)
4. Wang, Y., Shen, Y., Jin, H.: A Bi-model based RNN semantic frame parsing model for intent detection and slot filling. In: Proceedings of the Conference of the North American Chapter of the Association for Computational Linguistics: Human Language Technologies, vol. 2, pp. 309–314 (2018)
5. Wang, Y., Tang, L., He, T.: Attention-based CNN-BLSTM networks for joint intent detection and slot filling. In: Sun, M., Liu, T., Wang, X., Liu, Z., Liu, Y. (eds.) CCL/NLP-NABD -2018. LNCS (LNAI), vol. 11221, pp. 250–261. Springer, Cham (2018). https://doi.org/10.1007/978-3-030-01716-3_21
6. Kim, Y.: Convolutional neural networks for sentence classification. In: Proceedings of EMNLP 2014 Conference, pp. 1746–1751 (2014)
7. Kim, Y., Lee, S., Stratos, K.: OneNet: joint domain, intent, slot prediction for spoken language understanding. In: Automatic Speech Recognition and Understanding Workshop IEEE, pp. 547–553. IEEE (2017)
8. Zhou., C., Sun, C., Liu, Z., Lau, F.C.M.: A C-LSTM neural network for text classification. arXiv:1511.08630 (2015)
9. Hochreiter, S., Schmidhuber, J.: Long short-term memory. Neural Comput. 9(8), 1735–1780 (1997)
10. Lai, S., Xu, L., Liu, K., Zhao, J.: Recurrent convolutional neural networks for text classification. In: AAAI, vol. 333, pp. 2267–2273 (2015)
11. Liu, P., Qiu, X., Huang, X.: Recurrent neural network for text classification with multi-task learning. arXiv:1605.05101 (2016)
12. Huang, Z., Xu, W., Yu, K.: Bidirectional LSTM-CRF models for sequence tagging. arXiv:1508.01991 (2015)
13. Hakkani-Tür, D., Tur, G., Celikyilmaz, A., Chen, Y.N., Deng, L., Wang, Y.-Y.: Multi-domain joint semantic frame parsing using Bi-directional RNN-LSTM. In: Interspeech (2016)
14. Kurata, G., Xiang, B., Zhou, B., Yu, M.: Leveraging sentence-level information with encoder LSTM for semantic slot filling. arXiv:1601.01530 (2016)

15. Shi., Y., Yao, K., Chen, H., Yu, D., Pan, Y.-C., Hwang, M.-Y.: Recurrent support vector machines for slot tagging in spoken language understanding. In: Proceedings of Conference of the North American Chapter of the Association for Computational Linguistics: Human Language Technologies, pp. 393–399 (2016)
16. Raymond, C., Riccardi, G.: Generative and discriminative algorithms for spoken language understanding. In: International Speech Communication Association (2007)
17. Pennington, J., Socher, R., Manning, C.: Glove: Global vectors for word representation. In: Proceedings of Conference on Empirical Methods in Natural Language Processing (EMNLP), pp. 1532–1543 (2014)
18. Devlin, J., Chang, M.-W., Lee, K., Toutanova, K.: Bert: pre-training of deep bidirectional transformers for language understanding. arXiv:1810.04805 (2018)

Evaluating the Accuracy and Efficiency of Sentiment Analysis Pipelines with UIMA

Nabeela Altrabsheh, Georgios Kontonatsios, and Yannis Korkontzelos[(⊠)]

Department of Computer Science, Edge Hill University, Ormskirk, UK
{altrabsn,Georgios.Kontonatsios,Yannis.Korkontzelos}@edgehill.ac.uk

Abstract. Sentiment analysis methods co-ordinate text mining components, such as sentence splitters, tokenisers and classifiers, into pipelined applications to automatically analyse the emotions or sentiment expressed in textual content. However, the performance of sentiment analysis pipelines is known to be substantially affected by the constituent components. In this paper, we leverage the Unstructured Information Management Architecture (UIMA) to seamlessly co-ordinate components into sentiment analysis pipelines. We then evaluate a wide range of different combinations of text mining components to identify optimal settings. More specifically, we evaluate different pre-processing components, e.g. tokenisers and stemmers, feature weighting schemes, e.g. TF and TFIDF, feature types, e.g. bigrams, trigrams and bigrams+trigrams, and classification algorithms, e.g. Support Vector Machines, Random Forest and Naive Bayes, against 6 publicly available datasets. The results demonstrate that optimal configurations are consistent across the 6 datasets while our UIMA-based pipeline yields a robust performance when compared to baseline methods.

Keywords: Sentiment analysis · Text processing · Interoperability · UIMA

1 Introduction

The Unstructured Information Management Architecture (UIMA) [4] is a software framework that facilitates the development of interoperable text mining applications. UIMA-enabled components can be freely combined into larger pipelined applications, e.g. machine translation [8] and information extraction, using UIMA's common communication mechanism and shared data type hierarchy, i.e. Type System. Recent studies has demonstrated that UIMA-based pipelines can efficiently address a wide range of different text mining tasks [2,8].

In this paper, we use the UIMA framework to develop efficient sentiment analysis pipelines. We focus on sentiment analysis, considering that automatic sentiment analysis systems are being increasingly used in a number of applications, such as business and government intelligence. The popularity of the task can largely be associated with the vast amount of available data, especially

© Springer Nature Switzerland AG 2019
E. Métais et al. (Eds.): NLDB 2019, LNCS 11608, pp. 286–294, 2019.
https://doi.org/10.1007/978-3-030-23281-8_23

in social media. For example, sentiment analysis on Twitter has been used to identify concerns in urban environments [19].

Despite the popularity of sentiment analysis and the wide applicability of UIMA to many text processing tasks, UIMA has been used for sentiment analysis by a few studies, only. Rodriguez et al. [13] developed UIMA-based pipelines for capturing the sentiment expressed in customers' reviews about hotels.

This study investigates sentiment analysis using the UIMA framework. Further than Rodriguez et al. [13], (a) we investigate the effect of different preprocessing components, features, and feature selection on the overall performance of a sentiment analysis system, and (b) we compliment evaluation results with the execution times of each combination of components and classifiers. Our results show that execution times vary widely and that high execution times do not always match high accuracies. To the best of our knowledge, this is the first work that considers execution times while evaluating UIMA pipelines. The execution time of a sentiment analysis system is particularly important for real-time applications, especially when monitoring social media.

2 Related Work

The Unstructured Information Management Architecture (UIMA) has been employed widely for developing text processing applications in various domains. Kontonatsios et al. [8] extended UIMA workflows to facilitate the creation of multilingual and multimodal NLP applications. In the medical domain, UIMA has been applied to detect the smoking status of patients [17]. UIMA has been used to analyse hotel customer reviews [13], where sentiment analysis is modelled as a classification task. UIMA was shown to be suitable for designing and implementing sentiment analysis systems due to the reusability components.

Several studies have explored the time that classifiers take to identify polarity. For instance, Greaves et al. [6], who researched sentiment analysis to analyse patients' experience, concluded that the Naive Bayes Multinomial classifier was faster than other classifiers by a short margin of 0.2 s. Of course, data size can affect the model's running time. Running large datasets using limited computational resources can cause out-of-memory errors, and distributing the training task across many machines was shown to decrease running time by 47% [7]. Apart from classifier training, other components the pipeline, parameters and feature types can also affect execution times [5].

3 Experiments

As any other UIMA application, our sentiment analysis pipeline implements three basic operations: read (*Collection Reader*), process (*Analysis Engine*) and write (*CAS Consumer*). We have conducted 6 large-scale experiments to investigate the optimal pipeline configuration. More specifically, we evaluated all combinations of the following components: (1) CoreNLP and Snowball Tartarus stemmers, (2) TF and TF-IDF feature weighting schemes, (3) feature

types: unigrams, bigrams, trigrams and combinations of them, (4) frequency
thresholds for feature filtering, i.e. feature removal, and (5) classification algorithms: Support Vector Machines, Random Forest and Naive Bayes, as implemented in the WEKA platform. It should be noted that different pipeline configurations were created by simply changing the UIMA XML descriptor file.

Table 1. Data sources

Dataset type	Amazon [9]	IMDB [9]	SemEval [14]	Senti-140 [10]	UMICH [18]	Yelp [9]
	Training (subset)	Training (subset)	Development set	Training (subset)	Training set	Training (subset)
size	1,000	1,000	20,632	1,048,575	7,086	1,000
positive	500	500	7059	554,470	3,995	500
negative	500	500	3,231	494,105	3,091	500
neutral	-	-	10,342	-	-	-

*SemEval refers to Task 4A of SemEval 2016.

All combinations of the components above are evaluated in terms of accuracy
(Acc), precision (P), recall (R) and F-score ($F1$) using 10-fold cross validation.
In addition, we measured the execution time of each pipeline configuration. We
used 6 publicly available datasets. Table 1 shows the source, name, size and
number of documents labelled as positive, negative or neutral in each dataset.
The neutral label is only available in the *SemEval* dataset and we did not include
it in our experiments. Amazon, IMDB, UMICH and Yelp experiments were run
on a HP laptop with Intel core i5-8250u, 1.80 GHz, on Windows. SemEval and
Senti-140 experiments were run on an HP ProLiant DL360 Gen9 server running
Linux.

The first experiment evaluates our sentiment analysis pipeline when using
different combinations of pre-processing components. We use UIMA to plug and
play pre-processing components into pipelines, while using the same type-system,
to identify the best configuration. Many studies explored the effect of preprocessing on sentiment analysis. Preprocessing can improve performance up to 20%,
while analysing sentiment in students' feedback [1]. We develop 4 pipelines by
combining 2 tokenisers and 2 stemmers, common in the literature: (1) Standard
tokeniser (T1): segments a document into its tokens using whitespace characters
as delimiter. This tokeniser was implemented in-house, (2) StringTokenizer (T2):
from the java.util package[1], (3) englishStemmer (S1): from the tartarus.snowball
package[2], and (4) PorterStemmer (S2): from the tartarus.snowball package[3].
The first experiment evaluates 120 configurations: 2 tokenisers x 2 stemmers x
1 ngrams (unigrams+bigrams+trigrams combined) x 6 datasets x 5 classifiers.
The remaining experiments use the best performing combination.

---

[1] docs.oracle.com/javase/7/docs/api/java/util/StringTokenizer.html.

[2] snowball.tartarus.org/algorithms/english/stemmer.html.

[3] snowball.tartarus.org/algorithms/porter/stemmer.html.

The second experiment considers two feature weighting schemes: Term Frequency (TF) and Term Frequency Inverse Document Frequency (TF-IDF). TF and TF-IDF are different ways of assessing feature importance by assigning different weights.

Choosing features that represent data instances accurately for a particular task can lead to more accurate predictions. The most common feature types used for sentiment analysis are n-grams, i.e. sequences of n textual units, which can be letters, syllables or words [1]. N-grams usually consider tokens and are of one, two or three tokens long, i.e. unigrams, bigrams or trigram, respectively. Sarker et al. [15] and Pal and Gosh [11] used n-gram features for developing sentiment analysis methods and evaluated their methods against the same datasets that we use in this work. Here, we explore the following n-gram combinations: unigrams only, bigrams only, trigrams only, unigrams and bigrams, unigrams and trigrams, bigrams and trigrams, and all n-grams combined.

Table 2. Best pipeline configurations in terms of both F-Score and execution time across the 6 evaluation datasets. The table also reports the highest and lowest F-Score and the slowest and fastest execution time obtained by the different pipeline configurations.

		Pipeline configuration	F1	Time (mm:ss)		Pipeline configuration	F1	Time (mm:ss)
Highest F-Score	Amazon	CNB-T1-S1	.831	00:01	IMDB	CNB-T1-S1	.787	00:02
Lowest F-Score		NB-T1-S1	.748	00:05		NB-T2-S1	.675	01:00
Slowest Time		RF-T1-S2	.777	06:09		RF-T2-S1	.699	05:13
Fastest Time		CNB-T2-S2	.829	00:01		CNB-T1-S2	.773	00:02
Best configuration		**CNB-T1-S1**	**.831**	**00:01**		**CNB-T1-S1**	**.787**	**00:02**
Highest F-Score	SemEval	CNB-T1-S1	.832	00:02	Senti-140	LIB-T1-S1	.798	02:03:42
Lowest F-Score		NB-T2-S2	.588	07:06		CNB-T2-S1	.768	27:04
Slowest Time		RF-T1-S1	.753	01:22:05		LIB-T2-S2	.796	02:07:00
Fastest Time		CNB-T1-S1	.808	00:01		CNB-T1-S2	.779	26:20
Best configuration		**CNB-T1-S1**	**.832**	**00:02**		**CNB-T1-S2**	**.779**	**00:25**
Highest F-score	UMICH	RF-T2-S1	.998	17:46	Yelp	CNB-T1-S2	.798	00:01
Lowest F-Score		NB-T2-S1	.807	04:11		NB-T2-S1	.665	01:02
Slowest Time		RF-T1-S2	.997	22:23		RF-T2-S2	.745	04:35
Fastest Time		CNB-T2-S2	.979	00:01		RF-T2-S2	.745	04:35
Best configuration		**SVM-T2-S1**	**.991**	**00:05**		**CNB-T1-S2**	**.798**	**00:01**

The fourth experiment evaluates our pipeline when filtering features using a frequency threshold. Considering a research objective is to scale text processing pipelines to big data collections, we are interested in reducing the computational resources needed to execute them without reducing the accuracy of the underlying text mining models. Equal thresholds were set for all ngram features, and we experimented with threshold values in the range of $[1, 30]$. We aim to remove infrequent features to eliminate potential noise in the datasets. Running times are expected to decrease as threshold values increase. If the performance of the models does not decrease significantly as threshold values increase, then high values can safely be adopted, leading in models of smaller size that are easier to transfer and work with, without loss in prediction accuracy.

The choice of a classifier substantially affects the performance of the sentiment analysis pipeline. We experiment with the following classifiers: SVM, NB, RF, CNB and LibLinear. CNB and LibLinear have not been previously evaluated on these datasets[4].

4   Results and Discussion

Table 2 shows the lowest and highest F-score and the slowest and fastest execution time achieved by the pipeline configurations. We further report the best configuration considering both the F-score performance and the execution time. As an example, we observe that SVM-T2-S1 achieves an F-score of 0.991 on the $UMICH$ dataset, which is only marginally lower than the overall highest F-score, 0.998, achieved by RF-T2-S1. However, SVM-T2-S1 is our preferred configuration because it is substantially faster than RF-T2-S1. Overall, the CNB classifier obtained both a high F-score performance and a fast execution time in 5 out of 6 datasets.

Preprocessing: We evaluate 4 combinations of pre-processing components. Table 3 shows the average performance of the 4 pipeline configurations when

Table 3. Average performance of our sentiment analysis pipeline, on combinations of pre-processing components. The results are averaged over 5 classifiers, as discussed in Sect. 3.

Metric		Pipeline configuration					Pipeline configuration			
		T1- S1	T1-S2	T2-S1	T2-S2		T1- S1	T1-S2	T2-S1	T2-S2
Accuracy	Amazon	.802	**.803**	.802	**.803**	IMDB	**.736**	.734	.719	.717
Precision		.807	.807	.807	.807		**.740**	.738	.725	.723
Recall		.802	**.803**	.802	**.803**		**.740**	.733	.719	.717
F-score		.801	**.802**	.801	**.802**		**.734**	.732	.717	.715
Accuracy	SemEval	**.791**	.789	.774	.777	Senti-140	**.786**	.785	.783	.783
Precision		**.789**	.785	.774	.779		**.789**	**.789**	.787	.787
Recall		**.760**	.758	.746	.745		**.788**	.787	.785	.784
F-score		**.757**	.756	.740	.741		**.785**	**.785**	.782	.782
Accuracy	UMICH	**.964**	.959	.954	.955	Yelp	**.765**	.764	.745	.750
Precision		**.968**	.967	.962	.962		**.770**	.769	.750	.755
Recall		.967	.958	.957	**.969**		**.770**	.764	.745	.750
F-score		**.964**	.958	.953	.955		**.770**	.764	.744	.749

---

[4] Only CNB and LIB were evaluated on Senti-140, as the other classifiers failed to run due to out-of-memory errors.

applied to the 6 datasets. The performance is computed in terms of accuracy, precision, recall and F-score, while the reported results are average values across the performance obtained by the 5 classifiers. It can be observed that the T1-S1 configuration performed best in most cases. The improvement over the remaining configurations are insignificant.

TF & TF-IDF: TF weighting achieved slightly higher classification performance than TF-IDF in 4 out of 6 datasets, as shown in Table 4. TF-IDF was faster than TF in 5 out of 6 datasets. A larger time margin, 9 s, was observed on the Senti-140 dataset.

Table 4. Scores and execution times of CNB-T1-S1 using TF and TF-IDF feature weighting.

	Amazon		IMDB		SemEval		Senti-140		UMICH		Yelp	
	TF	TF-IDF	TF	TF-IDF	TF	TF-IDF	TF	TF-IDF	TF	TF-IDF	TF	TF-IDF
Time (sec)	.022	**.018**	.067	**.064**	1.119	**.278**	34.277	25.061	**.055**	.577	.387	**.020**
Acc	**.835**	.833	**.782**	.774	**.839**	.810	.772	.772	**.982**	.974	.787	**.788**
P	**.839**	.835	**.791**	.778	**.801**	.780	.780	.780	**.982**	.973	**.791**	.790
R	**.835**	.833	**.782**	.775	**.815**	.806	.777	.777	**.981**	.975	**.787**	.788
F1	**.834**	.833	**.780**	.774	**.807**	.789	.772	.772	**.982**	.974	.786	**.788**

Table 5. Features: Performance of the best configuration (T1, S1) on all datasets, pre-processing techniques and classifiers for each of the features. U: Unigrams, B: Bigrams, T: Trigrams

Metric		U	B	T	U+B	U+T	B+T	All		U	B	T	U+B	U+T	B+T	All
Accuracy	Amazon	.816	.704	.608	.702	**.835**	.831	.831	IMDB	**.816**	.635	.575	.648	.718	.807	.788
Precision		.819	.727	.676	.728	**.837**	.833	.833		**.816**	.637	.607	.650	.786	.810	.792
Recall		.816	.704	.608	.702	**.835**	.831	.831		**.816**	.635	.575	.648	.781	.807	.788
F-score		.816	.696	.567	.693	**.835**	.831	.831		**.816**	.634	.540	.647	.780	.807	.787
Accuracy	SemEval	.837	.773	.512	.681	.840	.821	**.844**	Senti-140	.738	.732	.696	.749	.768	**.877**	.772
Precision		.827	.746	.664	.707	.819	.793	**.832**		.744	.740	.714	.750	.774	.774	**.780**
Recall		.782	.775	.638	.736	.802	**.815**	.793		.743	.737	.703	.745	.772	.771	**.777**
F-score		.798	.753	.518	.676	**.809**	.801	.808		.739	.732	.694	.740	.768	.766	**.772**
Accuracy	UMICH	.797	.680	.691	.681	.796	**.800**	.794	Yelp	.930	.952	.968	.969	.975	.978	**.980**
Precision		**.819**	.688	.663	.691	.798	.801	.796		.972	.950	.966	.967	.973	.977	**.980**
Recall		.797	.680	.601	.681	.796	**.800**	.794		.973	.955	.970	.971	.975	.979	**.979**
F-score		.797	.677	.560	.677	.796	**.800**	.794		.972	.952	.968	.969	.974	.978	**.980**

Features: Table 5 shows the performance of n-gram feature combinations, introduced in Sect. 3. The performance is computed for the best configuration (T1 and S1). Trigram features yielded the lowest performance in most cases, while the combination of all n-grams performed best in 3 out of the 6 datasets. Unigrams and trigrams together obtained the highest performance on Yelp. The performance margin between the different feature types is substantial in several occasions. For example, unigrams achieved an improved F-score of 27.6% over

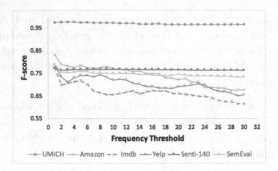

Fig. 1. F-score when using increasing frequency threshold values.

trigrams on the IMDB dataset. This suggests that careful feature selection can improve the performance of sentiment analysis pipelines.

Feature Selection: We filtered out features, i.e. n-grams, that occur less frequently than a pre-defined threshold. The results of applying threshold values in [1, 30], in Fig. 1, show that for smaller datasets, the performance decreases as we increase the threshold. For example, the F-score on Amazon, which consists of 1,000 reviews only, drops from 0.832 for a threshold of 1 to 0.676 for a threshold of 30. However, for larger datasets, e.g. Senti-140 that contains more than 1M documents, F-scores vary insignificantly.

Classifiers: CNB was the fastest and best. RF was the slowest, but performed best on UMICH. SVM and LIB performed competitively and quickly in all datasets.

Comparison with previous studies: We compare our pipeline with published results on the same datasets and classifiers, as shown in Tabl 6. Some published experiments used different parts of the datasets than what we used, thus we configured our experiments accordingly to compare fairly. For these comparisons, we used our best combination of pre-processing, feature extraction and selection methods and feature weighting. For SemEval, we used LibLinear instead of SVM and achieved marginally lower results than the published ones. Lastly, the method in [12] used 22,660 Senti-140 positive and negative instances. Since it is not mentioned which exaclty these instances were, we used the entire dataset with a frequency threshold of 100. We used the Liblinear classifier and the results were better by 2.6%.

Best performing model: CNB was the fastest classifier and often also performed best. It is beneficial for large datasets. The slowest classifier was RF. A combination of n-grams often performs best. The effect of frequency thresholding largely depends on the size of the data. Preprocessing matters and affects classification results. The best configuration, which achieved F-scores above 70% for all datasets, is the CNB model with tokeniser T1 and stemmer S1, all n-grams features and a frequency threshold of 6.

Table 6. Comparison between our sentiment analysis pipelines and state-of-the-art systems. Our scores have been computed using the same classifier, but different pre-processing and features (Sect. 3). Abbreviations - TOK: tokenisation, Ngr: Ngrams, BoW: Bag-of-Words, SL: stoplist, PR: punctuation removal, U: unigrams, ST: stemming, LC: lowercasing, B: bigrams, SL: sentiment lexicon, LW: elongated words, NEG: negation.

DataSet	Ref	Method			Published scores (%)				Our scores (%)			
		Classifier	Preprocessing	Features	Acc	P	R	F1	Acc	P	R	F1
Amazon	[11]	NB	TOK	Tokens, Ngr	82.4	-	-	-	75.0	76.0	75.0	74.8
	[16]	NB	-	BoW	-	78.9	-	-	75.0	76.1	75.0	74.8
IMDB	[11]	NB	TOK	Tokens, Ngr	78.6	-	-	-	72.3	72.7	72.3	72.1
	[16]	NB	-	BoW	-	78.9	-	-	72.3	72.7	72.3	72.1
Yelp	[11]	NB	TOK	Tokens, Ngr	82.7	-	-	-	70.5	70.7	70.5	70.4
	[16]	NB	-	BoW	-	60.3	-	-	70.5	**70.7**	70.5	70.4
UMICH	[3]	SVM	TOK, SL, PR	U	-	-	-	89	99.1	99.1	99	**99.1**
SemEval	[15]	SVM	ST, LC	Ngr	64.6	-	**63.7**	**63.2**	62.9	60.7	59.3	59.9
Senti-140	[12]	SVM	BoW, clustering	U, B, SL, LW, NEG	-	-	-	77.4	80.5	80.0	80.0	**80.0**

5   Conclusion

In this paper, we have investigated UIMA to optimise the accuracy and efficiency of sentiment analysis. We have demonstrated that UIMA can simplify the development of text-processing pipelines, wherein components can be freely combined using shared data types. We experimented with a wide range of pipeline configurations, considering various pre-processing components, classification algorithms, feature extraction methods and feature weighting schemes, to identify the best performing ones.

A potential limitation of our proposed sentiment analysis pipeline is that, like any other UIMA application, it is written as a sequential program, which limits its scalability. In the future we plan to leverage UIMA DUCC, i.e. the Distributed UIMA Cluster Computing platform, for scaling our sentiment analysis pipeline to big data collections. UIMA DUCC enables large-scale processing of big data collections by distributing a UIMA pipeline over a computer cluster while the constituent components of the pipeline can be executed in parallel across the different nodes of the cluster.

Acknowledgment. This research work is part of the TYPHON Project, which has received funding from the European Union's Horizon 2020 Research and Innovation Programme under grant agreement No. 780251.

References

1. Altrabsheh, N., Cocea, M., Fallahkhair, S.: Sentiment analysis: towards a tool for analysing real-time students feedback. In: ICTAI 2014, pp. 419–423. IEEE (2014)
2. Batista-Navarro, R., Carter, J., Ananiadou, S.: Argo: enabling the development of bespoke workflows and services for disease annotation. Database **2016** (2016)

3. Dridi, A., Recupero, D.R.: Leveraging semantics for sentiment polarity detection in social media. Int. J. Mach. Learn. Cybern., 1–11 (2017)
4. Ferrucci, D., Lally, A.: UIMA: an architectural approach to unstructured information processing in the corporate research environment. Nat. Lang. Eng. **10**(3–4), 327–348 (2004)
5. Go, A., Huang, L., Bhayani, R.: Twitter sentiment analysis. CS224N Project Report, Stanford (2009)
6. Greaves, F., Ramirez-Cano, D., Millett, C., et al.: Use of sentiment analysis for capturing patient experience from free-text comments posted online. J. Med. Internet Res. **15**(11), e239 (2013)
7. Khuc, V.N., Shivade, C., Ramnath, R., et al.: Towards building large-scale distributed systems for Twitter sentiment analysis. In: Proceedings of SAC, pp. 459–464. ACM (2012)
8. Kontonatsios, G., Thompson, P., Batista-Navarro, R.T., et al.: Extending an interoperable platform to facilitate the creation of multilingual and multimodal NLP applications. In: Proceedings of ACL 2013: System Demonstrations, pp. 43–48 (2013)
9. Kotzias, D., Denil, M., De Freitas, N., et al.: From group to individual labels using deep features. In: Proceedings of ACM SIGKDD 2015, pp. 597–606. ACM (2015)
10. Mohammad, S.M., Kiritchenko, S., Zhu, X.: NRC-Canada: Building the state-of-the-art in sentiment analysis of Tweets. arXiv preprint arXiv:1308.6242 (2013)
11. Pal, S., Ghosh, S.: Sentiment analysis using averaged histogram. Int. J. Comput. Appl. **162**(12) (2017)
12. Ren, Y., Wang, R., Ji, D.: A topic-enhanced word embedding for twitter sentiment classification. Inf. Sci. **369**, 188–198 (2016)
13. Rodrıguez-Penagos, C., Narbona, D.G., Sanabre, G.M., et al.: Sentiment analysis and visualization using UIMA and Solr. Unstructured Information Management Architecture (UIMA), p. 42 (2013)
14. Rosenthal, S., Farra, N., Nakov, P.: SemEval-2017 task 4: sentiment analysis in Twitter. In: Proceedings of SemEval-2017, pp. 502–518 (2017)
15. Sarker, A., Gonzalez, G.: HLP@UPenn at SemEval-2017 task 4A: a simple, self-optimizing text classification system combining dense and sparse vectors. In: Proceedings of SemEval-2017, pp. 640–643 (2017)
16. Sarma, P.K., Sethares, W.: Simple algorithms for sentiment analysis on sentiment rich, data poor domains. In: Proceedings of ACL 2018, pp. 3424–3435 (2018)
17. Sohn, S., Savova, G.K.: Mayo clinic smoking status classification system: extensions and improvements. In: AMIA Annual Symposium Proceedings, vol. 2009, p. 619. American Medical Informatics Association (2009)
18. UMICH: Dataset SI650 - sentiment classification (2011). https://goo.gl/Xfr8lI
19. Zavattaro, S.M., French, P.E., Mohanty, S.D.: A sentiment analysis of US local government Tweets: the connection between tone and citizen involvement. Gov. Inf. Q. **32**(3), 333–341 (2015)

Comparing Different Word Embeddings for Multiword Expression Identification

Aishwarya Ashok[1]([✉])(iD), Ramez Elmasri[1], and Ganapathy Natarajan[2](iD)

[1] University of Texas at Arlington, Arlington, TX 76019, USA
aishwarya.ashok@mavs.uta.edu, elmasri@cse.uta.edu
[2] Oregon State University, Corvallis, OR 97331, USA
gana.natarajan@oregonstate.edu

Abstract. The identification of Multi-Word Expressions (MWEs) is central to resolving ambiguity of phrases. Recent works show that deep learning methods outperform statistical and lexical based approaches. The deep learning approaches mostly use word2vec embedding; our paper aims at comparing the use of word2vec, GloVe, and a combination of the two word embeddings in identifying MWEs. GloVe, and the combination of word2vec and GloVe were marginally better in terms of F-score, identifying more unique words, and identifying words not seen in the train data. GloVe was marginally better at identifying Verbal Multi-Word Expressions (VMWEs) which tend to be the hardest group of MWEs because they can be gappy, which is caused by interleaving of words that are part of the MWE and words that are not part of the MWE. The major purpose of the paper is to compare the use of different word embeddings in identifying MWEs and not to suggest improvements to the state-of-the-art. Future work using different dimensions of word embedding vectors and use of fasttext are suggested.

Keywords: MWEs · Word2vec · GloVe

1 Introduction

Multi-Word expressions (MWEs) are combinations of words, which, when treated as a unit, have a different meaning than the individual words of the MWE. The part of speech (POS) tag composition alone can help identify certain MWEs like "Super Bowl" since the words are fixed in position and cannot be separated. A subgroup of MWEs, Verbal Multi-word Expressions (VMWEs) cannot be identified using only POS tag because the composition of POS tags and the number of words in VMWEs vary. Words that are part of the VMWE can be interleaved with words that are not part of the VMWE. This is often referred to as gappy and the number of words in the gap is not limited. A detailed explanation of different types of VMWEs may be found in [1].

MWEs have been referred to as "pain in the neck" by researchers [2]. The identification of MWEs is crucial to understanding semantics of a language.

© Springer Nature Switzerland AG 2019
E. Métais et al. (Eds.): NLDB 2019, LNCS 11608, pp. 295–302, 2019.
https://doi.org/10.1007/978-3-030-23281-8_24

Previous work on identification of MWEs has been done using syntactic parsers [3], statistical measures [4], and Conditional Random Fields (CRFs) [1]. There have been approaches to combine semantic features and statistical methods [5]. Machine learning methods such as Naive Bayes classifier [6] and Bayesian Network [7] have also been used.

Recent work has been done in representing words in d-dimensional vector spaces known as word embeddings; d is much smaller than the size of the vocabulary. The most commonly used word embeddings are word2vec [8] and GloVe [9]. Since the advent of word embeddings, deep learning methods have been widely used for various NLP tasks.

We model MWE identification as a supervised classification task with BIO as the class labels. 'B' tag identifies the beginning of a MWE, 'I' tag means inside the MWE and 'O' tag means outside the MWE or any word not part of the MWE. We want to study how using different word embeddings, along with Convolutional Neural Networks, affects the MWE identification. In particular, we would like to analyze which word embedding works better for gappy MWEs and VMWEs. We hypothesize that GloVe may work better than word2vec to identify MWEs generally and gappy MWEs specifically since GloVe is able to capture global information.

2   Description of Dataset

The dataset we used was the DiMSUM dataset [10] that was used in the 2016 Shared Task of SemEval 2016. The data is in CoNLL format consisting of the word position in a sentence, the word, the lemma of the word, POS tag (17 tags), BIO tag, offset of the MWE from the first word of the sentence, supersense label, and sentence ID. We used all the features except the supersense label.

We one-hot encoded categorical variables and included eight binary features, similar to [11] and [12], that give useful syntactic information such as includes single or double quotes, all upper case letters, word starts with an upper case letter, part of a URL, contains a number, contains a punctuation, made up of punctuation only, contains # or @. The contains # is useful to identify if it is a hashtag and the '@' is used to identify Twitter handles. We did some basic preprocessing to reduce the chance of unseen words in the word embeddings.

We built feature vectors using word2vec, GloVe, and by concatenating (appending) word2vec and GloVe feature vectors. To refer to these easily, we will use W2C, GloVe, and WG, for word2vec, GloVe, and word2vec+GloVe, respectively in the remainder of the paper.

3   Network Structure

We built a Convolutional Neural Network (CNN) with two convolutional layers, one fully connected (fc) layer, and the output layer. We varied the number of filters [100, 125, 150, 200, 250] and window sizes [1, 2, 3]. We used Stochastic Gradient Descent (SGD) optimizer with a learning rate of 0.01. We used tanh activation function at the convolutional layers and fc layer and softmax activation function at the output layer. We used 0.5 dropout rate after each convolution and 0.5 dropout after the fc layer; the dropout is to reduce overfitting.

For word2vec, we used the 2015 wikidump (~15 GB) to train the vectors and for GloVe, we used the pre-trained model from [9]. We used 100 dimension for both since we are restricted by GPU computing power. We performed 5-fold leave-one-out cross validation. We ran the setup for 600 epochs and only saved the best model based on the lowest error rate. We have reported the average F-score for the test set and generalization in percentage.

4 Related Work

Gharbieh et al. [12] made one of the first attempts to identify MWEs using deep learning. They use SVM as the baseline and implemented a variety of neural networks – Feed Forward, Recurrent, and Convolutional. They found that the Convolutional Neural Network worked best among the neural network models and they outperformed SVM and other methods used earlier. However, this work only used word2vec.

Another recent approach using Neural Networks is by [13]. They used a recurrent neural network and handled MWE detection as a supervised task. They did not attempt to work on the English dataset.

A paper on sentence classification [14], that used neural networks, made us consider a combination of word2vec and GloVe for the features. We adopted some commonly used hyperparameters explained in the paper for our experiments.

5 Evaluation and Results

We use the F-score evaluation method from [11] which is calculated by grouping the words together and using transitivity. The score penalizes errors that are at the beginning of the MWE ('B' tag) which is the hardest tag to find. Due to space constraints, a detailed explanation of the calculation is not provided. Readers are encouraged to refer to Schneider et al. [15] p. 553, Figure 4 for a detailed explanation.

We calculate generalization as the ratio of F-scores of the test data to the validation data. Generalization is a measure of how well our model was able to generalize on the test data based on the weights learned during training. Our models were not able to differentiate between idiomatic and literal meanings of certain expressions which lead to a decrease in the F-score.

5.1 Filters

We look at the test F-scores and compare the results across the number of filters shown in Fig. 1. It can be seen that W2C never had the highest F-score. GloVe performs better than the other two for all combinations of window sizes for a filter size of 100. For all other filter sizes, either GloVe or WG performed better. Generalization shown in Fig. 2 follows a trend similar to the F-score, with one exception.

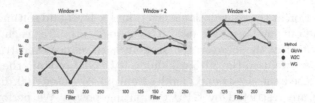

Fig. 1. Test F score for different filters and window sizes by embedding used

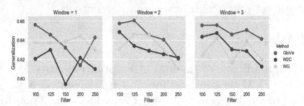

Fig. 2. Generalization for different filters and window sizes by embedding used

5.2   Window Size

In terms of window size we are only comparing GloVe and WG since the section on Filters makes it clear that W2C never had the highest F-scores. With window size 3, GloVe performs better than WG for all filter sizes. This is because GloVe can easily capture co-occurrence probabilities and with window size 3, it has access to more context. W2C uses a predictive model that performs well on small window sizes.

5.3   All Tags

Table 1 shows the percentage of 'B', 'I', and 'O' tags correctly identified in the test data at least 50% of the time. Table 2 looks at how many of the tags correctly identified by at least 50% of the models were identified correctly by all models. WG has higher percentage of tags correctly identified by all the models. On the other hand, with GloVe certain combinations of filter and window size do not consistently classify the tags, resulting in lower numbers. WG resolves some of the inconsistencies from using just W2C or GloVe, making WG more consistent and stable.

Table 1. Percentage of BIO tags classified correctly by at least 50% of the models

	W2C	GloVe	WG
B	14.59	14.86	**15.14**
I	22.41	22.52	**23.02**
O	97.4	**97.59**	97.44

Table 2. Percentage of BIO tags classified correctly by all models based on Table 1

	W2C	GloVe	WG
B	19.44	15.45	**22.32**
I	28.96	23.42	**30.4**
O	92.59	92.31	**93.41**

5.4   B Tags

The 'B' tag can occur anywhere and the 'I' tag depends on the 'B' tag. First, we filtered out only words that were correctly tagged 'B' by at least half the models. In this subset, we further look to see if each of those words and their POS tags were present in the train data to evaluate how well each of the three embeddings perform in identifying words that were never encountered in the train data.

If a word was classified as 'B' tag and it had the same POS tag in train and test data, we call it "'B' tag Best Words" due to the exact agreement. In the train data, for words that had more than half the BIO tags as 'B' and within 'B', at least half the B-tagged words had the same POS tag as the one correctly classified in the test data, we called it "'B' tag ≥50% Words". For example the word 'keep' had 'B' tag 7 times and 'O' tag 5 times in the train data; within 'B' tag the word 'keep' had VERB as POS tag 6 times and NOUN 1 time. Now 'keep' has at least half 'B' tag and VERB tag occurring 6 times which is the same combination as the correctly classified in test data. "'B' tag not matching" is when the word was classified correctly as 'B' but the POS tags did not match across train and test data.

Table 3 shows the resulting counts and POS tags for the different categories explained above. In summary, WG is able to identify more words (better learning), and identify more unseen words (good generalization). GloVe is able to classify more verbs as the 'B' tag - something that will help in identifying VMWEs.

5.5   I Tags

The 'I' tagged words following 'B' tag were not identified properly and they were mostly given 'O' tag. We looked at the train data and found that the words that got misclassified had almost all BIO tags to be 'O'. This was a common problem with all three word embeddings. For example, the MWE "customer", followed by one or two words that were part of the MWE or not, occurred 14 times and the models did well in tagging "customer" as 'B' but could never tag "service" in "customer service" as 'I'. This was because "service", as a word by itself, was tagged 'O' more than 80% of the time in the train data. We looked at the 'I' tags associated with MWEs for which the models classified 'B' tag correctly shown in Table 4.

Table 3. Counts of different 'B' tagged words using the different embeddings

Embedding		Total	POS Tag						
			ADJ	ADP	ADV	DET	NOUN	PROPN	VERB
	Total Words	107							
	Unique Words	73	4	1	1	1	15	27	24
	Words in Train	54	4	1	1	1	13	12	22
W2C	Words not in Train	19					2	15	2
	'B' tag Best Words	9					2	2	5
	'B' tag >=50% Words	27	3				5	6	13
	'B' tag not matching	18	1	1	1	1	6	4	4
	Total Words	109							
	Unique Words	72	1	1	1	1	14	27	27
	Words in Train	52	1	1	1	1	10	13	25
GloVe	Words not in Train	20					4	14	2
	'B' tag Best Words	8					1	2	5
	'B' tag >=50% Words	28	1				4	8	15
	'B' tag not matching	16		1	1	1	5	3	5
	Total Words	111							
	Unique Words	76	3	1	1	1	16	29	25
	Words in Train	54	3	1	1	1	12	13	23
WG	Words not in Train	22					4	16	2
	'B' tag Best Words	10			1		2	2	5
	'B' tag >=50% Words	29	3				5	7	14
	'B' tag not matching	15		1		1	5	4	4

Table 4. Counts of two-word MWEs and VMWEs.Gappy VMWE counts are indicated in () next to the VMWE counts

		W2C	GloVe	WG
MWEs	<50%	15	**17**	15
	≥50%	18	18	**20**
VMWEs	<50%	21 (5)	**22 (4)**	21 (5)
	≥50%	19 (5)	**25 (6)**	25 (4)

Both Words Same POS Tag. The MWE "San Antonio" occurred only thrice in the train data but all combinations classified it correctly. "San Francisco" is similar to "San Antonio" and does not occur in train data but "San Francisco" was classified correctly by W2C and WG models but GloVe could do well only more than half the time. GloVe had the problem of classifying "Francisco" as 'B' tag because the starting letter is upper case. GloVe did well with tagging 'B' for words whose starting letter was upper case irrespective of whether it was seen or not during training. "Charlie Sheen" was not in the train data but more than half the models classified it correctly because the train data has "Charlie Rose"; GloVe had the same problem, as discussed above for "Francisco".

VMWEs. The non-gappy VMWE "hung up" occurs only twice in the train data but all the models except one in GloVe was able to correctly classify it. For the MWE "picked up", all the models were able to identify it correctly; it occurred thrice in the train data with one of them being gappy. In general, GloVe was able to identify more forms of verbs as 'B' tag which led to the marginally better performance than the other two.

We looked at the 'I' tags associated with VMWEs for which the 'B' tags were classified correctly. GloVe and WG performed similarly in terms of accuracy percentage but GloVe could tag more gappy VMWEs than WG. This is because GloVe works with co-occurrence probabilities and hence can learn MWE patterns that are spread out.

5.6   MWEs Which Were Not Seen During Training

Even though "Pick-up" did not occur in the train data, "Pick up" occurred in the train data. Most of the models were able to tag 'Pick' as 'B' and 'up' as 'I'. Although '-', a punctuation usually tagged as 'O', was tagged 'I' in the Gold Standard, the models tagged it appropriately as 'O'. This would mean that given enough number of gappy MWEs in the train data for different lengths of gaps, the models would be able to identify the MWEs.

There were a few MWEs that did not occur in the train set and these were incomplete in the test set. A few of them were "ve been" which did not have the ' before ve to indicate abbreviation. In the MWE "back & forth", '&' never occurred in the train data as 'I' tag in the 50 times it was seen. The system was not able to understand that 'and' and '&' are equivalent to correctly classify '&' with an 'I' tag. Internet slang such as "bruh bruh" and "Guhhh deeeh" were a problem since they are out of vocabulary and the lemmas were same as the words.

6   Conclusions and Future Work

WG models were consistent in tagging over all combination of filter and window sizes. GloVe is marginally better at identifying VMWEs including those that are gappy. In terms of generalization, GloVe provides marginally better generalization from train to test data. The combination of GloVe and word2vec also provides better performance than word2vec. Word2vec seems to identify fewer MWEs; however, we would like to test the performance of word2vec and GloVe on other datasets. We would also like to increase the word embedding dimension size to 300 and run our experiments to study the effect. We are also interested in applying this to our Question Answering system to study the impact of MWEs in question understanding.

References

1. Maldonado, A., et al.: Detection of verbal multi-word expressions via conditional random fields with syntactic dependency features and semantic re-ranking. In: Markantonatou, S., Ramisch, C., Savary , Savary , A., Vincze, V. (eds.) Proceedings of the 13th Workshop on Multiword Expressions (MWE 2017). pp. 114–120. Association for Computational Linguistics, Valencia, Spain (Apr 2017)

2. Sag, I.A., Baldwin, T., Bond, F., Copestake, A., Flickinger, D.: Multiword expressions: a pain in the neck for NLP. In: Gelbukh, A. (ed.) CICLing 2002. LNCS, vol. 2276, pp. 1–15. Springer, Heidelberg (2002). https://doi.org/10.1007/3-540-45715-1_1

3. Nagy T., I., Vincze, V.: Vpctagger: detecting verb-particle constructions with syntax-based methods. In: Proceedings of the 10th Workshop on Multiword Expressions (MWE), pp. 17–25. Association for Computational Linguistics, Gothenburg, Sweden, April 2014

4. Fazly, A., Cook, P., Stevenson, S.: Unsupervised type and token identification of idiomatic expressions. Comput. Linguist. **35**(1), 61–103 (2009). https://doi.org/10.1162/coli.08-010-R1-07-048

5. Piao, S.S., Rayson, P., Archer, D., McEnery, T.: Comparing and combining a semantic tagger and a statistical tool for MWE extraction. Comput. Speech Lang. **19**(4), 378–397 (2005). https://doi.org/10.1016/j.csl.2004.11.002

6. Komai, M., Shindo, H., Matsumoto, Y.: An efficient annotation for phrasal verbs using dependency information. In: Proceedings of the 29th Pacific Asia Conference on Language, Information and Computation: Posters, pp. 125–131 (2015)

7. Tsvetkov, Y., Wintner, S.: Identification of multi-word expressions by combining multiple linguistic information sources. In: Proceedings of the Conference on Empirical Methods in Natural Language Processing. EMNLP 2011, pp. 836–845. Association for Computational Linguistics, Stroudsburg (2011)

8. Mikolov, T., Chen, K., Corrado, G., Dean, J.: Efficient estimation of word representations in vector space. arXiv preprint arXiv:1301.3781 (2013)

9. Pennington, J., Socher, R., Manning, C.D.: Glove: global vectors for word representation. In: Empirical Methods in Natural Language Processing (EMNLP), pp. 1532–1543 (2014)

10. Johannsen, A., Schneider, N., Hovy, D., Carpuat, M.: Dimsum 2016 shared task data (2015). Accessed 10 Aug 2018

11. Schneider, N., Danchik, E., Dyer, C., Smith, N.A.: Discriminative lexical semantic segmentation with gaps: running the MWE gamut. Trans. Assoc. Comput. Linguist. **2**, 193–206 (2014)

12. Gharbieh, W., Bhavsar, V., Cook, P.: Deep learning models for multiword expression identification. In: Proceedings of the 6th Joint Conference on Lexical and Computational Semantics (* SEM 2017), pp. 54–64 (2017)

13. Klyueva, N., Doucet, A., Straka, M.: Neural networks for multi-word expression detection. In: Proceedings of the 13th Workshop on Multiword Expressions (MWE 2017), pp. 60–65 (2017)

14. Zhang, Y., Wallace, B.: A sensitivity analysis of (and practitioners' guide to) convolutional neural networks for sentence classification. arXiv preprint arXiv:1510.03820 (2015)

15. Schneider, N., Hovy, D., Johannses, A., Carpuat, M.: SemEval-2016 task 10: detecting minimal semantic units and their meanings (DiMSUM). In: Proceedings of SemEval-2016, pp. 546–559 (2016)

Analysis and Prediction of Dyads in Twitter

Isa Inuwa-Dutse, Mark Liptrott, and Yannis Korkontzelos(✉)

Edge Hill University, Liverpool, UK
{dutsei,Mark.Liptrott,Yannis.Korkontzelos}@edgehill.ac.uk

Abstract. Social networks are useful for linking *micro* and *macro* levels of sociological theory by enabling the analysis of various forms of relationships. In social science, a taxonomy of social relationships is described as a function of closeness among users. The closer the users are, the more cohesive and trustworthy. Identifying dyadic ties, pairs of fully connected users, on Twitter is challenging due to the flexible and eccentric underlying connection patterns. The ability to follow anyone results in many unidirectional connections between socially disconnected users and ultimately affects clustering users and, in turn, the veracity of online content. Major challenges towards effective user clustering are the low number of dyads and efficient methods to identify more. In this study, we query over 17M *verified* and *unverified* Twitter user accounts and retrieve dyadic ties. In the collected data, 55% and 21% of *unverified* and *verified* profiles, respectively, participate in dyadic ties. We describe the importance of dyads in the detection of cohesive user groups and how they may be used to validate trustworthiness. We demonstrate how identifying and using dyadic ties will improve Twitter analysis, in the future. Finally, we develop a deep learning model for dyad prediction.

Keywords: Social networks · Twitter · Dyadic tie · Clustering

1 Introduction

Online socialisation, facilitated by platforms such as Twitter and Facebook, attracts much research interest and poses many questions. For a long time, *social networks* have been considered useful tools for linking *micro* and *macro* levels of sociological theory [6]. Many forms of social relationships have been analysed at various levels. Understanding social interactions today would be incomplete without taking online social relationships into account. Sufficient understanding of the structural properties of online platforms is important in designing a more *human-centric* internet [2], in the future.

However, the growing complexity and heterogeneity of connections makes the task of identifying communities and relationships at the micro levels challenging. Twitter allows every user to follow anyone, resulting in many unidirectional connections, which may not correspond to a social connection. This makes it

E. Métais et al. (Eds.): NLDB 2019, LNCS 11608, pp. 303–311, 2019.
https://doi.org/10.1007/978-3-030-23281-8_25

difficult to extract dyads. Thus, dyadic ties are usually overlooked in tasks such as clustering and in the authentication of online content posted by trusted users.

The definition of network community detection as a task varies in the literature and ground-truth evaluation data are rare and difficult to collect [11]. According to the *ego network model*, which is based on Dunbar's classification of social relationships [4], a *social support clique* consists of a few fully connected users with the strongest relationship in the network [2]. We opined that the level of trust is stronger among users that share dyadic ties and it is highly unlikely for a user in the group to misuse the network e.g. spread fake news or spam. However, acquiring large amounts of tweets sufficient to identify such cohesive groups is challenging and time consuming.

We analyse a large collection of dyadic and non-dyadic ties[1] and explore their potential contribution to online clustering and content authentication. In the collected data, 55% and 21% of *unverified* and *verified* profiles, respectively, are involved in dyadic ties. Despite this large proportion of dyads, a random collection of tweets corresponds to far fewer users in dyadic ties. We analyse the cohesiveness of cliques, i.e. fully connected groups, in terms of size. Finally, we propose a deep learning method for dyadic tie prediction, to avoid the time-consuming search for dyads on Twitter. The model achieves a promising performance when trained on real data. Checking Twitter users for dyadic ties can limit spurious content and allow content collection from legitimate users.

2   Related Work and Background

Networks and Online Social Networks: Relationships and structural properties of networks have been extensively studied at different levels of granularity and sophistication, ranging from the structure of microscopic organisms to complex networks, such as the internet [1,5,15,17]. While many properties are common across various networks, social networks are different with respect to the degree of correlation and tendency for clustering. The formation of clusters is easier and the correlation degree between users is positive [13].

Homophily, the tendency for humans to connect with people of similar characteristics, is central to human's social interaction [12]. Users in reciprocal relationship discuss similar topics [18]. Homophily has been investigated in the context of geolocation and popularity [10]. In social networks, the concept of dyad, or reciprocity, has been viewed from various perspectives and often with contradicting results. With respect to how popular users follow other similarly popular users, [10] reports low-level reciprocity and a high proportion of directed connections in Twitter. However, [18] reports high reciprocity by computing the ratio of follower/following. The probability of a user reciprocating a relationship, i.e. by following back, and how users of varying influence on Twitter reciprocate their followers have been investigated in [3]. We extend this by proposing a method to predict the likelihood of reciprocity between users. Dyadic ties in Twitter are rare due to the prevalence of directed ties. Previous studies collected datasets

---

[1] See github.com/ijdutse/dyads_in_Twitter for details about the data of the study.

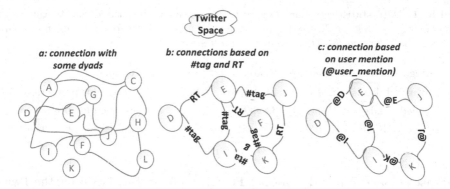

Fig. 1. Connections on Twitter manifest by sharing a link, *re-tweeting (RT)*, using the same *hashtags* and *user mentions (@)* or by *following*. The concept allows multiple connections among many diverse users and limits the chances of dyads.

from various social networks [11,19,20]. For example, the Twitter data collected in [11] is freely available but contains mainly directed ties.

Connection in Twitter: Online social media platforms, such as Twitter and Facebook, enable the empirical quantification and evaluation of social relationships among users to an unprecedented scale. Theories and analytical methods can be validated using real social data. We argue that the presence of random connections among some users on Twitter (see Fig. 1) contributes to the limited overall cohesiveness and the growing proportion of fake and spam contents. Users openly engaging with other users in a bidirectional manner will curtail the circulation of spurious information. Users with dyadic ties are more likely to be genuine, trustworthy and will probably result in a more cohesive clusters.

3   Method

Definition 1: **dyadic tie²** – a relation R over a set D is *dyadic iff* $aRb = 1, \forall a, b \in D$. In the context of this study, a follows b is a directed relationship. If b follows a back, then it is undirected and is called dyad (see Fig. 2).

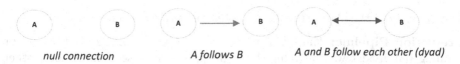

Fig. 2. Relations between two users A and B in Twitter: no relationship (*null connection*), directed relationship ($A \longrightarrow B$) and dyadic or pairwise relationship ($A \longleftrightarrow B$).

---

² Dyadic tie, pairwise or binary relations are used interchangeable in this work.

Table 1. Data statistics: Many *unverified* users had to be visited due to the large number of $1 - edge$ or directed connections, occurring when followers are not being followed back, i.e. $\exists a, b \in D, a \longrightarrow b = 1$ and $b \longrightarrow a = 0$.

Category	Seed size	Visited users	Retrieved	Remark
Unverified dyads	2,023	13,409,661	8,715	utilised for prediction
Verified dyads	1,999	3,893,075	–	not used for prediction
1-edge and null tie	1,700	–	7,014	utilised for prediction

Dataset Collection and Training Features: We collect data using the Twitter API and criteria that satisfy the definition of dyadic ties. We begin with 4022 seed users[3] from *verified* and *unverified* accounts. The profile of each user's network G, i.e. their list of *friends* and *followers*) was searched by a crawler to determine pairs of users that follow each other. Essentially, for each user network, $G = \{u | \exists u' \in G\}$ such that $u \cap u' = 1$, i.e. dyadic tie. Table 1 shows basic statistics of users visited by the collection crawler. In particular, it shows the counts of directed ($1 - edge$) connections and dyadic ties. Similarly, Fig. 4 summarises dyads in the *verified* and *unverified* user category.

We considered the following feature groups to train the prediction model:

- **Network features** f_n: followers, friends, account category
- **Text feature** f_t: account description

Features consist of a rich set of meta-data information describing users based on their behaviour and the textual part of their account description.

We use a *Convolutional Neural Network (CNN)* to extract textual features. This is essential because if the users comprising a potential dyad have conflicting ideologies expressed in their profile descriptions, the likelihood of dyadic tie is minimal. According to the collected data (Table 1) and insights from our empirical analysis, we can estimate the likelihood of a dyadic tie between users. B is likely to follow A back:

- if A and B are both in the unverified user's category
- if both A and B have low or relatively large number of followers or network size, i.e. based on the average of those metrics in the users' categories
- if A has more followers than B or if A is a verified user.

The opposite of the above statements holds for verified users.

Prediction Pipeline: The set of network and text features $F = \{f_n, f_t\}$ for training our model was introduced in the previous subsection. Among other intrinsic factors, these are the likely features a user can easily access in making a decision to follow back a request or not. Each user U_i is represented by the following vector of reciprocal relationships $U_r^i = [u_{i,j}, u_{i,k}, ..., u_{i,n}]$ where users

---

[3] These are genuine users devoid of spammers or social bots collected based on the SPD filtering technique [8].

$j, k..n$ have dyadic ties with user u_i. Features from the *account description text* are learned by applying a CNN on the *n-dimensional* embedding of tokens[4] in text. CNNs has been applied to various domains and many successful studies in NLP have used them [9,16,21]. In this study, the CNN is used as a textual feature extraction engine (Fig. 3), whose output is encoded using *Long Short-Term Memory (LSTM)*. The encoded vector is merged with the main features for training the prediction model.

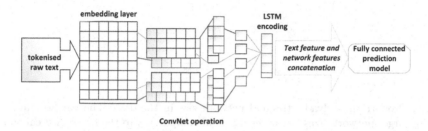

Fig. 3. The *embedding layer* accepts tokenised text and encodes each token in a dense 100-dimensional vector to be used by the *ConvNet* part. The *LSTM* layer transforms the output to a lightweight vector that is merged with the *network features* for training.

4 Dyads: Results Analysis

Network Topology: Firstly, we analyse the data to measure the depth of user relationships. The huge number of *visited users* for the number of *dyads* in Table 1 and Fig. 4 reveal a high proportion of *null connections* and *1-edge connections*. Subsequent analysis will focus on ordinary *unverified users* (Fig. 5).

Proportion of Nodes and Reciprocity: *Verified users* have more network neighbours than their *unverified counterparts*, but there is a higher proportion of dyadic ties in the *unverified category*, as shown in Fig. 4.

Automatic Detection of Dyads: Noting the flexibility of connections on Twitter and the lack of real connectivity (Fig. 1), large scale dyadic ties are rare and difficult to locate due to the *curse of dimensionality*. We aim to predict *the likelihood that user A who follows user B will be followed back*. The task can be modelled as binary classification. Given two users A and B connected with one edge connection, the goal is to predict whether a pairwise relationship will be established. We build a deep learning classifier that predicts the probability of a dyadic tie between two users on Twitter and then we compare the results with actual dyads collected for evaluation. Figure 6 show some results from the prediction model. Although the performance in sub-figures a and b is good, it is unstable and seems to be prone to overfitting, noting the proportional

---

[4] We utilise Glove word embeddings [14], pre-trained on tweet collections.

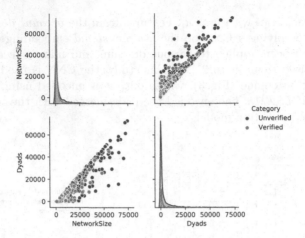

Fig. 4. Proportion of dyadic ties and network size in the data. The *verified* category exhibit larger network sizes but fewer dyads in comparison to the *unverified* category.

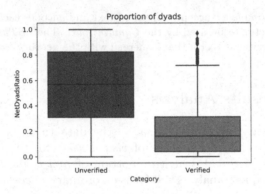

Fig. 5. Dyads proportions in *verified* and *unverified* profiles

relationship between the training accuracy and the validation loss, i.e. both are increasing. We increase the training epochs to 200 and add more layers to the network for stability (sub-figure c). There is room for improvement when using larger amounts of data and historical tweets from users.

Utility of Dyads in Clustering and Content Veracity: Phenomena in real life are associated with numerous network structures and embedded communities. The social media ecosystem enables various forms of interactions among diverse users at various levels. The high dominance of online content from influential users in Twitter makes it difficult to detect low level communities of average users [7]. Low-level communities are a better reflection of true connectivity with strong social cohesion. In this study, we show how the dyadic tie is widespread among users in the unverified category; regarded as proxy for average users on Twitter. The proposed approach can be extended to other social media

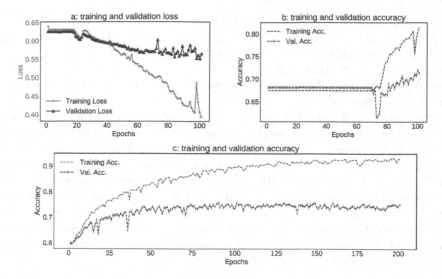

Fig. 6. Performance of the proposed model on the training and the validation set. The performance remains stable after the first 100 epochs.

platforms that support automatic reciprocal tie once one party accepts the other party as a friend, e.g. Facebook. In those platforms, users with dyadic ties subscribe to various groups and identifying users with common subscription will add additional layer of social cohesion.

A user with many dyadic ties can be a resourceful representation of a microcosm. Such a user can be regarded as a differential entity for deriving a set of related users. For instance, if U^3 denotes a user with many dyads of order 3, $3U^2$ and $6U$ are directly related to the user. The *constants* and *powers* relate to the user's network size and closeness to the original user, respectively. The group can be viewed as *microcosm* and can be exploited in tasks such as clustering. In the context of content veracity, a *microcosm* can be useful for analysing user groups with common online traits. Following the old adage, *birds of a feather flock together*, users who spread rumours or spam are likely to be connected together. In the future, we will explore these aspects from the perspective of dyadic ties.

5   Conclusion

Many relevant theories on various networks and social networks have been proposed and validated analytically or experimentally. Modern social media platforms, such as Twitter and Facebook, enable the empirical quantification and evaluation of social relationships among users on an unprecedented scale. Social network theories and analytical solutions can now be tested using real social data. We conducted an empirical analysis to understand dyads on Twitter, where connections among users are porous, and the composition of communities is not sufficiently cohesive. We began by collecting and analysing a large number of datasets consisting of pairwise users. This deeper insight into the underlying mechanisms in dyadic ties on Twitter will be beneficial to studies involving tweets, as the recognition of dyads can improve clustering and content validation. We demonstrated how the recognition of dyads can improve clustering and content validation tasks.

Due to the challenging and time-consuming task of collecting dyads on Twitter, we proposed an effective deep learning prediction method that returns the likelihood of two users engaging in a pairwise relationship. The fundamental conclusion is that dyadic ties can be accurately predicted (if the pair of users are socially active) enabling the identification of cohesive groups of users on Twitter. This approach can also be applied to detect cohesive communities of users on Twitter. Employing this strategy can limit the danger of spurious content and allow the collection of content from legitimate users. In the future we will employ the concept of *transitivity* to extend this research to include transitive users and model how to predict those users' behaviour.

Acknowledgements. This research work is part of the CROSSMINER Project, which has received funding from the European Union's Horizon 2020 Research and Innovation Programme under grant agreement No. 732223.

References

1. Albert, R., Barabási, A.L.: Statistical mechanics of complex networks. Rev. Mod. Phys. **74**(1), 47 (2002)
2. Arnaboldi, V., Guazzini, A., Passarella, A.: Egocentric online social networks: analysis of key features and prediction of tie strength in facebook. Comput. Commun. **36**(10–11), 1130–1144 (2013)
3. Cha, M., Benevenuto, F., Haddadi, H., et al.: The world of connections and information flow in twitter. IEEE Trans. Syst. Man Cybernet. Part A Syst. Hum. **42**(4), 991–998 (2012)
4. Dunbar, R.I.: The social brain hypothesis. Evol. Anthropol. Issues News Rev. Issues, News Rev. **6**(5), 178–190 (1998)
5. Erdős, P., Rényi, A.: On the evolution of random graphs. Publ. Math. Inst. Hung. Acad. Sci **5**(1), 17–60 (1960)
6. Granovetter, M.S.: The strength of weak ties. In: Social Networks, pp. 347–367. Elsevier (1977)
7. Inuwa-Dutse, I.: Modelling formation of online temporal communities. In: Proceedings of WWW, pp. 867–871. International WWW Conferences Committee (2018)
8. Inuwa-Dutse, I., Liptrott, M., Korkontzelos, I.: Detection of spam-posting accounts on twitter. Neurocomputing **315**, 496–511 (2018)
9. Kim, Y.: Convolutional neural networks for sentence classification. arXiv preprint arXiv:1408.5882 (2014)
10. Kwak, H., Lee, C., Park, H., Moon, S.: What is twitter, a social network or a news media? In: Proceedings of WWW, pp. 591–600. ACM (2010)
11. Leskovec, J., Mcauley, J.J.: Learning to discover social circles in ego networks. In: Proceedings of NIPS, pp. 539–547 (2012)

12. McPherson, M., Smith-Lovin, L., Cook, J.M.: Birds of a feather: homophily in social networks. Ann. Rev. Sociol. **27**(1), 415–444 (2001)
13. Newman, M.E., Park, J.: Why social networks are different from other types of networks. Phys. Rev. E **68**(3), 036122 (2003)
14. Pennington, J., Socher, R., Manning, C.: Glove: global vectors for word representation. In: Proceedings of EMNLP, pp. 1532–1543 (2014)
15. Scott, J.: Social network analysis. Sociology **22**(1), 109–127 (1988)
16. Wang, W.Y.: "liar, liar pants on fire": a new benchmark dataset for fake news detection. arXiv preprint arXiv:1705.00648 (2017)
17. Watts, D.J., Strogatz, S.H.: Collective dynamics of "small-world" networks. Nature **393**(6684), 440 (1998)
18. Weng, J., Lim, E.P., Jiang, J., et al.: Twitterrank: finding topic-sensitive influential twitterers. In: Proceedings of WSDM, pp. 261–270. ACM (2010)
19. Yang, J., Leskovec, J.: Defining and evaluating network communities based on ground-truth. Knowl. Inf. Syst. **42**(1), 181–213 (2015)
20. Yoshida, T.: Toward finding hidden communities based on user profile. J. Intell. Inf. Syst. **40**(2), 189–209 (2013)
21. Zhang, X., Zhao, J., LeCun, Y.: Character-level convolutional networks for text classification. In: Proceedings of NIPS, pp. 649–657 (2015)

Mathematical Expression Extraction from Unstructured Plain Text

Kulakshi Fernando[✉], Surangika Ranathunga, and Gihan Dias

Department of Computer Science and Engineering, University of Moratuwa,
Katubedda 10400, Sri Lanka
{kulakshif,surangika,gihan}@cse.mrt.ac.lk

Abstract. Mathematical expressions are often found embedded inline with unstructured plain text in the web and documents. They can vary from numbers and variable names to average-level mathematical expressions. Traditional rule-based techniques for mathematical expression extraction do not scale well across a wide range of expression types, and are less robust for expressions with slight typos and lexical ambiguities. This research employs sequential, as well as deep learning classifiers to identify mathematical expressions in a given unstructured text. We compare CRF, LSTM, Bi-LSTM with word embeddings, and Bi-LSTM with word and character embeddings. These were trained with a dataset containing 102K tokens and 9K mathematical expressions. Given the relatively small dataset, the CRF model out-performed RNN models.

Keywords: Mathematical expression extraction · Sequential tagging · Information extraction

1 Introduction

Simple mathematical expressions in questions and answers of types arithmetic, linear algebra, etc can be typed inline with text. When such mathematical expressions are added in structured format to documents, for example using Tex or XML, extracting them out from the ordinary text is a trivial process. However, sometimes text containing mathematical expressions in the web are unstructured. For example, a math related content in an email, or a mathematical problem submitted into an educational forum may contain mathematical expressions embedded in plain text. '*If A and B are disjointed sets, how can you find n (A union B)?*' is such a problem submitted to the well-known educational forum, Quora.com[1]. Therefore, any system (e.g. intelligent search engines) that needs to understand the math in a document, answer generation systems for mathematical problems expressed in natural language (math word problems), or answer grading systems in mathematics should be able to recognize unstructured mathematical expressions appearing inline with text.

---

[1] https://www.quora.com/If-A-and-B-are-disjointed-sets-how-can%2Dyou-find-n-A-union-B.

© Springer Nature Switzerland AG 2019
E. Métais et al. (Eds.): NLDB 2019, LNCS 11608, pp. 312–320, 2019.
https://doi.org/10.1007/978-3-030-23281-8_26

There is much research conducted to answer and grade math word problems in elementary and secondary level mathematical domains such as arithmetic, algebra, and geometry. However, most of the time the input to these systems is expected to be annotated for mathematical expressions [9,14]. When a lay person such as a teacher or a high school student is preparing the content, this is an overhead. If user interfaces are present, these systems [3] require all mathematical expressions to be typed using tools based on Tex or XML. This is an extra effort for simple mathematical expressions that can be typed inline as other text. Such difficulties can be avoided by automatically identifying mathematical expressions from other text in math problems.

Some systems read mathematical expressions as plain text and use regular expressions (regexes) to extract them [4,15]. Fernando et al. [4] focus on automatically solving math word problems related to set theory, and show that 75% of the errors in their system occurs due to the incapability of capturing unexpected expression formats. Therefore, in the presence of unseen expression types, typing errors, and lexical ambiguities, regexes can be less accurate and require more rules to identify relevant text. Tian et al. [16] use a heuristics and vocabulary based filtering mechanism to separate expressions from other text and an Hidden Markov Model (HMM) to verify text as expressions prior to extraction. They show that HMM perform well with large amounts of data, at the expense of a long training time. Therefore, it is worth exploring alternative techniques; especially techniques that do not require predefined rules or states, to extract mathematical expressions.

This research presents a mechanism to extract simple mathematical expressions that appear in-line with other non-mathematical (natural language) text[2]. The task of extracting mathematical expressions is mapped into a sequence tagging task, where space-separated tokens are tagged as expressions and other text using IOB (Inside-Outside-Beginning) format. Experiments were conducted using Conditional Random Fields (CRF), Long-Short Term Memory (LSTM) networks with word embeddings (we refer to this setting as W-LSTM hereafter), Bidirectional-LSTM (Bi-LSTM) with word embeddings (we refer this as W-Bi-LSTM), and Bi-LSTM with both word and character embeddings (we refer to this model as W-CH-Bi-LSTM). Experiments with CRF showed that character level properties of the text contribute noticeably to increasing the accuracy of expression extraction.

2   Challenges in Extracting Mathematical Expressions in Plain Text

The main challenge in identifying mathematical expressions from other text is semantic level ambiguities between expressions and non-expression text. One of the main semantic level ambiguities is variable names used in mathematical expressions. For example, the token 'a' in '*Let a be a positive integer*' is both

---

[2] The code and data is available here https://github.com/Kulakshi/math-expression-extraction.

a mathematical expression and a stop word. Another ambiguity is the use of a sequence of numbers in contrast to a list of some numbers. For example in the text *"In the sequence 7, 14, 28, x, 112.. what is the value of x?"*, the mathematical expression is the sequence:' *7, 14, 28, x, 112..'* whereas in *"Find arithmetic mean of the numbers in the list 8 − a, 8, 8 + a"*, mathematical expressions are individual expressions separated by commas.

Text copied from examination papers or tutorials often include question numbers as digits, roman numbers or letters that can be misinterpreted as mathematical expressions. Years, table or figure labels, and abbreviated names for irrelevant entities are few other occasions that can be misinterpreted as mathematical expressions.

When tokenizing text combined with mathematical expressions, they may get split into different combinations when spaces are considered as the delimiter. For example, if the expression is *'x + 1'*, we get three tokens. If the same is written as *'x + 1'* we get only two tokens. Another important fact is the start and end of a mathematical expression. An expression that contains words such as *'A = total area of five circles of radius r'*, or an expression with typos like *'set A = {Students of grade five and set B = {Girls in grade five}'* do not have a clear lexical separation from the usual text.

3   Related Work

Fernando et al. [4] extracted set theory related mathematical expressions from the unstructured plain text using regexes. However, they claim that regexes fail when unexpected expression formats are met and it is the main reason for reducing the accuracy of the system. Seo et al. [15] also extracted geometry related mathematical expressions using regexes.

Work of Tian et al. [16] is the most relevant research we could find that focused on extracting mathematical expressions in general from unstructured plain text. They eliminate other text in a document using heuristics and vocabulary based filtering. An HMM with 8 hidden states denoting parts of mathematical expressions is then used to verify filtered expressions before extraction. They mapped mathematical symbols that are observable in the text into these hidden states. They trained the model with 13,423 expressions and achieved over 89% accuracy and 77% recall. However, the dataset is not available to be used to compare their results empirically.

HMMs, CRFs [10], Convolution networks and Recurrent Neural Networks (RNNs) are well-known for sequence classification for decades. LSTM networks [7] reduce the vanishing gradient problem present in standard RNNs and perform well for long sequences of data. Bi-LSTM increases the accuracy of sequence tagging by considering both past and future inputs as used in the work of Graves et al. [6]. There is plenty of research [2,8] that successfully used LSTM networks with different varieties and combinations for the sequential classification task. Many research including the work of Ling et al. [12] shows that using character level information with LSTM networks is effective in increasing the performance of language modeling. In the presence of well defined features, Nikola [13] shows that CRF can perform as well as Bi-LSTM models in a sequential classification task.

4 Methodology

4.1 Dataset and Pre-processing

An adequately annotated dataset for mathematical expressions was not available to use in our research. Thus we adapted the dataset provided by Task 10 of SemEval-2019, "Math Question Answering"[3] and the dataset of Fernando et al. [4]. The former dataset includes mathematical problems that belong to closed-vocabulary algebra, open-vocabulary algebra, geometry, probability, and data representation. Some questions that belong to multiple domains are not categorized. Most of the mathematical expressions in this dataset were given using LaTeX. Such expressions were converted into inline mathematical expressions. For example, '8\times(2 ∧ 4a) = \frac{2 ∧ 2b}{2 ∧ 3}' was converted into $8 \times (2 \wedge 4a) = 2 \wedge 2b/2 \wedge 3$. For now, mathematical expressions that cannot be written along the mean-line; text that includes special scripts such as superscripts and subscripts are not handled in this work. The dataset of Fernando et al. [4] consists of elementary set theory problems that include both text and expressions in set notation. Both the datasets were adapted for this research by tokenizing text and annotating tokens as described next.

Text with inline expressions was then tokenized and tagged in IOB format for expressions and other texts to prepare the dataset. For example, the problem *If $a - 5 = 0$, what is the value of $a + 5$?* is tagged as *O B-EXP I-EXP I-EXP I-EXP I-EXP O O O O O B-EXP I-EXP I-EXP O*.

Dataset statistics are shown in Table 1 ('elementary set theory' category contains problems of the dataset of Fernando et al. [4]).

Table 1. Statistics of problems in the dataset

Category	#problems	#expressions	#tokens	#expression tokens
Closed-algebra	1088	3832	29541	10886
Open-algebra	360	1059	16332	1372
Geometry	702	2351	22677	4288
Other	86	124	3634	156
Uncatagorized	528	1717	22796	3219
Elementary set theory	487	2419	31380	16308

Cetintas et al. [1] show that traditional text pre-processing tasks such as stemming and stop word removal affect negatively in mathematical text categorization tasks since math related information are lost with such pre-processing

---
[3] https://github.com/allenai/semeval-2019-task-10.

approaches. For example, words such as *'a'*, *'than'* and words with suffixes such as *'rd'*, *'th'* are important to identify math related text. This is relevant to mathematical expression extraction as well, given the ambiguities (see Sect. 2). Therefore we avoided such text pre-processing methods.

Some expressions in the dataset had typos. Specially equations converted from LATEX to plain text included syntactical errors. They were kept intact since the purpose of this research includes identifying expressions that might contain errors.

4.2 Experiment Setup

All models were trained using 10-fold cross validation. The CRF model was trained by using unigrams as the baseline feature. Other features in Table 2 were added incrementally to evaluate the effectiveness of each feature set. Features that contribute effectively to the model were selected using the validation dataset. Some features (shown in non-italic letters in Table 2) were adapted from the work of Finkel et al. [5] and Huang et al. [8]. These features were originally used for normal NER tagging. A few more features were added that seems relevant to mathematical expressions. These features are distinguished from aforementioned features by the italic font in Table 2).

Table 2. Features used for CRF divided into categories. (The left most column contains a label for each set of features)

Context features	
A	Uni-grams, bi-grams, tri-grams and their frequencies
Token level features	
B	*Whether the token is a single character*
C	Case related features (all upper case, all lower case, starts with capital, contains non-initial capital letters)
D	Features related to character type (*contains only digits, contains math operators, contains bracket delimiters,* contains only letters, is a mix of digits and letters, contains punctuation marks, word shape, word shape summarization [8])
E	Last two and last three suffixes
Semantic level features	
F	POS tag of the token and surrounding tokens

This being the first research conducted for expression extraction using RNNs, we used vanilla LSTM and Bi-LSTM model architectures for first two settings (W-LSTM and W-Bi-LSTM), which comprised an word embedding layer where the 50-length output is subjected to a 10% dropout to avoid overfitting, one LSTM layer and an output layer with softmax normalization. The third setting

(W-CH-Bi-LSTM) comprises of an additional embedding layer with an encoder for characters, concatenated with word embeddings [11]. The model architecture is shown in Fig. 1. Prior to the training, we experimented with different optimizers (standard SGD, Adam and RMSProp), dropout rates (10%, 20%, 50%), different batch sizes (10, 32, 50 & 100) and epochs (10, 20, 50, 100, 500 and 1000), and selected batch size of 10, 10% dropout and RMSProp optimizer with 0.001 learning rate to train the models.

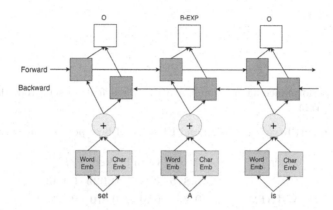

Fig. 1. High level architecture of W-CH-Bi-LSTM model. Source: [8,11]

5   Evaluation and Results

We need to calculate the accuracy of extracting complete mathematical expressions. Given the set of expected expressions E, and the set of predicted expressions P, true-positives (TP), false-positives (FP) and false-negatives (FN) were calculated for each model as follows.

$$TP = \{\,Expressions\ in\ both\ E\ and\ P\,\}$$
$$FP = \{\,Expressions\ in\ P\ but\ not\ in\ and\ E\,\}$$
$$FN = \{\,Expressions\ in\ E\ but\ not\ in\ P\,\}$$

Figure 2 shows the results of the CRF model for cumulatively added features. When considering the CRF model, case related features and character-based features such as whether there are digits in the token, whether only letters contain in the token, and the suffix of the token increased the performance of expression extraction in a significant rate.

Table 3 shows the comparison of the best results of all the models. The W-Bi-LSTM model performs better than other RNN models. Since the dataset is small-sized, the CRF model performs better than RNN models.

When analysing errors, the words such as 'Mickey-Mouse' had been predicted as expressions by RNN models due to non alphanumeric symbols. In addition,

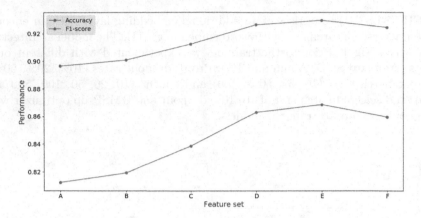

Fig. 2. CRF performance against cumulatively added feature sets from the set A to set G listed in Table 2

Table 3. Accuracy, Recall, Precision and F1-score of best performance of all models

Model	Acc.	Recall	Prec.	F1
CRF	**0.875**	**0.937**	**0.929**	**0.933**
W-LSTM	0.805	0.914	0.870	0.892
W-Bi-LSTM	0.824	0.917	0.890	0.903
W-CH-Bi-LSTM	0.811	0.916	0.875	0.895

the models exhibited poor performance in differentiating related numbers to solve the problem and unrelated numbers such as years in the text. The errors suggest that more semantic level features such as POS could help in increasing the accuracy.

6   Conclusion

In this research, the problem of extracting mathematical expressions from the unstructured plain text was modeled as a sequential text classification problem, an empirical evaluation was carried out on the state-of-the-art classifiers and a manually annotated dataset suitable to identify mathematical expressions in the text is presented. Given the dataset is small-sized, the achieved results are justifiable. While CRF performed the best, Bi-LSTM networks performed better than LSTM network.

LSTM with a CRF in the output layer helps to increase the accuracy of a sequential tagging task since it considers the possible transitions between output labels [8]. We hope to experiment with this in the future. After extracting mathematical expressions, it is useful to identify syntactically correct expressions separately. We plan to develop a rule-based parser to recognize any errors in filtered expressions, which can be used to give feedback to the user in end-user applications such automatic math problem solvers and e-learning systems.

Acknowledgment. This research was funded by a Senate Research Committee (SRC) Grant of the University of Moratuwa and LK Domain Registry.

References

1. Cetintas, S., Si, L., Xin, Y.P., Zhang, D., Park, J.Y.: Automatic text categorization of mathematical word problems. In: FLAIRS Conference (2009)
2. Chen, T., Xu, R., He, Y., Wang, X.: Improving sentiment analysis via sentence type classification using BiLSTM-CRF and CNN. Expert. Syst. Appl. **72**, 221–230 (2017)
3. Erabadda, B., Ranathunga, S., Dias, G.: Automatic identification of errors in multi-step answers to algebra questions. In: 2017 IEEE 17th International Conference on Advanced Learning Technologies (ICALT), pp. 215–219. IEEE (2017)
4. Fernando, K., Ranathunga, S., Dias, G.: Answer generation for set type math word problems. In: Proceedings of the 2018 International Conference of Advances in ICT for Emerging Regions (2018)
5. Finkel, J.R., Grenager, T., Manning, C.: Incorporating non-local information into information extraction systems by gibbs sampling. In: Proceedings of the 43rd Annual Meeting on Association for Computational Linguistics, pp. 363–370. Association for Computational Linguistics (2005)
6. Graves, A., Mohamed, A.r., Hinton, G.: Speech recognition with deep recurrent neural networks. In: 2013 IEEE International Conference on Acoustics, Speech and Signal Processing (ICASSP), pp. 6645–6649. IEEE (2013)
7. Hochreiter, S., Schmidhuber, J.: Long short-term memory. Neural Comput. **9**(8), 1735–1780 (1997)
8. Huang, Z., Xu, W., Yu, K.: Bidirectional LSTM-CRF models for sequence tagging. arXiv preprint arXiv:1508.01991 (2015)
9. Kadupitiya, J., Ranathunga, S., Dias, G.: Automated assessment of multi-step answers for mathematical word problems. In: 2016 Sixteenth International Conference on Advances in ICT for Emerging Regions (ICTer), pp. 66–71. IEEE (2016)
10. Lafferty, J., McCallum, A., Pereira, F.C.: Conditional random fields: Probabilistic models for segmenting and labeling sequence data (2001)
11. Lample, G., Ballesteros, M., Subramanian, S., Kawakami, K., Dyer, C.: Neural architectures for named entity recognition. arXiv preprint arXiv:1603.01360 (2016)
12. Ling, W., et al.: Finding function in form: compositional character models for open vocabulary word representation. arXiv preprint arXiv:1508.02096 (2015)
13. Ljubešić, N.: Comparing CRF and LSTM performance on the task of morphosyntactic tagging of non-standard varieties of South Slavic languages. In: Proceedings of the Fifth Workshop on NLP for Similar Languages, Varieties and Dialects (VarDial 2018), pp. 156–163 (2018)
14. Matsuzaki, T., Ito, T., Iwane, H., Anai, H., Arai, N.H.: Semantic parsing of pre-university math problems. In: Proceedings of the 55th Annual Meeting of the Association for Computational Linguistics (Volume 1: Long Papers), vol. 1, pp. 2131–2141 (2017)

15. Seo, M., Hajishirzi, H., Farhadi, A., Etzioni, O., Malcolm, C.: Solving geometry problems: Combining text and diagram interpretation. In: Proceedings of the 2015 Conference on Empirical Methods in Natural Language Processing, pp. 1466–1476 (2015)
16. Tian, X., Bai, R., Yang, F., Bai, J., Li, X.: Mathematical expression extraction in text fields of documents based on HMM. J. Comput. Commun. 5(14), 1 (2017)

A Study on Self-attention Mechanism
for AMR-to-text Generation

Vu Trong Sinh$^{(\boxtimes)}$ and Nguyen Le Minh$^{(\boxtimes)}$

Japan Advanced Institute of Science and Technology (JAIST), Nomi, Japan
{sinhvtr,nguyenml}@jaist.ac.jp

Abstract. Introduced by Vaswani *et al.*, transformer architecture, with
the effective use of self-attention mechanism, has shown outstanding performance in translating sequence of text from one language to another.
In this paper, we conduct experiments using the self-attention in converting an abstract meaning representation (AMR) graph, a semantic
representation, into a natural language sentence, also known as the task
of AMR-to-text generation. On the benchmark dataset for this task, we
obtain promising results comparing to existing deep learning methods in
the literature.

Keywords: Abstract meaning representation · Self attention ·
Text generation

1 Introduction

Abstract Meaning Representation (AMR) [5] is defined as a semantic representation language that encodes the core meaning of a sentence into a graph structure.
This graph is rooted, directed, edge-labeled and leaf-labeled. Every vertex and
edge of the graph are labeled according to the sense of the words in a sentence.
AMRs can be represented in several ways: graph structure for the computer to
store in its memory, or Penman notation for human to read and write with ease.
We give an example of AMR annotation for the sentence *"From among them,
pick out 50 for submission to an assessment committee to assess"*. As shown
in Table 1, the nodes (e.g. "thing", "pick-out-03", "access-01") represent concepts, and the edges (e.g. ":arg0", ":quant") represent relations between those
concepts.

AMR has been applied as an intermediate meaning representation for solving
various tasks in natural language processing (NLP) including machine translation [3], text summarization [6], event extraction [11], machine comprehension
[12]. To gain more success in those applications, the problem of AMR parsing
and AMR generation has to be solved effectively. While many approaches have
been proposed for the text-to-AMR parsing task, the number of published AMR-
to-text generation is comparably small. The generation task is non-trivial since
AMR graphs abstracted away tense, number as well as functional words such
as prepositions, articles. Recent methods for generating text from AMR based

E. Métais et al. (Eds.): NLDB 2019, LNCS 11608, pp. 321–328, 2019.
https://doi.org/10.1007/978-3-030-23281-8_27

on the success of deep learning encoder-decoder architecture, in which the input for the encoder side could be a sequence of AMR Penman notation or a graph structure.

Table 1. Abstract meaning representation for the sentence *"From among them, pick out 50 for submission to an assessment committee to assess"*

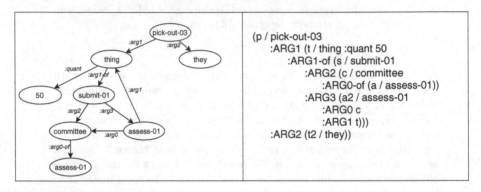

Inspired by the transformer network introduced by Vaswani *et al.* [17] that achieved outstanding performance on machine translation, we investigate the use of its core component, the self-attention mechanism, applying as a incorporated component to the encoder-decoder architecture in AMR generation problem. Our experimental results show comparative BLEU score on the newest release AMR dataset LDC2017T10.

2 Related Works

In this section, we give a short summary of previous methods in AMR generation using deep learning approaches. Pourdamghani *et al.* [10] was the first one who apply machine translation approach in AMR generation. They converted AMR graphs, which were written in Penman notation form, to a sequence of text through a linearization process. With these pairs of linearized AMRs and corresponding sentences, they considered AMR generation task as a machine translation problem and implemented a phrase-based model to obtain the final text.

Following this approach, Konstas *et al.* [4] proposed the first neural model for both the AMR parsing and AMR generation problem (NeuralAMR). The authors used an encoder-decoder model built upon a long short term memory (LSTM) neural network. As this architecture required a large set of training data to achieve good results, Konstas *et al.* (2017) used their own AMR parser to automatically annotate millions of unlabeled sentences before training their main system; the obtained AMR graphs are then used as additional training data. To deal with the problem of data sparsity addressing in Peng et al.

work [8], NeuralAMR adopt anonymization algorithm. In detail, they first replaced the subgraphs that represent open-class tokens (such as *"country :name name :op1 United :op2 States"*) with predefined placeholders (such as) before decoding, and then recovered the corresponding surface tokens (such as "United States") after decoding.

Different from the rule-based anonymization algorithm above, Song *et al.* [16] incorporated a char-level LSTM over character of input tokens and a copy network [2] on top of the decoder side. This architecture also helps generate the named entities, dates and numbers effectively. Song *et al.* also proposed a novel graph to sequence model (Graph2Seq), in which the authors encoded the AMR graph with a bidirectional LSTM encoder, performing through a graph-state transition. This graph encoder helped prevent information loss through the linearization process, especially when the graph become large. With the same amount of training data as NeuralAMR, this graph to sequence model achieved the state of the art result on a benchmark test set.

3 Incorporating the Self-attention Mechanism

3.1 The Baseline Model

We take both the sequence to sequence and graph to sequence model in Song *et al.* work [16] as our baseline model. The input for these models could be either an AMR graph or its linearized sequence from Penman notation. We keep using the char-level LSTM over input tokens as well as the copy network to tackle the data sparsity problem.

3.2 Self Attention Sequence to Sequence Model

Given an AMR graph in Penman notation, we use the linearization algorithm of Konstas *et al.* [4] to obtain a sequence of tokens $v_1, \ldots, v_n$, where n is the number of tokens. All variables and verb senses are removed from the annotation so that all vertices in the original graph could be considered as a normal word. For instance, the AMR notation shown in Table 1 can be linearized as follow: *"pick-out :arg1 (thing :quant 50 :arg1-of (submit :arg2 (committee :arg0-of assess) :arg3 (assess :arg0 committee :arg1 thing))) :arg2 they"*.

We follow the transformer architecture in Vaswani *et al.* work [17] to build the lower layers for our sequence to sequence model. Specifically, after processing the source tokens with two sub-layers: self-attention followed by a position-wise feed forward layer; and the target tokens with three sub-layers: self-attention followed by vanilla attention, followed by a position-wise feed forward layer. The self-attention outputs are sent to a bidirectional LSTM similar to the model designed in Song *et al.* work. The self-attention layers in the decoder side uses masking to prevent a given output position from incorporating information about future output positions during training.

In both the encoder and decoder side, self-attention sub-layers employ h attention heads. To form the sub-layer output, results from each head are concatenated and a parameterized linear transformation is applied. Each attention head operates on an input sequence of tokens, $v = (v_1, ..., v_n)$ of n elements where $v_i \in R^{d_v}$, and computes a new sequence $z = (z_1, ..., z_n)$ of the same length where $z_i \in R^{d_z}$.

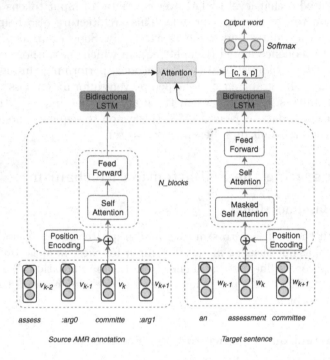

Fig. 1. AMR Sequence to sequence model incorporated with self-attention mechanism

The whole model architecture is represented in Fig. 1. When generating the t-th word, the decoder relies on the attention memory, the previous hidden state from the LSTM layers, the probability distribution to decide copying the word from source tokens or generating a new one (the value c, s and p in Fig. 1, respectively).

3.3 Self Attention Graph to Sequence Model

Dealing with the graph structure from AMR, we adopt the graph encoder in Song *et al.* [16]. For a graph $G = \{V, E\}$, We represent each node $v_i \in V$ by a hidden state vector h_i. The state of the graph can thus be represented as $g = \{h_i\}$. Information exchange between a current node v_i and all nodes connected to it are captured by a sequence of state transitions $\{g_0, g_1, ..., g_k\}$. In particular, the transition from g_{t-1} to g_t consists of a hidden state transition for

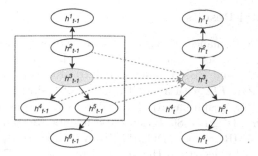

Fig. 2. Transition from graph state g_{t-1} to g_t, where information from the current node h_{t-1}^3, its incoming node h_{t-1}^2 and outgoing nodes h_{t-1}^4, h_{t-1}^5 are captured and transferred to h_t^3

each node h_{t-1}^i to h_t^i. Figure 2 shows a demonstration of graph state transition, detailed formulas can be found in the original paper. After k iterations, we obtain the last hidden state of the graph, containing all the hidden vectors of nodes in it, where k is the maximum graph diameter in the dataset. These node hidden vectors are then passed through the self-attention encoder similar to Sect. 3.2.

4   Experiments and Results

4.1   Dataset and Hyper-parameter Settings

We use the latest release AMR corpus (*LDC2017T10*) as our experimental dataset, which contains 36,521 instances for training, 1,368 for development and 1,371 for test. Each instance contains an English sentence and an AMR graph in Penman notation. Because of lacking hardware resources, we do not conduct our experiments on silver data sampled from external corpus (like NeuralAMR and Graph2Seq using Gigaword corpus).

In the experiments with sequence to sequence model, the vocabulary is shared between the encoder and the decoder. The word embeddings are initialized from Glove pretrained word embeddings [9] with embedding size is set to 300. For the graph to sequence experiments, we extract the edge label vocabulary to be used in the graph encoder. All the baseline hyper-paramenters for graph to sequence and sequence to sequence model are kept the same as in [16].

Table 2. Hyper-parameter settings

Word vocab size	27,876
Edge vocab size	119
Edge label dimension	100
d_{model}	300
N_{heads}	6
N_{blocks}	6
Feed forward dimension	1200

Other hyper-parameters for self-attention mechanism as well as dataset information can be found in Table 2. Following existing work, we evaluate the results with the BLEU metric [7].

4.2 Experimental Results

We compare the performance of self-attention incorporated models with the baseline model as well as other works in the literature. JAMR-generator (alignment-based) [1] and PBMT (phrase-based) [10] were trained on the old AMR corpus *LDC2014T12* with a small number of training samples. TSP [15] (graph-based), SNRG (graph-based) [14], NeuralAMR [4] and Graph2Seq [16] use a newer version, LDC2015E86. In our experiments, we train our models with the newest release AMR corpus, LDC2017T10, which keeps the same test set as LDC2015E86, but a bit less than the splitted test set in LDC2014T12 (1,400 samples). Since Song *et al.* did not publish their results of sequence to sequence model trained on gold data only, we use the default configuration to train a new one on *LDC2017T10* (Seq2Seq). Table 3 shows the BLEU scores of all the models on the test set.

From the result table, it can be recognized that Selfatt+Seq2Seq outperforms the baseline sequence to sequence model with nearly 3 BLEU score increased. This helps prove the effectiveness of incorporating the self-attention mechanism with the basic sequence to sequence model in the AMR generation task. However, combining the default transformer architecture with the graph to sequence approach does not bring high quality performance when the Selfatt+Graph2Seq obtains the slightly lower score than original Graph2Seq model. This result is probably due to the position encoding used in our architecture, which is naturally suitable for sequence of words rather than graph nodes.

Comparing to the full Graph2Seq model, NeuralAMR model as well as other traditional methods, our BLEU scores are still lower by a large margin. An experiment with hugh amount of training data and a fine tuning strategy must be conducted to have a more significant improvement (e.g Graph2Seq obtains more than 10 BLEU scores improvement after this process).

Table 3. BLEU scores on test set

Model	Corpus	Number of Training samples	BLEU score
Seq2Seq	LDC2017T10	36,521	15.49
Selfatt + Seq2Seq (Ours)	LDC2017T10	36,521	**18.36**
Graph2Seq	LDC2017T10	36,521	20.76[1]
Selfatt + Graph2Seq (Ours)	LDC2017T10	36,521	19.45
Graph2Seq	LDC2015E86 + Gigaword	16,833 + 2M	**33.0**
NeuralAMR	LDC2015E86 + Gigaword	16,833 + 2M	32.3
TSP	LDC2015E86	16,833	22.4
SNRG	LDC2015E86	16,833	25.6
JAMR-generator	LDC2014T12	10,000	22.0
PBMT	LDC2014T12	10,000	26.9

[1] We train the Graph2Seq model with the same setting as the publish source code, but the BLEU score is not as high as reported in the original paper (20.76 vs 22.7)

5   Conclusions and Future Works

In this paper, we investigated the use of self-attention mechanism in generating natural language from AMR graphs. We incorporated this mechanism with both sequence to sequence and graph to sequence baseline model. Our models obtained promising results compared to other deep learning approaches, but still far from the state of the art model due to the lack of training time.

For the future work, we would like to explore the use of pretrained model such as BERT, ELMO to have a better embedding representation for the input tokens. We also aim to build an graph to sequence transformer model for AMR generation by applying the relative position encoding introduced by Shaw *et al.* recently [13] as a replacement to the current positional encoding.

References

1. Flanigan, J., Dyer, C., Smith, N.A., Carbonell, J.: Generation from abstract meaning representation using tree transducers. In: Proceedings of the 2016 Conference of the North American Chapter of the Association for Computational Linguistics: Human Language Technologies, pp. 731–739. Association for Computational Linguistics (2016). https://doi.org/10.18653/v1/N16-1087, http://aclweb.org/anthology/N16-1087

2. Gu, J., Lu, Z., Li, H., Li, V.O.: Incorporating copying mechanism in sequence-to-sequence learning. In: Proceedings of the 54th Annual Meeting of the Association for Computational Linguistics, Long Papers, vol. 1, pp. 1631–1640. Association for Computational Linguistics (2016). https://doi.org/10.18653/v1/P16-1154, http://aclweb.org/anthology/P16-1154

3. Jones, B., Andreas, J., Bauer, D., Moritz Hermann, K., Knight, K.: Semantics-based machine translation with hyperedge replacement grammars. In: 24th International Conference on Computational Linguistics - Proceedings of COLING 2012: Technical Papers, pp. 1359–1376, December 2012

4. Konstas, I., Iyer, S., Yatskar, M., Choi, Y., Zettlemoyer, L.: Neural AMR: Sequence-to-sequence models for parsing and generation. In: Proceedings of the 55th Annual Meeting of the Association for Computational Linguistics, Long Papers, vol. 1, pp. 146–157. Association for Computational Linguistics (2017). https://doi.org/10.18653/v1/P17-1014, http://aclweb.org/anthology/P17-1014

5. Banarescu, L., et al.: Abstract meaning representation for sembanking. In: Proceedings of the 7th Linguistic Annotation Workshop and Interoperability with Discourse, pp. 178–186 (2013)

6. Liu, F., Flanigan, J., Thomson, S., Sadeh, N., Smith, N.A.: Toward abstractive summarization using semantic representations. In: NAACL, pp. 1077–1086 (2015)

7. Papineni, K., Roukos, S., Ward, T., Zhu, W.J.: Bleu: a method for automatic evaluation of machine translation. In: Proceedings of 40th Annual Meeting of the Association for Computational Linguistics, pp. 311–318. Association for Computational Linguistics, Philadelphia, July 2002. https://doi.org/10.3115/1073083.1073135, https://www.aclweb.org/anthology/P02-1040

8. Peng, X., Wang, C., Gildea, D., Xue, N.: Addressing the data sparsity issue in neural AMR parsing. In: Proceedings of the 15th Conference of the European Chapter of the Association for Computational Linguistics, Long Papers, vol. 1, pp. 366–375. Association for Computational Linguistics (2017), http://aclweb.org/anthology/E17-1035

9. Pennington, J., Socher, R., Manning, C.: Glove: global vectors for word representation. In: Proceedings of the 2014 Conference on Empirical Methods in Natural Language Processing (EMNLP), pp. 1532–1543. Association for Computational Linguistics (2014). https://doi.org/10.3115/v1/D14-1162, http://aclweb.org/anthology/D14-1162

10. Pourdamghani, N., Knight, K., Hermjakob, U.: Generating English from abstract meaning representations. In: Proceedings of the 9th International Natural Language Generation Conference, pp. 21–25. Association for Computational Linguistics (2016). https://doi.org/10.18653/v1/W16-6603, http://aclweb.org/anthology/W16-6603

11. Rao, S., Marcu, D., Knight, K., Daumé III, H.: Biomedical event extraction using abstract meaning representation. BioNLP **2017**, 126–135 (2017)

12. Sachan, M., Xing, E.: Machine comprehension using rich semantic representations. In: Proceedings of the 54th Annual Meeting of the Association for Computational Linguistics, Short Papers, vol. 2, pp. 486–492. Association for Computational Linguistics (2016). https://doi.org/10.18653/v1/P16-2079, http://www.aclweb.org/anthology/P16-2079

13. Shaw, P., Uszkoreit, J., Vaswani, A.: Self-attention with relative position representations. In: Proceedings of the 2018 Conference of the North American Chapter of the Association for Computational Linguistics: Human Language Technologies, (Short Papers), vol. 2 pp. 464–468. Association for Computational Linguistics (2018). https://doi.org/10.18653/v1/N18-2074, http://aclweb.org/anthology/N18-2074

14. Song, L., Peng, X., Zhang, Y., Wang, Z., Gildea, D.: AMR-to-text generation with synchronous node replacement grammar. In: Proceedings of the 55th Annual Meeting of the Association for Computational Linguistics, Short Papers, vol. 2, pp. 7–13. Association for Computational Linguistics (2017). https://doi.org/10.18653/v1/P17-2002, http://aclweb.org/anthology/P17-2002

15. Song, L., Zhang, Y., Peng, X., Wang, Z., Gildea, D.: AMR-to-text generation as a traveling salesman problem. In: Proceedings of the 2016 Conference on Empirical Methods in Natural Language Processing, pp. 2084–2089. Association for Computational Linguistics (2016). https://doi.org/10.18653/v1/D16-1224, http://aclweb.org/anthology/D16-1224

16. Song, L., Zhang, Y., Wang, Z., Gildea, D.: A graph-to-sequence model for AMR-to-text generation. In: Proceedings of the 56th Annual Meeting of the Association for Computational Linguistics, Long Papers, vol. 1, pp. 1616–1626. Association for Computational Linguistics (2018). http://aclweb.org/anthology/P18-1150

17. Vaswani, A., et al.: Attention is all you need. In: Advances in Neural Information Processing Systems, pp. 5998–6008 (2017)

PreMedOnto: A Computer Assisted Ontology for Precision Medicine

Noha S. Tawfik[1,2]([✉]) and Marco R. Spruit[2]

[1] Computer Engineering Department, College of Engineering,
Arab Academy for Science, Technology, and Maritime Transport (AAST),
Alexandria 1029, Egypt
noha.abdelsalam@aast.edu
[2] Department of Information and Computing Sciences, Utrecht University,
3584 CC Utrecht, The Netherlands
{n.s.tawfik,m.r.spruit}@uu.nl

Abstract. This paper proposes an ontology learning framework that combines text mining, information extraction and retrieval. The proposed model takes advantage of existing structured knowledge by reusing terms and concepts from other ontologies. We further apply the methodology to create a detailed ontology for the emerging precision medicine (PM) domain by collecting a corpus of relevant articles and mapping its frequent terms to existing concepts. The resulting ontology consists of 543 annotated classes. The ontology was also tested for effectiveness by applying two evaluation frameworks to validate its design and quality. The results demonstrate that the ontology learning system is able to capture and represent the semantics of the PM domain with high precision and significance. Moreover, the computer-assisted construction process reduced dependency on expert knowledge. The developed *PreMedOnto* ontology could be further used to enhance the potentials of other NLP applications in the PM domain.

Keywords: Precision medicine · Data mining · Ontology reuse

1 Introduction

Ontologies are data models that transform domain's data into machine-readable representations to describe how a domain's information is organized. We adopt its original definition by Gruber as "An explicit specification of a conceptualization" [13]. By definition, they capture a wide variety of rich semantics by organizing knowledge into a hierarchy of concepts and relationships. It is considered one of the most reliable data representation models in today's semantic world, however, manual ontology development is an expensive task, both in terms of time and money. Ontology learning is the process of creating new ontologies from scratch whereas ontology population is concerned with augmenting existing ontologies with instances and properties. Both tasks require deploying efficient techniques to automatically process enormous amounts of domain-specific,

© Springer Nature Switzerland AG 2019
E. Métais et al. (Eds.): NLDB 2019, LNCS 11608, pp. 329–336, 2019.
https://doi.org/10.1007/978-3-030-23281-8_28

unstructured resources. While the latter task is hard, the former task is particularly challenging as computer models must closely mimic domain experts in interpreting meanings for constructing the ontology [7] and are usually accompanied by efficiency and precision issues. An alternative to overcome such limitations is to take advantage of existing knowledge bases, as not only it would minimize the human factor, but it would potentially achieve better precision and reduce redundancy [6]. Reusing contents would also guarantee a consistent representation of domain knowledge given the quality of the source ontology. The practice is quite established as part of the Web Ontology Language (OWL) specification and is also supported by the Open Biological and Biomedical Ontology (OBO) Foundry [17]. This study focuses on building an ontology for the precision medicine (PM) domain. The PM approach seeks to identify the best and the most effective practices for patients based on their genetic, environmental, and lifestyle factors. Although the concept has been around for many years, recently there has been an increase of public research funding and dedication to adopt the concept into practice versus the 'one-size-fits-all' method. Accordingly, there has been a substantial increase in the number of publications related to the PM concept [22]. However, the PM domain lacks a clear and organized hierarchy of its general, investigations, diagnostics and treatments' terminologies. The main contribution of this research is the compilation and development of the precision medicine ontology (*PreMedOnto*). Such an ontology helps in defining and shaping the precision medicine domain and its related vocabulary which improves the understanding of the field.

2   Related Work

In the recent years, ontology has become a preferable way to represent biological data [2]. There is a great amount of published research in the ontology engineering field, however, our survey is only limited to ontology engineering models built for the medical domain. Casteleiro et al. was able to build an ontology for the sepsis disease from an unannotated biomedical corpus. Their model used Latent Semantic Analysis (LSA) and Latent Dirichlet Allocation (LDA), as well as the neural language models Continuous Bag-of-Words (CBOW) and Skip-grams [5]. They also exploited the same model to enrich the cardiovascular diseases ontology (CVDO) from PubMed articles. A reuse-based method was proposed by Gedzelman et al. to construct another ontology for cardiovascular diseases [12] using UMLS and MeSH thesaurus. Cahyani and Wasito investigated the use of Ontology Design Patterns (ODP) to construct an Alzheimer's Disease ontology. Their model uses existing vocabulary and glossary to extract terms and relations from published articles and match them against the patterns [8]. Another Alzheimer's disease ontology was developed by Drame et al. [9], they cluster bilingual terms from English and French corpora, according to the UMLS thesaurus, and align them by integrating new concepts. In [16], the authors propose a framework for updating existing medical ontologies. Their approach consists of 4 steps: extract relevant terms, apply machine learning techniques to infer polysemy, detect the concepts related to the term using clustering algorithms and finally, link terms to the exact positions in the ontology. Gao, Chen and Wang also suggested a model for extending ontologies [11] and applied it to the PHARE ontology. Their research took advantage of PMC repository to train a word2Vec model and uses random indexing to enrich ontology labels. In [15], Kang et al. attempted to tailor the general adverse event ontology to build specific diseases ontology (DSOAE). They used design patterns and addressed the specifications needed for the chronic kidney disease by adding new classes, relations and properties. Another model was proposed in [14], where the authors reused the existing GALEN ontology to build a specific ontology for the juvenile rheumatoid arthritis disease. Their semi-automatic approach relies on extracting relevant parts of the old ontology and refine them to ensure consistency and safety so that the semantics of imported concepts are not changed. Amato et al. [4] populated an ontology constructed by a domain expert with RDF templates extracted from medical records. Sanchez and Moreno [19] suggest a web based approach for building medical ontologies from scratch. It uses a set of user query words to collect web documents. Documents with the highest web search hit counts are considered valid taxonomic specialization for the domain. Named entities and verbs are then extracted to generate one-level taxonomy with general terms. The next stage is non-taxonomic learning where the extracted verbs are used as domain patterns and again used as web queries. Finally, the verb phrase is used to link each pair of concept. In [3], Alobaidi et al. combined UMLS thesaurus and Linked Open Data (LOD) classes to identify medical concepts and associate them to their corresponding formal semantics. Shah et al. constructed a framework based on MetaMap and SemRep to reuse terms from SNOMED-CT ontology. They applied the framework to construct an ontology that combines the dental and medical domain to allow better reasoning over common knowledge [20].

3   Methods

3.1   Proposed Model

Our ontology learning methodology is based on the concept of ontology reuse, where we adapt content from existing ontologies to model the PM domain. The model also relies on the assumption that the concepts that must be included in the ontology are mapped from the frequently mentioned terms present in the domain-specific data. And their co-occurrences frequency depicts the relations among them. To successfully achieve this goal, our proposed framework consists of 5 phases, Fig. 1 illustrates the overall learning process overview.

Knowledge Acquisition. In our work, we used a publicly available list of PM keywords and synonyms constructed by conducting a systematic search through multiple web resources, including: academic, news and health websites. As this list is manually compiled and verified, we refer to it as the PM vocab. The list is divided into three categories: keywords and synonyms for personalized medicine,

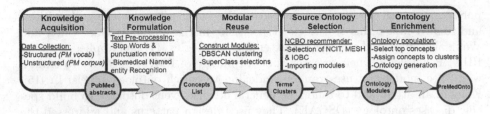

Fig. 1. Overview of the ontology learning framework.

keywords and synonyms for personal genomics and keywords and synonyms for diagnostics, biomarkers and testing. More details on the creation of the vocabulary could be found in [1]. In this paper we only use the last category since we aim at modelling the PM domain from a clinical and scientific point of view. In addition, we collected all titles and abstracts included in the PubMed repository discussing the PM concept. All articles included in PubMed are associated with Medical Subject Headings (MeSH) terms used for indexing articles. The search query used was "precision medicine" [Majr], adding the [Majr] term next to the original query restricts the search engine to return citations where the PM concept is the major focus of the article. In scientific literature, medical terminology is usually used interchangeably to describe the same concept. The MeSH entry terms or cross-references ensure that closely related terms and synonyms are all included when querying a certain term. In our case, the entry list has other terms such as Personalized Medicine and Individualized Medicine. The collection process was conducted through the Bio Python package that connects to NCBI E-utilities to retrieve and download articles. The results are then filtered so that all records with missing or incomplete abstract texts or in a foreign language other than English are excluded. This resulted in a total of 5,206 articles that serve as the *PM corpus*.

Knowledge Formulation. We preprocess all abstracts in the *PM corpus* to filter out stop words, symbols and punctuation. Due to the ambiguity of reporting biological or clinical results, MetaMap[1] was used for medical entity recognition. The output at this stage is a set of 6,832 distinct terms and concepts from the corpus. To guarantee precision, we do not map all terms extracted as this could lead to ambiguity and inconsistency in representing the domain knowledge. All terms mentioned more than once are ranked in descending order of their occurrence frequencies. Extracted terms are included only if their mention count exceeded a threshold. The threshold value is calculated as the weighted average occurrences of terms in documents to ensure that less significant words are removed.

Modular Reuse. In this stage, the *PM vocab* is used to create seed ontology modules where terms are mapped to a set of disjoint clusters. We started by analyzing the terms included in the *PM vocab* according to their relevance

---

[1] https://metamap.nlm.nih.gov/.

and commonness. We built a symmetric matrix of cosine similarity scores for every pair of word vectors that exist in the vocabulary. The word embeddings model was pretrained over a set of over 10 million biomedical articles from PubMed. The matrix was fed to a density-based spatial clustering of applications with noise (DBSCAN) clustering algorithm implemented through the Scikit library. We opted for the DBSCAN clustering algorithm since it allows unsupervised learning over data and does not require the number of clusters a priori. This process created a total of 5 clusters. Following the creation of clusters, we rank all terms included according to their centrality and create one module per cluster. The top ranked concepts per cluster serve as the ontology super-classes. The original *PM vocab* set contained 100 terms that refer to diagnostic and testing procedures. Out of the 100 terms, only 73 were correctly clustered while 27 terms were regarded as noise by the clustering algorithm. Among the top candidate terms for each cluster, 25 were mapped as parent and child classes. Finally, we add all the non-used terms from the PM vocab to the list of concepts extracted from the *PM corpus*.

Source Ontology Selection. It is critical to determine the correct ontology that can serve as the base of the newly developed *PreMedOnto*. The criteria of choosing the ontology include coverage, acceptance and semantic language used. The NCBO ontology recommender is employed to suggest the best ontology for each module over all 895 existing ontologies. To maximize the coverage factor, we opted for the ontology set option which returns the best set of combined ontologies. The weights configuration for the recommender scoring function was set to the default settings. The final ranking of ontologies to be reused was: National Cancer Institute Thesaurus (NCIT)[2] , Medical Subject Headings (MeSH)[3] and Interlinking Ontology for Biological Concepts (IOBC)[4]. From the selected ontologies, we import all candidate classes with their ancestors, and verify that all remaining concepts per cluster are included in the module as child nodes. All redundant concepts in the *PM vocab* are removed by checking synonyms of each imported class.

Ontology Enrichment. In the final stage, each module is enriched by assigning relevant concepts extracted from the *PM corpus* in the knowledge formulation phase. We first extract the Uniform Resource Identifier (URI) corresponding to each concept. The ontofox [21] tool supports efficient ontology reuse by extending the Minimum Information to Reference an External Ontology Term MIREOT concept. The MIREOT approach favors selective class imports instead of importing the ontology as a whole. The ontofox web tool takes as input the base ontology, source terms URIs and parent classes URIs. It also allows users to choose the appropriate settings of the import process such as importing or omitting intermediate classes between input child and parent or deciding which annotation properties to return.

---

[2] https://bioportal.bioontology.org/ontologies/NCIT.
[3] https://bioportal.bioontology.org/ontologies/MESH.
[4] https://bioportal.bioontology.org/ontologies/IOBC.

3.2 Evaluation

Assessing the ontology output is a key factor in all ontology learning techniques. Not only to ensure the ontology quality before referencing and adopting it in other semantics-aware applications, but also to highlight errors and shortcomings. There are two different evaluative perspectives: ontology quality and ontology correctness. In this research, we carried out a two-fold evaluation process to measure the effectiveness of the constructed ontology: the first experiment assesses the ontology design whereas the second computes multiple quality features. To detect any design error in *PreMedOnto*, we use OntOlogy Pitfall Scanner (OOPS) online tool [18]. OOPS evaluates an OWL ontology against a catalogue of common mistakes in ontology The tool produces a summary of all pitfalls found within the ontology with extended information on each and a label indicating its importance level. We also apply the ontology quality evaluation framework (OQauRE) [10] to validate the quality of classes and axioms in *PreMedOnto*. OQauRe is a quantitative method based on the original software product quality requirements and evaluation concept. The framework computes multiple quality characteristics including structure, quality in use, reliability, compatibility, maintainability, operability, functional adequacy, transferability, performance efficiency. The generated metrics are mapped to quantitative values ranging from 1 to 5 with 3 is the minimum score and considered as accepted.

4 Results

The final output of the ontology learning process is the *PreMedOnto* in the standard OWL format. A total of 543 classes imported from 3 medical ontologies. Table 1 provides a brief summary of some of its metrics. The ontology can be accessed, viewed and downloaded from http://bioportal.bioontology.org/ontologies/PREMEDONTO.

Table 1. Summary of the *PreMedOnto* metrics generated by the Protégé framework.

Metric		Metric	
Classes	543	Classes with a single child	111
Average number of children	3	Maximum number of children	90
Properties	10	Maximum depth	7

The obtained results of evaluating *PreMedOnto* against the 41 pitfalls included in OOPS's catalogue, show that the ontology is free from critical and important pitfalls while there exist 3 cases of minor pitfalls. The former finding ensures the consistency and sustainability of the ontology, while the later might suggests corrections for better organization. The pitfalls detected are related to missing annotations, lack of connectivity and inverse relationship declaration. However, we find them irrelevant, as they do not threaten the functionality of the ontology. The second experiment provides quantitative indicators of the quality of *PreMedOnto*. The computed scores for structure, compatibility and maintainability metrics were 3.5, 4.2 and 4.5 respectively. The ontology has successfully passed the minimal level required and is considered above average in most characteristics. It is worthy to mention that each quality measure is also associated to multiple sub-characteristics and hence indicates multiple quality aspects.

5   Conclusions

PreMedOnto is an application ontology built for the precision medicine domain on top of gold standard biomedical ontologies. The ontology learning process involves mining the PubMed repository to extract domain specific abstracts and vocabulary as sources of data. The information gathered is clustered and outlined to determine main modules. It reuses terms and concepts from NCIT, MeSH and IOBC to construct the ontology hierarchy. The evaluations demonstrate that the ontology content is reliable and consistent. We also plan to add a possible extra experiment to validate the ontology utility and applicability in the PM domain. The intended experiment involves human validation of the ontology by medical experts through a survey of questions.

References

1. Ali-Khan, S., Kowal, S., Luth, W., Gold, R., Bubela, T.: Terminology for personalized medicine: a systematic collection terminology for personalized medicine. Technical report (2016)
2. Alobaidi, M., Malik, K.M., Hussain, M.: Automated ontology generation framework powered by linked biomedical ontologies for disease-drug domain. Comput. Methods Programs Biomed. **165**, 117–128 (2018). https://doi.org/10.1016/j.cmpb.2018.08.010
3. Alobaidi, M., Malik, K.M., Sabra, S.: Linked open data-based framework for automatic biomedical ontology generation. BMC Bioinform. **19**(1), 319 (2018). https://doi.org/10.1186/s12859-018-2339-3
4. Amato, F., Santo, A.D., Moscato, V., Picariello, A., Serpico, D., Sperli, G.: A lexicon-grammar based methodology for ontology population for e-health applications. In: 2015 Ninth International Conference on Complex, Intelligent, and Software Intensive Systems. pp. 521–526. IEEE, July 2015. https://doi.org/10.1109/CISIS.2015.76
5. Arguello Casteleiro, M., et al.: Deep learning meets ontologies: experiments to anchor the cardiovascular disease ontology in the biomedical literature. J. Biomed. Semant. **9**(1), 13 (2018). https://doi.org/10.1186/s13326-018-0181-1
6. Bontas, E.P., Mochol, M., Tolksdorf, R.: Case Studies on Ontology Reuse. Technical report
7. Buitelaar, P., Cimiano, P., Magnini, B.: Ontology learning from text: methods. Eval. Appl. (2005). https://doi.org/10.1162/coli.2006.32.4.569
8. Cahyani, D.E., Wasito, I.: Automatic ontology construction using text corpora and ontology design patterns (ODPs) in Alzheimer's disease. Jurnal Ilmu Komputer dan Informasi **10**(2), 59 (2017). https://doi.org/10.21609/jiki.v10i2.374

9. Dramé, K., et al.: Reuse of termino-ontological resources and text corpora for building a multilingual domain ontology: an application to Alzheimer's disease. J. Biomed. Inform. **48**, 171–182 (2014). https://doi.org/10.1016/J.JBI.2013.12.013

10. Duque-ramos, A., Duque-ramos, A., Fernández-breis, J.T., Stevens, R., Aussenac-gilles, N.: OQuaRE: a SQuaRE-based approach for evaluating the quality of ontologies. J. Res. Pract. Inf. Technol. **43**, 159 (2011)

11. Gao, M., Chen, F., Wang, R.: Improving Medical Ontology Based on Word Embedding (2018). https://doi.org/10.1145/3194480.3194490

12. Gedzelman, S., Simonet, M., Bernhard, D., Diallo, G., Palmer, P.: Building an ontology of cardio-vascular diseases for concept-based information retrieval. In: Computers in Cardiology, 2005, pp. 255–258. IEEE (2005). https://doi.org/10.1109/CIC.2005.1588085

13. Gruber, T.R.: A translation approach to portable ontology specifications. Knowl. Acquisition **5**(2), 199–220 (1993). https://doi.org/10.1006/KNAC.1993.1008

14. Jiménez-Ruiz, E., Cuenca Grau, B., Sattler, U., Schneider, T., Berlanga, R.: Safe and Economic Re-Use of Ontologies: A Logic-Based Methodology and Tool Support. Technical report

15. Kang, Y., Fink, J.C., Doerfler, R., Zhou, L.: Disease specific ontology of adverse events: ontology extension and adaptation for chronic kidney disease. Comput. Biol. Med. **101**, 210–217 (2018). https://doi.org/10.1016/J.COMPBIOMED.2018.08.024

16. Lossio-Ventura, J.A., Jonquet, C., Roche, M., Teisseire, M.: A Way to Automatically Enrich Biomedical Ontologies. https://doi.org/10.5441/002/edbt.2016.82

17. Ochs, C., Perl, Y., Geller, J., Arabandi, S., Tudorache, T., Musen, M.A.: An empirical analysis of ontology reuse in BioPortal. J. Biomed. Inform. **71**, 165–177 (2017). https://doi.org/10.1016/J.JBI.2017.05.021

18. Poveda-Villalón, M., Carmen Suárez-Figueroa, M., Ángel García-Delgado, M., Gómez-Pérez, A.: OOPS! (OntOlogy Pitfall Scanner!): supporting ontology evaluation on-line. Technical report (2009)

19. Sánchez, D., Moreno, A.: Learning medical ontologies from the Web. Technical report

20. Shah, T., Rabhi, F., Ray, P., Taylor, K.: A guiding framework for ontology reuse in the biomedical domain. In: 2014 47th Hawaii International Conference on System Sciences, pp. 2878–2887. IEEE January 2014. https://doi.org/10.1109/HICSS.2014.360

21. Xiang, Z., Courtot, M., Brinkman, R.R., Ruttenberg, A., He, Y.: OntoFox: web-based support for ontology reuse. BMC Res. Notes **3**(1), 175 (2010). https://doi.org/10.1186/1756-0500-3-175

22. Yates, L.R., et al.: The european society for medical oncology (ESMO) precision medicine glossary. Ann. Oncol. **29**(1), 30–35 (2018). https://doi.org/10.1093/annonc/mdx707

An Approach for Arabic Diacritization

Ismail Hadjir[1,2(✉)], Mohamed Abbache[3(✉)],
and Fatma Zohra Belkredim[4(✉)]

[1] Mathematics and Computing Department, Faculty of Sciences,
Dr. Yahia Fares University of Medea, Medea, Medea Province, Algeria
hadjir.ismail@univ-medea.dz
[2] Linguistics Department, Faculty of Literature,
Algiers2 University of Bouzareah, Algiers, Algeria
[3] icrOKids CEO, Tianjin, China
m.abbache@yahoo.fr
[4] Mathematics and Its Applications Laboratory, Faculty of Exact Sciences
and Computing, Hassiba Ben Bouali University of Chlef,
Ouled Fares, Chlef Province, Algeria
f.belkredim@univ-chlef.dz

Abstract. Modern Standard Arabic (MSA) contains optional diacritical marks (diacritics, in Arabic harakat), which became less used in Arabic books, newspapers and other written media. Diacritics are very important for readability and understandability of texts. Their absence causes critical problems that add to the lexical, morphological and semantic ambiguities. In this paper, we present an automatic diacritization system of the Arabic language, using Hidden Markov Models with the Viterbi's algorithm, based on probabilities based on learning on diacritized Arabic texts. The corpus used was mostly composed of religious texts. Our results were satisfactory, achieving a precision of up to 80% at the word level.

Keywords: Diacritization · Modern Standard Arabic ·
Hidden Markov Models · Viterbi algorithm

1 Introduction

Arabic writing has appeared in the form of letters without diacritics, which is what the Arab person currently reads in books, newspapers, advertisements and on the Internet. The diacritic marks in Arabic calligraphy represent the tones in the Chinese language or vowels in the French and English languages, that is to say, they make it possible to specify the pronunciation and the meaning of a word (e.g., if we compare the two words porte and portée, they have different pronunciations and meanings, but if we exclude the vowels (e and ée), the two words will become (port), the result is another word having a different pronunciation and meaning from the initial two words. This is what happens in MSA. Diacritics denote "dhamaó", "fathaó", "kassraǫ"", "souk-ounó", "chaddaó" and "tanwinǫóó". Changing a diacritics of the letters composing an Arabic word, changes the meaning of the word, for example the word "عَلَ" means "to

© Springer Nature Switzerland AG 2019
E. Métais et al. (Eds.): NLDB 2019, LNCS 11608, pp. 337–344, 2019.
https://doi.org/10.1007/978-3-030-23281-8_29

know" if we put "kasra" for the last character and "soukoun" for the second one, it becomes "عِلْم" which mean "science". The lexical, morphological and semantic ambiguities caused by no-diacritized Arabic texts become a challenging problem in Arabic NLP, since several companies and researchers have been involved in this field and have proposed multiple solutions to solve it. They tried to develop an automatic diacritization system of Arabic texts, which can be used in translation systems, speech recognition, and so on.

The remaining of this paper is organized as follows: Sect. 2 describes the related works done in the field of automatic diacritization systems; Sect. 3 presents the modelization of our automatic system; Sect. 4 presents implementation of the system and the results evaluation. Finally, we conclude with Sect. 5.

2 Related Works

Lots of efforts have been made on Arabic diacritization (diacritics restoration or vowelization) by using rule-based, statistical, and hybrid approaches.

In 1998, Mustafa [16] used four algorithms to search for Arabic words diacritized or not diacritized. In 2000, Chelba et al. [4] proposed a method for automatic analysis of components in an English text, using sequence probability. In 2001, Goweder et al. [8] performed statistical analysis on the words repetition on texts extracted from the "Al-Hayat" Newspaper; they constructed an 18.5 million word corpus, with articles tagged as belonging to one of 7 domains. they outlined the profile of the data and how they assessed its representativeness. Furthermore, Kontrovich et al. [10] used diacritized texts of Hebrew texts; their system was based on an independent dictionary, the function of the word and HMM Models. In 2002, Gal [6] used a Hidden Markov Model (HMM) for analyzing the Arabic diacritized texts, after that he builds a model for automatic Arabic text diacritization. Smrz et al. [15] developed a research in which the topics discussed include linguistic data retrieval, morphology and morphotactics modeling using n-gram models, and description of the language on the analytical level. In 2012, Hamdi [1] used the morphosyntactic analyzer MAD for diacritization and other works and Khorsheed [18] presented a system for Arabic language diacritization using Hidden Markov Models (HMMs). The system employs the renowned HMM Tool Kit (HTK). Each single diacritic was represented as a separate model. The concatenation of the output models was coupled with the input character sequence to form the fully diacritized text; The data corpus used, includes more than 24000 sentences. In 2014, Bebah et al. [11] used a statistical approach based on two hidden Markov Models (HMM) by Viterbi algorithm and Al Khalil Morphosys. In 2015, Abandah et al. [7] presented a sequence transcription approach for the automatic diacritization of Arabic text and Hadj Ameur et al. [20] presented a new approach to restore Arabic diacritics using a statistical language model and dynamic programming. Their system was based on two models: a bi-gram-based model which was first used for vocalization and a 4-gram character-based model which was then used to handle the words that remain non vocalized (OOV words). The optimal vocalized word sequence was selected using the Viterbi algorithm from Dynamic Programming. In addition, Azmi et al. [3] produced a survey of the recent algorithms developed to solve the diacritization problems. In 2017,

Zerrouki et al. [17] created a corpus for Arabic diacritized texts, called "Tashkeela". Alnefaie et al. [13] created a system of diacritization that restores the diacritical markings. Furthermore, Alansary [14] presented an approach to Arabic automatic diacritisation called "Alserag" and Darwish et al. [5] presented a new and fast state-of-the-art Arabic diacritizer that guesses the diacritics of words and then their case endings. In addition, Fashwan et al. [2] created an automatic diacritization system for Standard Modern Arabic texts called "Shakkil"; it was based on a hybrid approach and they obtained 1.88% as diacritic error rate (DER), and 9.36% as Word error rate (WER) and finally Diab et al. [12] investigate the impact of Arabic diacritization on statistical machine translation (SMT). In 2018, Darwish and al. [19] presented their research and benchmark results on the automatic diacritization of two Maghrebi sub-dialects, namely Tunisian and Moroccan, using Conditional Random Fields (CRF). Aside from using character n-grams as features, they also employed character-level Brown clusters. They achieved word-level diacritization errors of 2.9% and 3.8% for Moroccan and Tunisian respectively. Also Jarrar et al. [21] proposed the Subsume knowledge-based algorithm, the Imply rule-based algorithm, and the Alike machine-learning-based algorithm. They evaluated the soundness, completeness, and accuracy of the algorithms against a large dataset of 86,886 word pairs.

3 Diacritization Processing

The diacritization of texts need corpus. The Corpus used is Tashkeela [17]. The corpus is freely available; it contains 75 million of fully vocalized words, mainly 97 books from classical and modern Arabic language. The corpus is collected from manually vocalized texts using a web crawling process, and is mostly composed of Islamic classical books. From this corpus we took 26 Books for the learning operation.

3.1 System Overview

The proposed solution is based on the Hidden Markov models (HMM), on the automatic learning and the algorithm of Viterbi. A Markov chain is useful when we need to compute a probability for a sequence of observable events. Hidden Markov model (like words that we see in the input) and hidden events (like part-of-speech tags) that we think of as causal factors in our probabilistic model.

The most common decoding algorithms for HMMs is the Viterbi algorithm. It is a kind of a dynamic programming algorithm that makes uses of a dynamic programming trellis. The idea is to process the observation sequence left to right, filling out the trellis. The result given is the most probable path by taking the maximum over all possible previous state sequences [9].

The system described in this paper and summarized in Fig. 1, receives a no-diacritized Arabic text as an input data and returns the same text diacritized as an output result. We noticed that a sentence in Arabic language is a structure, that each word's diacritics represent a state, The passage from one state to another is a transition and each word represents the observation which is generated by the state. After computing the transitions and generation of probabilities, we can simulate the obtained structure to an HMM, by applying the Viterbi's Method to find the states sequences as shown in Fig. 2.

340   I. Hadjir et al.

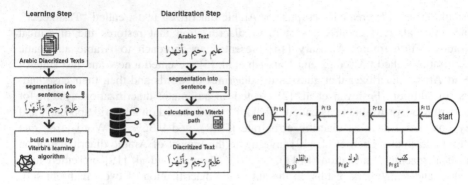

Fig. 1. Diacritization system   **Fig. 2.** Diacritized sentence modeling.

We create an automatic learning system based on a rich corpus for generating the HMM, which will be used in the next steps by our automatic diacritization system in order to know the most probable diacritization for the input text.

3.2 Pre-processing

To make the system faster, we create a method for converting the string to a number. In diacritics case we convert the mark to the mark position as shown in Table 1.

Table 1. Diacritics position

1	2	3	4	5	6	7	8	9
	ٗ	ٗ	ٗ	ٗ	ٗ	ٗ	ٗ	ٗ

For Arabic words, we use Eq. (1), to get the word's position by computing an index to each Arabic character (e.g. كتب = 37952) (Table 2).

$$f(x) = \sum_{i=1}^{l(x)} pos(x_i) * 36^{l(x)-i} \tag{1}$$

Table 2. Alphabetical character position

1	2	3	4	5	6	7	8	9	10	11	12
ء	آ	أ	ؤ	إ	ئ	ا	ب	ة	ت	ث	ج
13	14	15	16	17	18	19	20	21	22	23	24
ح	خ	د	ذ	ر	ز	س	ش	ص	ض	ط	ظ
25	26	27	28	29	30	31	32	33	34	35	36
ع	غ	ف	ق	ك	ل	م	ن	ه	و	ى	ي

x: is the word; i: is the count variable; $l(x)$: is the character number of a word; *pos:* is the character position; and x_i: is character at the position i.

3.3   The Learning Step

This method consists of filling the sets of States, Observations, Transitions and Generations, which permit the system to learn. We have done a learning on the obtained structure, using the Viterbi's algorithm. The learning process is divided into 04 principle steps:

1. Reading of documents from the given corpus.
2. Each line from the previous documents is segmented into sentences.
3. Each sentence is sent to the learning method.
4. All sets of states, observations, transitions and generations are saved in our Database.

For the learning operation, we used 26 documents (see Table 3) from our corpus. The learning operation lasted 169 s. The size of the corpus used was 292Mo; it contained 18 667 588 words with repetitions. At the end of the learning operation we obtained: 25 640 states, 469 532 generations, 272 681 words (without repetitions) and 1 546 404 transitions.

Table 3. Learning corpus

شرح البهجة الوردية.txt	شرح السير الكبير.txt	شرح حدود ابن عرفة.txt
شرح مختصر خليل للخرشي.txt	شرح منتهى الإرادات.txt	شرح ميارة.txt
صحيح مسلم.txt	طرح التثريب.txt	طلبة الطلبة.txt
غذاء الألباب في شرح منظومة الآداب.txt	غمز عيون البصائر في شرح الأشباه والنظائر.txt	فتاوى الرملي.txt
فتاوى السبكي.txt	فتح القدير.txt	كشاف القناع عن متن الإقناع.txt
مجمع الأنهر في شرح ملتقى الأبحر.txt	مجمع الضمانات.txt	مشكل الآثار.txt
مطالب أولي النهى في شرح غاية المنتهى.txt	معالم القربة في طلب الحسبة.txt	معين الحكام فيما يتردد بين الخصمين من الأحكام.txt
مغازي الواقدي.txt	مغني المحتاج إلى معرفة ألفاظ المنهاج.txt	منح الجليل شرح مختصر خليل.txt
نهاية الرتبة الظريفة في طلب الحسبة الشريفة.txt	نهاية المحتاج إلى شرح المنهاج.txt	

3.4   Diacritization Step

Segmentation and Conservation of Symbols: In this step, we browse the input texts, character by character. At the end of the process we have an output board that guaranteed that each word is in a case and each symbol is in another case.

Global Treatment of Diacritisation.

It used the principle of Viterbi to calculate the most probable sequences. But we have modified the Algorithm to be more compatible with our needs to resolve the problem.

If the maximum probability is not zero, so the word is diacritized, otherwise we will use a second method which diacritize the word according to the diacritization of another word (e.g. كتب is considered as a nearest word "عمل") (see Fig. 3). In the case of an Interruption to find a transition (S ----->S') we estimate two possibilities:

1. The transition do not exist, but it is possible that the term exists, so we diacritize it.
2. Usually Arabic terms take the same diacritization (e.g. the term " كتب " could take the same diacritization as: " رسم ، عرف ، وجد ..... ").

Otherwise, the system cannot diacritize this term. The diacritization interface looks the figure shown in Fig. 3.

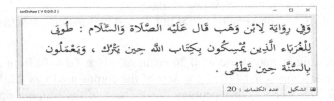

Fig. 3. Diacritization interface

4   System Evaluation

In order to evaluate our system, we create another program, which compare a random Arabic diacritized text (not existing in the learning corpus) with the output of the diacritization system and calculate the precision at the character level (2), and at the word level (3), as following:

$$Precision\ Char = 100 * \frac{\sum_{i=1}^{length(phrase)} R(i)}{Nb\_Characters} \tag{2}$$

$$Precision\ Word = 100 * \frac{Nb_{correctwords}}{Nb\_Words} \tag{3}$$

4.1   Results

Our system diacritize three religious texts, we obtained a precision of up to 80% at the word level word and 90% at the character level. We also compared our results with the Mishkal[1] system, which is a free online Arabic texts diacritizer, based on the Arabic complex rules.

---

[1] https://tahadz.com/mishkal/.

4.2 Comparison of Our System with "Mishkal"

In order to compare our system with Mishkal (see Footnote 1), we took randomly three (03) documents from the corpus used, that do not belong to the (26 documents chosen in the learning step). The results of the comparison are summarised in Table 4.

Table 4. Comparison between our system and Mishkal.

	Book	المبسوط	الانصاف	رد المحتار
	Text Index	28/134	4/193	16/368
	Characters	578 chars	362 chars	928 chars
	Words	144 words	77 words	222 words
Our system	Char prec.	96%	94%	95%
	Word prec.	86%	84%	87%
Mishkal	Char prec.	92%	89%	93%
	Word prec.	73%	67%	75%

5 Conclusion

In this paper we have presented two essential steps in the statistical model, which are well known in the field of automatic processing of the Arabic language: learning and diacritization that can significantly improve the quality of diacritisation. Our study revealed that our system does not take into account syntactic, morphological and semantic processing. The system designed and realized in this project is based on Hidden Markov Model (HMM) by Viterbi's method.

Our research project acknowledges these results and takes into account both a Viterbi Learning and a treatment approach. This is the reason for which we have chosen this approach, in order to make the diacritization of Arabic text more effective. We note more the learning is higher in a domain more the diacritization is best in that domain.

References

1. Hamdi, A.: Apport de la diacritisation dans l'analyse morphosyntaxique de l'Arabe. In: JEP-TALN-RECITAL 2012, Volume 3: RECITAL (2012)
2. Fashwan, A., Alansary, S.: SHAKKIL: an automatic diacritization system for modern standard Arabic texts. Phonetics and Linguistics Department, Faculty of Arts, Alexandria University, Alexandria, Egypt (2017)
3. Azmi, Almajed: Survey much of the literature on MSA diacritization (2015)
4. Chelba, C., Jelinek, F.: Structured language modeling. Comput. Speech Lang. 14(4), 283–332 (2000)
5. Darwish, K., Mubarak, H., Abdelali, A.: Arabic diacritization: stats, rules, and hacks. In: Proceedings of The Third Arabic Natural Language Processing Workshop (WANLP), Valencia, Spain, pp. 9–17 (2017)

6. Gal, Y.: An HMM approach to vowel restoration in Arabic and Hebrew (2002)
7. Abandah, G., Graves, A., Al-Shagoor, B., Arabiyat, A., Jamour, F., Al-Taee, M.: Automatic diacritization of Arabic text using recurrent neural networks. Int. J. Doc. Anal. Recognit. **18** (2), 183–197 (2015)
8. Goweder, A., de Roeck, A.: Assessment of a significant Arabic corpus. In: Arabic NLP Workshop at ACL/EACL, Toulouse, France (2001)
9. Jurafsky, D., Martin, J.H.: Speech and language processing. In: Draft Chapters in Progress (2018)
10. Kontrovich, L., Lee, D.D.: Learning semitic languages with Hidden Markov Models. In: NIPS 2001 Workshop on Machine Learning Methods for Text and Images (2001)
11. Bebah, M., Amine, C., Azzeddine, M., Abdelhak, L.: Hybrid approaches for automatic vowelization of Arabic texts. Int. J. Nat. Lang. Comput. (IJNLC) **3**, 53–71 (2014). https://doi.org/10.5121/ijnlc.2014.3404
12. Diab, M., Ghoneim, M., Habash, N.: Arabic diacritization in the context of statistical machine translation (2007)
13. Alnefaie, R., Azmi, A.M.: Automatic minimal diacritization of Arabic texts. In: 3rd International Conference on Arabic Computational Linguistics, Dubai, United Arab Emirates, 5–6 November 2017
14. Alansary, S.: Alserag: an automatic diacritization system for Arabic. In: Hassanien, A.E., Shaalan, K., Gaber, T., Azar, A.T., Tolba, M.F. (eds.) AISI 2016. AISC, vol. 533, pp. 182–192. Springer, Cham (2017). https://doi.org/10.1007/978-3-319-48308-5_18
15. Smrž, O., Zemánek, P.: Sherds from an Arabic treebanking mosaic. Bull. Math. Linguist. **78**, 63–76 (2002)
16. Mustafa, S.H.: Arabic string searching in the context of character code standards and orthographic variations. Comput. Stand. Interfaces **20**(1), 31–51 (1998)
17. Zerrouki, T., Balla, A.: Tashkeela: novel corpus of Arabic vocalized texts, data for auto-diacritization systems. Data Brief **11**, 147–151 (2017)
18. Khorsheed, M.S.: A HMM-based system to diacritize arabic text. J. Softw. Eng. Appl., 124–127 (2012). https://doi.org/10.4236/jsea.2012.512b024
19. Darwish, K., Abdelali, A., Mubarak, H., Samih, Y., Attia, M.: Diacritization of Moroccan and Tunisian Arabic Dialects: A CRF Approach (2018)
20. Hadj Ameur, M.S., Moulahoum, Y., Guessoum, A.: Restoration of Arabic diacritics using a multilevel statistical model. In: Amine, A., Bellatreche, L., Elberrichi, Z., Neuhold, Erich J., Wrembel, R. (eds.) CIIA 2015. IAICT, vol. 456, pp. 181–192. Springer, Cham (2015). https://doi.org/10.1007/978-3-319-19578-0_15
21. Jarrar, M., Zaraket, F., Asia, R., Amayreh, H.: Diacritic-based matching of Arabic words. In: ACM Transactions on Asian and Low-Resource Language Information Processing, vol. 18, no. 2, Article 10, December 2018

A Novel Approach Towards Fake News Detection: Deep Learning Augmented with Textual Entailment Features

Tanik Saikh[1]([✉]), Amit Anand[2], Asif Ekbal[1], and Pushpak Bhattacharyya[1]

[1] Indian Institute of Technology Patna, Bihta, India
{1821cs08,asif,pb}@iitp.ac.in
[2] Indian Institute of Information Technology Kalyani, Kalyani, India
amitanand@iiitkalyani.ac.in

Abstract. The phenomenal growth in web information has nourished research endeavours for automatic fact checking, or fake news and/or misinformation detection. This is one of the very emerging and challenging problems in Natural Language Processing (NLP), Machine Learning (ML) and Data Science. One such problem relates to estimating the veracity of a news story, which is a complex and deep problem. The very recently released Fake News Challenge Stage 1 (FNC-1) dataset introduced the benchmark FNC stage-1: stance detection task. This task could be an effective first step towards building a robust fact checking system. In this paper, we correlate this stance detection problem with Textual Entailment (TE). We present the systems which are based on statistical machine learning (ML), Deep Learning (DL), and a combination of both. Empirical evaluation shows encouraging performance, outperforming the state-of-the-art system.

Keywords: Fake news · Stance detection · Deep learning ·
Machine learning · Textual entailment

1 Introduction

In recent years, people are very communicative with the advent of the Internet. A lot of communications and conversations are happening through text, image, audio and video etc. This generates a lot of data everyday. The proliferation of these data/information in social media, online news feeds and tweets etc. demand for checking the truthfulness of these data/information. It is a tedious job even for the human being to do it manually. Hence, it is imperative to build the automated system which should be able to perform the tasks of detecting fake or misinformation, false claim detection, judging the veracity of a textual content made by a person etc.

Detecting veracity of information is a very challenging and demanding problem in Artificial Intelligence (AI), difficult even for a human being to understand the news contents all the time. Lately, [12] organized a shared task to

© Springer Nature Switzerland AG 2019
E. Métais et al. (Eds.): NLDB 2019, LNCS 11608, pp. 345–358, 2019.
https://doi.org/10.1007/978-3-030-23281-8_30

investigate how AI and Natural Language Processing (NLP) techniques could be promoted to combat fake news, entitled as Fake News challenge stage-I (FNC-I): Stance Detection. It could be a valuable first step towards helping human fact checkers to identify the false claims. Basically, to check the veracity of a claim/headline/report, it is important to see what other news agencies are saying about that particular claim/headline/report. There are multiple reportings available for a particular claim/headline/report produced by the different news agencies. Sometimes the document (body texts) agrees/supports the claim, sometimes it contradicts, sometimes discusses, or sometimes it remains completely unrelated to the claim. This is called stance, i.e. the relation between the *headline* and the *body* text. This is exactly what is defined in the dataset released in the shared task, FNC-I. The dataset contains <*Headline, Body Text, Stance*> triples. An example from the dataset is shown in Table 1. For this experiment, we assume the titles as claim/fact and the documents related to a particular title as body text. So if a particular title generally agrees with one and/or many of the body texts, then that particular title/claim could be most probably legitimate, otherwise, if there is no supporting body text to that claim, then that claim might be most probably fake. In this way, we can detect the truthfulness of a claim/report through stance detection. The shared task gained a lot of responses, with 50 teams from both academia and industry submitted their systems. Briefly, input to the system is a claim and the output corresponds to determining whether it is fake or genuine. We pose the problem as a classification problem, i.e. stance classification. The problem is conceptually very similar to a very well-known problem in NLP, namely TE [9] or Natural Language Inference (NLI) [3,15,16]. The definition of which is as follows: Given two pieces of texts, one is the *Premise(P)* and the other one is *Hypothesis(H)*, the system has to decide whether H is the logical consequence of P or not and/or H is true in every circumstance (possible world) in which P is true. For example, *P: "John's assassin is in jail"* entails *H: "John is dead"* and *P: "Mary shifted to France three years back."* entails *H: "Mary lives in France"*. Indeed, in both the above examples *H* is the logical consequence of *P*. We correlate the problem of stance detection to TE as follows: If a body text entails a claim, then it corresponds to actually support or agree or discuss; if it contradicts, then it corresponds to refute/disagree and if it does not provide any information related to the claim then it is completely unrelated (to the claim). We propose two approaches which are based on *viz. i. Statistical/Traditional ML and ii. DL*. The first approach makes use of a conventional set of features which are typically used for the task of TE. The second approach is an end-to-end deep learning approach and is based on the prior work [20]. We consider their model as the baseline in our experiments. The task described in [6] has shown how external knowledge could be helpful for DL based NLI models. Motivated by this we incorporate the ML features into our proposed DL architecture.

Contributions of our current work are two-fold, *viz* (i). We relate the problem to TE and propose various ML based models. We exploit the TE-based features and show the effect of TE for stance classification and further for fake news

Table 1. Headline and text snippets from documents and respective stances from the FNC training dataset

Headline: Hong Kong protesters go Ferguson style: 'Hands up, don't shoot'	
Stance	Body text
Agree	Hong Kong protesters have "emulated" the Ferguson gesture in their recent protests
Disagree	Photographs of Hong Kong protests have been discussed in the context of Ferguson....
Discuss	HONG KONG—Thousands of pro-democracy demonstrations in Hong Kong have....
Unrelated	A Russian fisherman says that Justin Bieber saved his life...

detection. (ii). We merge the ML feature values and the features extracted from the DL network, and feed into a feed-forward neural network. In this way we provide the external knowledge to neural network based model. This system outperforms the state-of-the art reported in the literature for the problem on this particular dataset. The paper is organized as follows. Section 2 describes brief overview of the related works followed by proposed methodologies (Sect. 3), dataset (Sect. 4), the experiments, results along with proper analysis (Sect. 5), and conclude (Sect. 6).

2 Related Work

Automatic fake news detection has recently gained attention to the researchers and developers. The papers [7,26] defined fact checking problem and they correlated this problem with the problem of TE. We also correlate, and make use of different TE based features. The work defined in [27] first released a large dataset for fake news detection and proposed a hybrid model to integrate the statement and speaker's meta data and performed classification. The task of [11] also posited a novel dataset called *Emergent*, which was driven from the digital Journalism project, namely Emergent [22]. They additionally proposed a logistic regression model for the stance detection, where features are extracted from the headline and news body pairs. The dataset that we employ in this experiment is an extended version of this Emergent dataset.

The task defined in [1] made use of conditional encoding network with two Bi-LSTMs to detect stance of tweets with some targets. They nurtured two separate LSTM networks, one for the tweet and another one for the target. The first hidden state of the LSTM for the target was initialized with the final hidden state of the LSTM for the tweet. The work described in [19] also utilized the stance detection dataset. They proposed four models which are based on *Bag of word (BoW), basic LSTM, LSTM with attention, and condition encoding LSTM with attention* and showed that the model with condition encoding LSTM with attention mechanism yielded the highest result among the results produced by all these models, which demonstrated the efficiency of attention technique in extracting from a long sequence (news body) of information relevant to a small query (article title). They reported the highest accuracy of 80.8%.

The task defined in [23] presented a novel hierarchical attention model for stance detection. Especially they fostered a model to represent the document and their linguistic features with attention technique. Additionally, on the top of document representation, they made use of attention mechanism to estimate the importance of different linguistic features and learnt overlapping attention between the document and the linguistic information. The work described in [12] performed deep analysis of the three best participating systems of FNC-1. They showed that, the class wise and macro-averaged F1 score is the best way for validating the model for stance detection, as the shared task's standard evaluation metric is severely affected by the imbalanced class distribution of the dataset. We also followed these two metrics in addition to the standard metric provided by fake news challenge to evaluate our systems. Apart from these, the tasks on stance detection for fake news detection which made use of Fake news dataset could be found in [12,14,17,18]. It has been studied in other languages too like Arabic which could be found in [10].

3   Proposed Method

As stated earlier, We use both traditional supervised Machine learning and the deep learning approaches.

3.1   Feature Based Machine Learning Approach

We propose a supervised machine learning approach based on Support Vector Machine (SVM) [5,24] and Multilayer Perceptron (MLP) [2,8] to detect the stance between the headline and the body text. This model aims to develop a machine learning based system where different TE-based features are employed. The features include *Synonyms, Antonyms, Hypernyms, Hyponyms, Overlapping Tokens, Longest Common Overlap, Modal verbs, Polarity, Numerals, Named Entities, and Cosine Similarity.* The following points elaborate all these features.

Synonyms: Presence of synonymous words in two pieces of text snippets reveal that they are semantically similar, like *X bought Y* implies *X acquired Z% of the Y's shares,* because *acquire* is the synonym of *bought.* For each word in title, we search for the synonym of that particular word in the body text. If it is present then the feature value of "1" is assigned otherwise "0".

Antonyms: This is also a vital feature for detecting TE, which is a pervasive form of entailment trigger, where a word is replaced by it's antonym. Sentences like *T: "Oil price is surging"* does not imply *T: "Oil price is falling down.".* The feature value is computed in the reverse direction to what was followed in the synonym feature.

Hypernyms: Sometime certain concepts are generalized from one text to another, which leads to entailment. Like T: *"Beckham plays football."* entails H: "Beckham plays game.". So if there was *football* in headline and *game* in the body then we assign "1" otherwise "0".

Hyponyms: It is also observed that sometimes concepts are specialized, which, in turn, lead to entailment. Like T: *"Reptiles have scale."* entails H: "Snakes have scale.". So if Hyponyms of a word in title is present in body text, then the value of "1" is assigned, otherwise "0".

Overlapping Tokens: Overlapping tokens between two comparing text snippets can help in deciding entailment. The number of overlapping tokens between the headlines and body texts become the feature value of this feature.

Longest Common Overlap: Longest matching between two texts also matters a lot in taking the decision of Entailment. The value of this feature is computed as the maximum overlapping length between two pair of texts normalized by the number of words present in the body text.

Modal Verbs: It represents the presence of modal auxiliary verbs (like: can, should, must etc) which denote the possibility or necessity and sometimes lead to wrong entailment. Like T: *"The govt. may approve anti-corruption bill."* does not entails H: "The govt. approved anti corruption bill.". This feature is important for predicting the classes (like agree and discuss) between title and body text pairs. So, if it is present in any of the title or body text then the value of "0" is assigned and if it is present or absent in both the headline and body text then the value of "1" is assigned.

Polarity Features: These features determine whether the fact asserted or it's negation is going to occur, like (not, never, deny etc) are the polarity features. If we fully rely on lexical matching, the presence of negation word might cause problem in taking the decision for entailment. For example, T: *"The watchman denied that he was sleeping."* does not entail H: "The watchman was sleeping.". We compute this feature's value following the procedure as described in [21] for computing this polarity feature value.

Numerals: In some cases certain level of numeric calculation affect the entailment decision. Like T: *"3 men and 2 women were found dead in the apartment."* entails H: "5 people were found dead in apartment.". We assign the value of "1", if we found such matching, otherwise "0" is assigned.

Named Entity Information: Named Entities (NEs) (like, person, location, organization) between two text snippets sometime affect in entailment decision. We search for any matching pair of NEs between the headline and body text. A value of "1" is assigned if NEs match, otherwise a value of "0" is assigned.

Cosine Similarity: This is very popular and a benchmark similarity metric, widely used among the researchers over the years to find similarity between two pieces of texts. It could be a feature for entailment also. We pass headline and body separately to Universal Sentence Encoder (USE). USE produce vector representation of headline and body. We compute the cosine similarity between these two vectors and assign as the value of this feature.

We apply different classifiers like SVM and MLP. The results obtained using these classifiers are shown in the results and discussion section (i.e in Sect. 5).

3.2 Deep Learning Based Approach

We propose two DL based approaches. One is based on the model defined in [20]. The difference from our propose model is in the representation layer. We apply the universal sentence encoder (USE) [4] to obtain the representations of titles and body texts, whereas they utilized Term Frequency-Inverse Document Frequency (tf-idf) for the same purpose. The another one is based on the first one but incorporated with ML based features values. The USE comes into two variants one exploiting the Transformer [25] architecture and the other one is based on the Deep Average Network (DAN) [13]. We make use of the Transformer based USE because it is observed that transfer learning from the transformer based sentence encoder performs better than transfer learning from the DAN encoder.

This model utilize the encoding sub-graph of the transformer architecture to produce the sentence/document's embedding. This kind of sub-graph provides context aware representation of words in a sentence by utilizing attention without hampering the ordering and the identity of other words. To obtain the fixed length sentence encoding vector, element-wise sum of the representations of each word is taken into account, which is further normalized by the square root of the length of the sentence.

The headline and body pairs are given to USE, which produces the representations for both headline and body, but separately. These representations are concatenated and subjected as inputs to feed-forward neural networks (dense layers) with ReLU activation function. Four such layers have been used, and this decision was taken in an empirical manner. We perform the experiments by taking the different number of layers. We obtain the highest performance with four layers. The outputs obtain from the fourth layer are given to a final layer with softmax activation function for final prediction. This layer predicts the class having the highest probability score. Architecture of the proposed model is shown in Fig. 1(a).

We modify our first approach to offer the second one. We incorporate the features values used in ML approach in the representation layer, as shown in the Fig. 1(b). We concatenate these values (computed for 11 features) with the representations obtained for headline and body from USE.

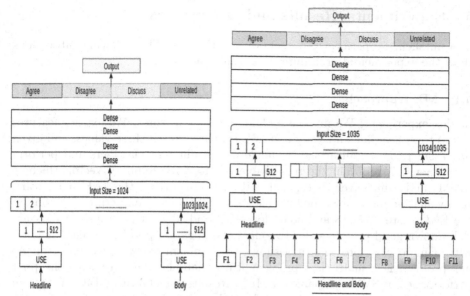

(a) Architectural diagram of the propose first DL system with Universal Sentence Encode

(b) Architectural diagram of the propose second DL system augmented with different ML features

Fig. 1. The architecture of the proposed two systems

4   Data

We make use of the benchmark dataset released in the shared task FNC-I for fake news detection through stance detection. The key statistics of the dataset are shown in Table 2. The dataset is highly imbalanced. So the task organizers[1] provide a standard metric to mitigate this problem. The metric is a weighted based evaluation system which comprises of two levels. In the first level, 25% weight is given for classifying headline and body text as related or unrelated and in the second level, 75% weight is given for classifying related pairs as agrees, disagrees, or discuss. The justification behind this is: classifying agrees, disagrees, or discusses is more difficult and relevant to fake news detection rather than just classifying headline–body pairs as related and unrelated.

Table 2. Number of instances, distribution of classes and average length of title and body in training and test set of FNC-1 dataset

Dataset	Example pairs	Classes				Avg. Length	
		Unrelated	Discuss	Agree	Disagree	Body	Title
Training	49972	0.73131	0.17828	0.0736012	0.0168094	369	11
Testing	25413	0.722032	0.17466	0.074833	0.027427	347	11

[1] http://www.fakenewschallenge.org/.

5   Experiments, Results and Discussions

In a nutshell, we perform three sets of experiments. The following subsections show the experimental procedures and results obtained.

5.1   ML Approach

In this experiment, We make use of 11 different features. We extract features values from headline and body text. We concatenate all these values, and given to classifier for classification. We make use of different classifiers and perform experiments. We obtain the remarkable results with Support Vector Machine (SVM) and Multi-layer Perceptron (MLP). We compute the FNC score using the evaluation metric provided in Fake News Challenge Competition. We obtain the FNC score of 72.13 and 56.04 for MLP and SVM, respectively. SVMs are well known good performer for two-class classification problem, even if it plays with a multi-class problem, it assumes the problem as two class problem. As our problem is a multi-class problem, this might be the reason for the poor performance of SVM compared to MLP. Results are shown in Table 4. Due to space constraints we are unable to show the confusion matrices for all of our proposed models. However, we show the confusion matrix for the best performing model.

Sensitivity Analysis of the Features: We perform feature ablation study to understand the contribution of each feature. The F1 scores are obtained by removing one feature after another. Results are shown in the Table 3. It shows that cosine similarity followed and Named Entities (because news titles/documents are full of different names) are the most contributing features in our experiment.

Table 3. Feature sensitivity analysis and effect of each feature on F1

Features removed	F1	Increment/decrement
None	**0.4777**	0
Synonyms	0.4757	−0.0020
Antonyms	0.4756	−0.0021
Longest common overlap	0.4679	−0.0098
Hypernym	0.4701	−0.0076
Hyponym	0.4724	−0.0053
NER	0.4653	−0.0124
Modality	0.4731	−0.0046
Overlapping tokens	0.4729	−0.0048
Numerals	0.4700	−0.0077
Polarity	0.4763	−0.0014
Cosine similarity	0.4364	**−0.0413**

5.2   Deep Learning

We propose two models which utilize the DL platform. The first one is based on USE and another one is where we incorporate the ML features values into USE based Model.

Universal Sentence Encoder Model: All the modern ML techniques fully rely on the vector representation of words, phrases and sentences. We obtain the embedding of title and body by utilizing transformer based USE. It takes lowercased Pen Tree Bank (PTB) tokenized[2] string of any length as input and produces the representation of fixed (512) dimensional embedding vector as output. We concatenate the representations of title and body text. The concatenated vector further send to four feed forward neural network layers. The representation obtained from the fourth feed forward neural network is further fed into a final layer for classification. The final layer predicts appropriate labels (Agree, Disagree, Discuss and Unrelated) having the maximum probability score. The architecture of this approach is shown in Fig. 1(a). We obtain the FNC score of 76.9 in this experiment.

Universal Sentence Encoder Model Incorporated with ML Features: In this experiment we inject the ML based features in the previous model. We concatenate the 11 features values with the vector representation for headline and body text. So the representation become a vector of 1035 dimension. This representation is further subjected as input to four feed forward neural network layers, placed one after another. The output obtained from the fourth feed forward neural network is given to a final layer with softmax activation function for final prediction. The architecture of this model is shown in the Fig. 1(b). We obtain the FNC score of 82.54 in this experiment.

Hyperparameters: We tune the hyperparameters in this experiment and mark the results and freeze the model having the hyperparameters which produces the best result. For example, the hidden layer size is tuned from 64 units to 256 units, batch size input from 64 to 256, dropout from 0.2 to 0.3. For all the experiments Rectified linear Unit (ReLU) activation function is used in all the feed-forward neural networks. The loss function and optimizer are cross entropy and ADAM respectively. The training iterations i.e. epoch was 50 for all the experiments and also we used checkpoint, to check the model's accuracy get increased or not, if it get increased only then the weights get updated. The final layer for the output prediction is with softmax activation function.

5.3   Comparison with the State of the Art and Other Prior Models

We perform an exhaustive comparison with previous three best participating systems on this dataset. The comparison is shown in Table 4. Apart from the FNC, we also compute the performance of our model using different modalities of evaluation metrics like *"overall F1"*, *"FNC"*, *"per class F1"* (for Agree, Disagree,

---

[2] https://nlp.stanford.edu/software/tokenizer.shtml.

Discuss and Unrelated). The DL model augmented with TE based features i.e. the third one has achieved the highest FNC score which outperforms the state-of-the-art reported in the literature by the FNC score of 0.5 margin. This model also beats the official baseline provided by the shared task organizers and also the score of the system [20] which we assumed as the baseline in this experiments. The result of this system is shown in the 3rd row (UCLMR system) in all formats. We also obtain the overall F1 score of 63.6%, and also the F1 score of 61.1% for agree class which is the highest among all the prior models. We also obtain the highest F1 score of 59.54% in disagree class with SVM classifier which is also the highest F1-score among all the previous system's score. However, we are not able to overcome the performance of human which is shown in row no 12 of the Table 4. This indicates there are lots of room that are available for improvement. The first participating system obtained an FNC score of 0.8204. The system is an ensemble of two 2D CNNs on word embedding of headline and body respectively. The resulting output is then fed into an MLP of three hidden layers and a decision tree based system composition of 5 features. Our two deep learning systems are based on the UCLMR system [20] with some modifications *viz: i. at the representation layer and ii. at hidden layer (that model was one feed-forward neural network, and we have four)*. In the third model, in addition to these we inject TE based ML features.

Table 4. The prior six best results and the results obtained by our proposed models on the dataset

SN	System	FNC-1	F1	Agree	Disagree	Discuss	Unrelated
	Previous Models						
	TALOSCOMB(TREE+CNN)	0.8204	0.582	0.539	0.035	0.760	0.994
2	ATHENE	0.8197	0.604	0.487	0.151	**0.780**	**0.996**
3	UCLMR	0.8172	0.583	0.479	0.114	0.747	0.989
4	featMLP	0.825	0.607	0.530	0.151	0.766	0.982
5	stackLSTM	0.821	0.609	0.501	0.180	0.757	0.995
6	MAJORITY VOTE	0.394	0.210	0.0	0.0	0.0	0.839
	Proposed Models						
7	SVM	0.5604	0.4150	0.0073	**0.5954**	0.1084	0.9489
8	MLP	0.7213	0.4777	0.3462	0.0	0.6328	0.9315
9	Univ_Sen_Enc	0.769	0.570	0.436	0.187	0.712	0.944
10	Univ_Sen_Enc_Features	**0.8254**	**0.636**	**0.611**	0.214	0.746	0.972
11	Official Baseline	**0.7520**	X	X	X	X	X
12	HUMAN UPPER BOUND	0.859	0.754	0.588	0.667	0.765	0.997

5.4   Error Analysis

Every system has some pros and cons. Our system has some disadvantages too. We perform error analysis of our best performing system. We take miss-classified

Table 5. Confusion matrix obtained by the best performing DL approach on the test set

	Agree	Disagree	Discuss	Unrelated
Agree	1162	55	590	96
Disagree	233	149	258	57
Discuss	804	154	3323	180
Unrelated	92	33	395	17829

instances into account. We make a rigorous analysis of those instances and try to analysis why our model fails. The Table 5 shows the confusion matrix.

Our observations could be as follows:

• The dataset is enriched with Named Entities, phrasal verbs, and Multi-word expressions. The bodies are having multiple number of repetitive words, and sentences too which we need to take care separately in future. • The length variation between the title and the body is very high. • It is observed that the model is performing badly where headlines and body texts are of question answer type, i.e. Headline is question and the body text explaining it like answer. We need to investigate this in future.

6   Conclusion and Future Work

Detection of misinformation/fake news and fact checking is a very challenging and utmost task these days to mankind. In this paper, we try to mitigate this problem. The dataset released in Fake News Challenge for detecting fake news through stance detection serves this purpose. We relate this problem to TE as they are conceptually similar. We offer the systems which are based on ML, DL and combination of both. In ML, we foster the different TE-based features apply to different classifiers (SVM and MLP), and obtain remarkable results. In DL, we pose two models, one is USE based and the other one is the modified version of the USE model but augmented with TE based features. We make use of different performance measures i.e. *FNC, overall F1, per class F1 score* etc. Our proposed model outperforms the state-of-the-art system in *FNC* and *F1 score*, and *F1 score of Agree class* by the third DL model i.e. the model augmented with TE features. The system also outperforms the state-of-the-art *F1 score of Disagree class* by our SVM based model. In future we would like to: • enrich the propose models by incorporating many more lexical/syntactic/semantic based features and address the issues raised by the proposed models. • do more in-depth and rigorous error analysis of the previous three best participating systems to get more insights. • incorporate the external knowledge (i.e. world knowledge) into the existing system.

Acknowledgments. Asif Ekbal acknowledges Young Faculty Research Fellowship (YFRF), supported by Visvesvaraya PhD scheme for Electronics and IT, Ministry of Electronics and Information Technology (MeitY), Government of India, being implemented by Digital India Corporation (formerly Media Lab Asia).

References

1. Augenstein, I., Rocktäschel, T., Vlachos, A., Bontcheva, K.: Stance detection with bidirectional conditional encoding. In: Proceedings of the 2016 Conference on Empirical Methods in Natural Language Processing, Austin, Texas, pp. 876–885. Association for Computational Linguistics (2016)
2. Becerra, R., Joya, G., García Bermúdez, R.V., Velázquez, L., Rodríguez, R., Pino, C.: Saccadic points classification using multilayer perceptron and random forest classifiers in EOG recordings of patients with ataxia SCA2. In: Rojas, I., Joya, G., Cabestany, J. (eds.) IWANN 2013. LNCS, vol. 7903, pp. 115–123. Springer, Heidelberg (2013). https://doi.org/10.1007/978-3-642-38682-4_14
3. Bowman, S.R., Angeli, G., Potts, C., Manning, C.D.: A large annotated corpus for learning natural language inference. In: Proceedings of the 2015 Conference on Empirical Methods in Natural Language Processing, Lisbon, Portugal, pp. 632–642. Association for Computational Linguistics (2015)
4. Cer, D., et al.: Universal sentence encoder for English. In: Proceedings of the 2018 Conference on Empirical Methods in Natural Language Processing: System Demonstrations, Brussels, Belgium, pp. 169–174. Association for Computational Linguistics (2018)
5. Chang, C.C., Lin, C.J.: LIBSVM: a library for support vector machines. ACM Trans. Intell. Syst. Technol. **2**(3), 27:1–27:27 (2011)
6. Chen, Q., Zhu, X., Ling, Z.H., Inkpen, D., Wei, S.: Neural natural language inference models enhanced with external knowledge. In: Proceedings of the 56th Annual Meeting of the Association for Computational Linguistics (Volume 1: Long Papers), pp. 2406–2417. Association for Computational Linguistics (2018)
7. Ciampaglia, G.L., Shiralkar, P., Rocha, L.M., Bollen, J., Menczer, F., Flammini, A.: Computational fact checking from knowledge networks. PLoS One **10**(6), e0128193 (2015)
8. Costa, W., Fonseca, L., Körting, T.: Classifying grasslands and cultivated pastures in the brazilian cerrado using support vector machines, multilayer perceptrons and autoencoders. In: Perner, P. (ed.) MLDM 2015. LNCS (LNAI), vol. 9166, pp. 187–198. Springer, Cham (2015). https://doi.org/10.1007/978-3-319-21024-7_13
9. Dagan, I., Glickman, O., Magnini, B.: The PASCAL recognising textual entailment challenge. In: Quiñonero-Candela, J., Dagan, I., Magnini, B., d'Alché-Buc, F. (eds.) MLCW 2005. LNCS (LNAI), vol. 3944, pp. 177–190. Springer, Heidelberg (2006). https://doi.org/10.1007/11736790_9
10. Darwish, K., Magdy, W., Zanouda, T.: Improved stance prediction in a user similarity feature space. In: Proceedings of the 2017 IEEE/ACM International Conference on Advances in Social Networks Analysis and Mining 2017, Sydney, Australia, 31 July–03 August 2017, pp. 145–148 (2017)
11. Ferreira, W., Vlachos, A.: Emergent: a novel data-set for stance classification. In: Proceedings of the 2016 Conference of the North American Chapter of the Association for Computational Linguistics: Human Language Technologies, San Diego, California, pp. 1163–1168. Association for Computational Linguistics (2016)

12. Hanselowski, A., et al.: A retrospective analysis of the fake news challenge stance-detection task. In: Proceedings of the 27th International Conference on Computational Linguistics, Santa Fe, New Mexico, USA, pp. 1859–1874. Association for Computational Linguistics (2018)
13. Iyyer, M., Manjunatha, V., Boyd-Graber, J., Daumé III, H.: Deep unordered composition rivals syntactic methods for text classification. In: Proceedings of the 53rd Annual Meeting of the Association for Computational Linguistics and the 7th International Joint Conference on Natural Language Processing (Volume 1: Long Papers), Beijing, China, pp. 1681–1691. Association for Computational Linguistics (2015)
14. Thorne, J., Chen, M., Myrianthous, G., Pu, J., Wang, X., Vlachos., A.: Fake news stance detection using stacked ensemble of classifiers. In: Proceedings of the EMNLP Workshop on Natural Language Processing meets Journalism, Copenhagen, Denmark, pp. 80–83 (2017)
15. MacCartney, B., Grenager, T., de Marneffe, M.C., Cer, D., Manning, C.D.: Learning to recognize features of valid textual entailments. In: Proceedings of the Human Language Technology Conference of the NAACL, Main Conference (2006)
16. MacCartney, B., Manning, C.D.: Natural logic for textual inference. In: Proceedings of the ACL-PASCAL Workshop on Textual Entailment and Paraphrasing, RTE 2007, Stroudsburg, PA, USA, pp. 193–200. Association for Computational Linguistics (2007)
17. Mohtarami, M., Baly, R., Glass, J., Nakov, P., Màrquez, L., Moschitti, A.: Automatic stance detection using end-to-end memory networks. In: Proceedings of the 2018 Conference of the North American Chapter of the Association for Computational Linguistics: Human Language Technologies, Volume 1 (Long Papers), New Orleans, Louisiana, pp. 767–776. Association for Computational Linguistics (2018)
18. Pérez-Rosas, V., Kleinberg, B., Lefevre, A., Mihalcea, R.: Automatic detection of fake news. In: Proceedings of the 27th International Conference on Computational Linguistics, Santa Fe, New Mexico, USA, pp. 3391–3401. Association for Computational Linguistics (2018)
19. Pfohl, S., Triebe, O., Legros, F.: Stance detection for the fake news challenge with attention and conditional encoding (2017)
20. Riedel, B., Augenstein, I., Spithourakis, G.P., Riedel, S.: A simple but tough-to-beat baseline for the fake news challenge stance detection task. CoRR abs/1707.03264 (2017)
21. Saikh, T., Ghosal, T., Ekbal, A., Bhattacharyya, P.: Document level novelty detection: textual entailment lends a helping hand. In: Proceedings of the 14th International Conference on Natural Language Processing (ICON-2017), Kolkata, India, pp. 131–140. NLP Association of India, December 2017
22. Silverman, C.: Lies, damn lies and viral content (2015)
23. Sun, Q., Wang, Z., Zhu, Q., Zhou, G.: Stance detection with hierarchical attention network. In: Proceedings of the 27th International Conference on Computational Linguistics, Santa Fe, New Mexico, USA, pp. 2399–2409. Association for Computational Linguistics (2018)
24. Vapnik, V.N.: The Nature of Statistical Learning Theory. Springer, New York (1995). https://doi.org/10.1007/978-1-4757-2440-0
25. Vaswani, A., et al.: Attention is all you need. In: Guyon, I., et al. (eds.) Advances in Neural Information Processing Systems 30, pp. 5998–6008 (2017)

26. Vlachos, A., Riedel, S.: Fact checking: task definition and dataset construction. In: Proceedings of the ACL 2014 Workshop on Language Technologies and Computational Social Science, Baltimore, MD, USA, pp. 18–22. Association for Computational Linguistics (2014)
27. Wang, W.Y.: "Liar, liar pants on fire": a new benchmark dataset for fake news detection. In: Proceedings of the 55th Annual Meeting of the Association for Computational Linguistics (Volume 2: Short Papers), pp. 422–426. Association for Computational Linguistics (2017)

Contextualized Word Embeddings in a Neural Open Information Extraction Model

Injy Sarhan[1,2]([envelope]) and Marco R. Spruit[2]

[1] Computer Engineering Department, College of Engineering,
Arab Academy for Science, Technology and Maritime Transport (AAST),
Abukir, Alexandria 1029, Egypt
[2] Department of Information and Computing Sciences, Utrecht University,
Princetonplein 5, 3584 CC Utrecht, The Netherlands
{i.a.a.sarhan,m.r.spruit}@uu.nl

Abstract. Open Information Extraction (OIE) is a challenging task of extracting relation tuples from an unstructured corpus. While several OIE algorithms have been developed in the past decade, only few employ deep learning techniques. In this paper, a novel OIE neural model that leverages Recurrent Neural Networks (RNN) using Gated Recurrent Units (GRUs) is presented. Moreover, we integrate the innovative contextual word embeddings into our OIE model, which further enhances the performance. The results demonstrate that our proposed neural OIE model outperforms the existing state-of-art on two datasets.

Keywords: Open Information Extraction · Word embeddings · RNN · GRU · LSTM

1 Introduction

Natural Language Processing (NLP) techniques that facilitates the process of fetching important information from large data are highly demanded. With the ongoing development in the field of NLP, OIE gained a massive amount of attention in the past years. It is the process of extracting a relation tuple from a text corpus in the form of *<Entity1> <Relation> <Entity 2>* as seen in Table 1.

OIE plays a fundamental role in turning massive, unstructured text corpora into factual information, it can be used as a foundation to many NLP tasks, including, Information Extraction, Question Answering and Summarization.

Previously, OIE paradigms either utilized automatically assembled training data or hand-crafted heuristics. Nonetheless, after deep learning techniques paved their way in various NLP tasks researchers aimed their focus towards neural networks.

RNN is a robust class of artificial neural networks, contrary to Feed-Forward networks, RRNs can loop among nodes, thus it's capable of apprehending temporal behavior. This results in permitting information to persist in them, by selecting which information to keep and which to forget by taking into consideration the current input and the previous data it received.

© Springer Nature Switzerland AG 2019
E. Métais et al. (Eds.): NLDB 2019, LNCS 11608, pp. 359–367, 2019.
https://doi.org/10.1007/978-3-030-23281-8_31

Table 1. Open information extraction example

Sentence	Barack Obama born August 4, 1961 in Hawaii served as the 44th President of the USA
Extracted tuples	\<Barack Obama – Born - August 4\>
	\<Barack Obama – Born - Hawaii\>
	\<Barack Obama – Served as- President of the USA\>

In this paper we present an OIE model that employs RNNs to extract relation triples. Recently, RNNs proved their importance by achieving notable performance in various NLP tasks such as translation [1] and speech recognition [2], they are heavily applied in Google Home [3] and Amazon's Alexa [4]. The features that make RNNs a good fit for NLP applications are notable [5]. For instance, they take into consideration the order of the words, in addition, GPU can be utilized to carry out RNN computation therefore they perform well on large datasets. Also, RNNs can handle arbitrary input and output lengths. Furthermore, we demonstrate that contextual embedding enhances the overall performance of OIE task compared to non-contextual word embedding techniques.

The remainder of this paper is structured as follows; Sect. 2 reviews the existing OIE state-of-art models, while Sect. 3 presents the proposed OIE model, followed by the results and evaluation in Sect. 4. Finally, Sect. 5 concludes the paper and discusses future work.

2 Related Work

In this section we review existing OIE state-of-art architectures, a complete picture can be found in [6]. OIE can be categorized into two broad categories, approaches that requires automatically machine learning classifiers and approaches that utilizes hand-crafted rules [7]. Newly, deep learning techniques started paving their way towards OIE systems.

2.1 Machine Learning Classifiers

In 2007, Banko et al. [8] introduced TextRunner, the first OIE system is a fully implemented, highly adaptable, self-supervised system that relies on shallow syntactic analysis. It makes use of a domain-independent technique on a text corpus in order to extract relation tuples. TextRunner extracts all possible relation tuples by making a single pass over the corpus using Conditional Random Field classifier, tuples that are classified as trustworthy are reserved by the extractor.

Wikipedia-based Open Extractor (WOE) system [9], introduced by Wu and Weld, that operate in two modes: WOEPos and WOEParse. The WOEPos system employs a CRF extractor trained with shallow syntactic features, in contrast to WOEParse that makes use of a rich dictionary of dependency path patterns. Heuristically matching

Wikipedia info box values with corresponding text for automatic assembly of training examples is the primary idea behind WOE herby enhancing TextRunner's performance.

2.2   Hand-Crafted Rules

REVERB proposed by Fader et al. [10]. REVERB relies on the process of relation phrases that meet syntactic and lexical constraints, afterwards it extracts noun phrase argument pairs for each relation phrase. Logistic regression classifier is latter used to assign a confidence score for each extracted tuple. Subsequently, Etzioni et al. [11] presented the second generation of OIE, R2A2 by combing REVERB with an argument identifier - ARGLEARNER - to enrich argument extraction for the relation phrases.

Del Corro and Gemulla proposed ClausIE [12], a clause-based OIE system that expoilts the linguistic knowledge of the grammar of the English language to locate clauses in an input corpus. It determines the dependency parse of the input sentence to realize its syntactical structure. Then, the algorithm acquires a set of coherent derived clauses based on the dependency parse and small domain-independent lexica and generate one or more propositions for each clause. ClausIE fundamentally vary from the aforementioned OIE systems in the way that it doesn't utilize any training data in contrast to REVERB [10] and TextRunner [8].

2.3   Neural Approaches

A neural OIE paradigm was proposed by Cui et al. [13] that employs a Recurrent Neural Network (RNN) encoder-decoder framework. The encoder-decoder infrastructure is a method for text generation and has already been utilized in other NLP tasks successfully as illustrated in [13] The encoder inputs a variable length sequence and outputs a compressed representation vector, which is then passed to the decoder, resulting in the output sequence produced by the decoder. Both the encoder and decoder use a 3-layer Long Short-Term Memory (LSTM) [14]. Training data is obtained from high confidence binary extractions from state-of-the-art OIE system. Thus, the extraction of high-quality tuples.

In addition to the work of Cui et al., Stanovsky et al. [15] developed a Bidirectional LSTM transducer to extract OIE tuples, proving that supervised learning can have a strong impact on OIE performance. By extending the work made on deep semantic role labeling to extract OIE tuples authors of [15] were able to achieve notable results. Moreover, their work emphasis that research on Question Answering-Semantic Role Labeling paradigms can greatly benefit future OIE models.

3   Proposed Model

Our proposed model is built on the work of Stanovsky et al. [15] by treating OIE task as a sequencing labeling problem resulting in the extraction of multiple, overlapping tuples for each sentence.

The proposed neural network framework takes a fixed length vector of an embedded sentence as an input. In addition, predicates are the building blocks of any language, they denote strong actions which are considered extremely effective in extracting relations of interest. Thus, following the work of [15], we assume that the predicate in each sentence represents the relation that's associated with the tuple, therefore the predicate is sent to the network as a feature vector along with the Part of Speech (POS) tag of the sentence using NLTK [16].

3.1 Contextual Embedding

ELMo (Embedding from Language Models) [17] is a deep contextualized word representation that models both: complex syntactic and semantic features of a word and the way in which these words' uses differ throughout linguistic. The key idea behind ELMo is contextual embedding, thus the representation of each word differs according to its neighboring words. The generated word vectors are acquired from the functions of the internal states of a deep bidirectional language model, which is pre-trained on a large dataset. We integrated ELMo embedding in our OIE model, results proved that contextual embedding yield to a better performance. The aforementioned neural OIE methods utilized either GloVe [18] or Word2Vec [19], both are non-contextual word embeddings. Comparative results are demonstrated in the subsequent section.

3.2 GRU Model Architecture

RNNs are hard to train due the vanishing and the exploding gradient descent problems during the back-propagating process [20]. Efforts were made to overcome this complication, hence LSTMs and GRUs were developed. They both successfully dealt with the difficultly of training RNNs. Indeed, LSTM and GRU are considered very effective models for learning very long contexts. The way they are used in [21] allows to train on long word-contexts.

GRUs are comparatively new and employs fewer number of parameters than LSTMs which eventually entails that GRUs are both lighter and faster to train than LSTM. GRU merges LSTM's Input and Forget gate in the Update gate. In Addition, it merges the cell state and the hidden state which lowers the complexity of the model.

Contrary to LSTM, GRU has 2 gates instead of 3:

- Reset Gate: that decides how to integrate the previous memory with the current input.
- Update Gate: that determines the amount that it should keep from the prior memory.

For GRU, the hidden state Ht is computed as [22]:

$$Z_t = \sigma(X_t U_z + H_t - 1 W_z) \qquad (1)$$

$$R_t = \sigma(X_t U_r + H_t - 1 W_r) \qquad (2)$$

$$h_t = \tanh(X_t U_h + (R_t * H_t - 1) W_h) \qquad (3)$$

$$H_t = (1 - Z_t) * h_t - 1 + Z_t * h_t \tag{4}$$

Where Z and R denotes the update gate and the reset gate respectively. X represents the input vector, while U and W represent parameter vectors.

Our proposed OIE architecture is shown in Fig. 1. In our OIE model we implemented a 2-Layer Bidirectional GRUs. The default application of RNNs is to assess information in a single direction. However, it has been shown that modelling information in a bidirectional technique results in better performance [21, 23]. A Bidirectional GRU was employed to encapsulate forward and backward lexical semantics of each word in a given sentence. A bidirectional network can be generated in 2 different approaches; either by having 2 RNN operating in opposing directions or within the internal architecture of the RNN itself, in our model we employed the latter approach.

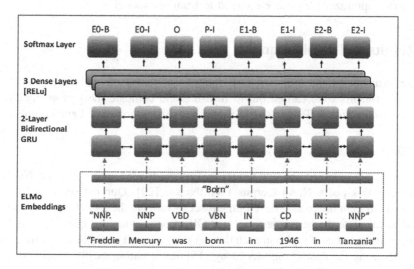

Fig. 1. Our OIE model architecture.

After encoding the 3 inputs using ELMo -the word, the POS tag of each word and the predicate as shown in Eq. (5)- they are all concatenated and passed as single feature vector to the Bidirectional GRU. Subsequently, the Bidirectional GRU outputs a tensor that's passed to 3-layer Time Distributed Dense layer which is later passed to the SoftMax layer for label prediction.

$$FeatureVector = ELMo(Word) \oplus ELMo(POS) \oplus ELMo(Predicate) \tag{5}$$

Eventually, SoftMax layer assigns a probability of each word belonging to a certain label. We used BIO tags (Begin – Intermediate – Outside) [24] that's demonstrates the location of each word in the sentence, and each label is later assigned accordingly as shown in the last layer in Fig. 1. A sentence might include more than one entity, each sentence may output more than one tuple as the example in Table 1; however, our model captures binary relations. If a sentence contains no relation between the words only the predicate is assigned as "*P-B*" and label "*O*" is allocated to the remaining the words in the sentence.

3.3   Hyperparameters Settings

Our neural OIE architecture was implemented using Keras framework [25] with TensorFlow backend [26]. Our model was trained on 10 epochs with the dropout rate set to 0.1 for regularization to avoid over-fitting. The data is divided into 100 batches. Moreover, we use early stopping to terminate training when the performance stops improving. Each Bidirectional GRU has 128 units, which is the same number of the hidden units in the subsequent 3 Time Distributed Dense layers. The activation function used in the 3 Time Distributed Dense layers is Rectified Linear Unit (ReLU) [27]. Adam optimizer [28] was employed to train our model.

4   Results and Evaluation

The performance of the proposed OIE model was tested on two different datasets. Three experiments were carried out to measure and compare the performance of the proposed Bidirectional GRU-based OIE approach using contextual embedding.

4.1   Dataset

The dataset we obtained for our model is further divided into two sets: Newswire corpus and Wikipedia News Corpus (WikiNews) [29]. Our dataset is split into a training set to train the model, development set to validate the model and a test set that is used to calculate the performance of our OIE proposed architecture. The number of sentences and number of tuples in each dataset can be found in Table 2. We tried to test our model using the dataset introduced by [15] that is automatically generated from a Question Answering dataset, but we couldn't obtain it.

Table 2.   Information on the datasets used.

Dataset	Train set # Sent/# Tuples	Dev set # Sent/# Tuples	Test set # Sent/# Tuples
Newswire	744/2173	249/727	248/737
WikiNews	1174/2906	392/946	393/993

4.2   Experimental Results and Analysis

Three evaluation metrics were used to measure the performance of our model: Recall (R), Precision (P) and F-measure (F). All the aforementioned measures were expressed as percentages throughout the experiments. With the F-measure being the breakthrough performance measure. Detailed results of the experiments can be found in Table 3.

Table 3. Summary of the results on both datasets.

Network architecture	Newswire			WikiNews		
	P	R	F	P	R	F
BiLSTM (GloVe)	41.1	45.1	43.0	47.3	46.9	47.1
BiLSTM (ELMo)	43.0	46.7	44.8	49.6	50.7	50.1
GRU (GloVe)	41.0	42.6	41.8	40.8	45.1	42.8
HAN (ELMo)	32.2	43.3	36.9	37.5	45.9	41.3
BiGRU (GloVe)	51.7	49.1	50.4	58.4	54.0	56.1
BiGRU (ELMo)	**53.0**	**51.4**	**52.1**	**60.1**	**57.2**	**58.7**

Experiment 1

In the first experiment we compare the results of employing ELMo embeddings against GloVe embeddings. As demonstrated in Table 3, when a Bidirectional GRU network is employed using ELMo instead of GloVe it yields to an increase in the F-Measure from 56.1% to 58.7% on WikiNews dataset and from 50.4% to 52.1% on Newswire dataset. An increase in the F-Measure by 3% can also be observed when a Bidirectional LSTM model that uses contextual embeddings is employed in contrast to non-contextual embedding. Hence, contextual embeddings have a notable effect on the performance of OIE task.

Experiment 2

Subsequently, in the second experiment We compare our OIE model (BiGRU (ELMo)) against the model proposed by Stanovsky et al. [15] (BiLSTM (GloVe)). Table 3 shows the effect of utilizing contextual word embedding in a Bidirectional GRU network on extracting relation triples. The proposed model achieved an F-Measure of 52.1% compared to 43.0% achieved by [15] on Newswire dataset. Results on Wiki-News dataset followed the same trend, our model increased the F-measure by 11.6%. It is observed that the proposed OIE system outperforms the model proposed by [15].

Experiment 3

In the final experiment, we illustrate the effect of implementing a Bidirectional GRU instead of single direction GRU network. As we previously mentioned in Sect. 3.2, unidirectional networks can only have access to past information, thus output is based on what the network have previously learned, unlike bidirectional networks that can capture both, past and future information. This elaborates the massive decrease in the F-Measure by of the GRU network by 8.6% and by 13.3% on Newswire and Wiki-News respectively, compared to our proposed Bidirectional GRU model.

It is note-worthy that we tested the effect of building a Hierarchical Attention Network (HAN) [30] over a RNN. HAN employs stacked RNN on word-level to capture the informative words in a sentence, it then combines the representation of those vital words to produce a sentence vector [30]. However, the OIE model under-performed using HAN.

5   Conclusion and Future Work

The Bidirectional GRU-based OIE model with contextual word embeddings presented here delivers higher performance than existing state-of-the-art algorithm. The impact that contextual embedding had over our OIE architecture is notable in our experiments. In addition, Bidirectional GRU enhanced the performance with less complexity when compared to Bidirectional LSTM.

We believe that there is still a room for development in the field OIE. OIE can't be regarded as a solved NLP task. For instance, little work has been done in extracting N-ary relations, the main focus has been directed towards the extraction of binary relations, omitting the importance of higher order relations. The presented work can be further extended to extract N-ary relation. In the future, we would like to test our model on a larger dataset and would like to test the adaptability of the model on other languages. Finally, this approach can be employed in other NLP tasks such as question answering and summarization.

References

1. Cho, K., et al.: Learning phrase representations using RNN encoder-decoder for statistical machine translation. arXiv preprint arXiv:1406.1078 (2014)
2. Graves, A., Mohamed, A., Hinton, G.: Speech recognition with deep recurrent neural networks. In: 2013 IEEE International Conference on Acoustics, Speech and Signal Processing. IEEE (2013)
3. Li, B., et al.: Acoustic modeling for Google Home. In: Interspeech (2017)
4. Chung, H., et al.: Alexa, can I trust you? Computer **50**(9), 100–104 (2017)
5. Yin, W., et al.: Comparative study of CNN and RNN for natural language processing. arXiv preprint arXiv:1702.01923 (2017)
6. Sarhan, I., Spruit, M.: Uncovering algorithmic approaches in open information extraction: a literature review. In: 30th Benelux Conference on Artificial Intelligence. Springer CSAI/JADS (2018)
7. Gamallo, P.: An over view of open information extraction (invited talk). In: OASIcs-Open Access Series in Informatics, vol. 38. Schloss Dagstuhl Leibniz Zentrum fuer Informatik (2014)
8. Banko, M., Cafarella, M.J., Soderland, S., Broadhead, M., Etzioni, O.: Open information extraction from the web. In: IJCAI, vol. 7, pp. 2670–2676 (2007)
9. Wu, F., Weld, D.S.: Open information extraction using Wikipedia. In: Proceedings of the 48th Annual Meeting of the Association for Computational Linguistics. Association for Computational Linguistics (2010)
10. Fader, A., Soderland, S., Etzioni, O.: Identifying relations for open information extraction. In: Proceedings of the Conference on Empirical Methods in Natural Language Processing. Association for Computational Linguistics (2011)
11. Etzioni, O., Fader, A., Christensen, J., Soderland, S., Mausam, M.: Open information extraction: the second generation. In: IJCAI, vol. 11, pp. 3–10 (2011)
12. Del Corro, L., Gemulla, R.: ClausIE: clause-based open information extraction. In: Proceedings of the 22nd International Conference on WWW, pp. 355–366. ACM (2013)
13. Cui, L., Wei, F., Zhou, M.: Neural open information extraction. arXiv:1805.04270 (2018)

14. Hochreiter, S., Schmidhuber, J.: Long short-term memory. Neural Comput. **9**(8), 1735–1780 (1997). https://doi.org/10.1162/neco.1997.9.8.1735
15. Stanovsky, G., et al.: Supervised open information extraction. In: Proceedings of the 2018 Conference of the North American Chapter of the Association for Computational Linguistics: Human Language Technologies, Volume 1 (Long Papers), vol. 1 (2018)
16. Loper, E., Bird, S.: NLTK: the natural language toolkit. arXiv preprint cs/0205028 (2002)
17. Peters, M.E., et al.: Deep contextualized word representations. arXiv preprint arXiv:1802. 05365 (2018)
18. Pennington, J., Socher, R., Manning, C.: Glove: Global vectors for word representation. In: 2014 Conference on Empirical Methods in Natural Language Processing (EMNLP) (2014)
19. Mikolov, T., Sutskever, I., Chen, K., Corrado, G., Dean, J.: Distributed representations of words and phrases and their compositionality. In: Advances in Neural Information Processing Systems, pp. 3111–3119 (2013)
20. Pascanu, R., Mikolov, T., Bengio, Y.: On the difficulty of training recurrent neural networks. In: International Conference on Machine Learning (2013)
21. Vukotic, V., Raymond, C., Gravier, G.: A step beyond local observations with a dialog aware bidirectional GRU network for Spoken Language Understanding. In: Interspeech (2016)
22. Cho, K., van Merrienboer, B., Gulcehre, C., Bougares, F., Schwenk, H., Bengio, Y.: Learning phrase representations using RNN encoder-decoder for statistical machine translation. In: EMNLP (2014)
23. Bansal, T., Belanger, D., McCallum, A.: Ask the GRU: multi-task learning for deep text recommendations. In: Proceedings of the 10th ACM Conference on Recommender Systems. ACM (2016)
24. Ramshaw, L.A., Marcus, M.P.: Text chunking using transformation-based learning. In: Armstrong, S., Church, K., Isabelle, P., Manzi, S., Tzoukermann, E., Yarowsky, D. (eds.) Natural Language Processing Using Very Large Corpora. Text, Speech and Language Technology, vol. 11, pp. 157–176. Springer, Dordrecht (1999). https://doi.org/10.1007/978-94-017-2390-9_10
25. Chollet, F.: Keras 2015. https://github.com/fchollet/keras. Accessed 20 Mar 2019
26. Abadi, M., et al.: Tensorflow: a system for large-scale machine learning. In: 12th Symposium on Operating Systems Design and Implementation (OSDI 2016) (2016)
27. Nair, V., Hinton, G.E.: Rectified linear units improve restricted boltzmann machines. In: Proceedings of the 27th International Conference on Machine Learning (ICML-10) (2010)
28. Kingma, D.P., Ba, J.: Adam: a method for stochastic optimization. arXiv preprint arXiv: 1412.6980 (2014)
29. Stanovsky, G., Dagan, I.: Creating a large benchmark for open information extraction. In: Proceedings of the 2016 Conference on EMNLP (2016)
30. Yang, Z., et al.: Hierarchical attention networks for document classification. In: Proceedings of the 2016 Conference of the North American Chapter of the Association for Computational Linguistics: Human Language Technologies (2016)

Towards Recognition of Textual Entailment in the Biomedical Domain

Noha S. Tawfik[1,2]([✉]) and Marco R. Spruit[2]

[1] Computer Engineering Department, College of Engineering, Arab Academy
for Science, Technology, and Maritime Transport (AAST), Alexandria 1029, Egypt
noha.abdelsalam@aast.edu
[2] Department of Information and Computing Sciences, Utrecht University,
3584 CC Utrecht, The Netherlands
{n.s.tawfik,m.r.spruit}@uu.nl
https://www.uu.nl/en/organisation/department-of-information-and-computing-sciences

Abstract. The medical literature suffers from disagreements among
authors discussing the same topic or treatment. With thousands of articles published daily, there is a need to detect inconsistent and often
contradictory findings. Natural language inference (NLI) gained a lot
of interest in the past years, however, domain-specific NLI systems are
yet to be examined in depth. In this paper, we conduct several experiments on sentence pairs extracted from PubMed abstracts, to infer
whether they express entailment, contradiction or neutral meanings.
The main focus of this research is to recognize textual entailment in
published evidence-based medicine findings. We explore popular NLI
models and sentence embeddings, adapted to the biomedical domain.
We further investigate improving the inference detection abilities of the
models by incorporating traditional machine learning (ML) features with
deep learning (DL) architecture. The proposed model serves in capturing
biomedical language's representations by combining lexical, contextual
and compositional semantics.

Keywords: Transfer learning · Textual entailment ·
Sentence embeddings

1 Introduction

In the last decade, the rate of conducting clinical and medical research has
changed dramatically, in terms of both quantity and quality. Subsequently, the
number of published results in forms of research papers, clinical trials and text-
books has witnessed a growth spurt. Catillon's synthesis [2] estimates that the
number of clinical trials has increased from 10 per day in 1975 to 55 and 95 in
1995 and 2015 respectively. In 2017, the PubMed repository contained around
27 million articles, 2 million medical reviews, 500,000 clinical trials and 70,000
systematic reviews. Contribution of medical research is evaluated according to

© Springer Nature Switzerland AG 2019
E. Métais et al. (Eds.): NLDB 2019, LNCS 11608, pp. 368–375, 2019.
https://doi.org/10.1007/978-3-030-23281-8_32

its applicability in the clinical practice and its ability to aid future research in the same field. It is then critical to assess and resonate with published findings specifically when there is more and more evidence on disagreements and contradiction between outcomes [8, 10].

Our work aims to improve the process of evaluating scientific contributions by detecting textual inference between results reported in biomedical abstracts. This paper proposes a model for labeling sentence pairs as entailed, contradictory or neutral. The model relics on linguistic and domain-specific hand-crafted features and recent state-of-the-art sentence encoders. The novelty of our approach is the integration of conventional machine learning features with an encoder-based deep neural network.

2   Related Work

Textual entailment has been widely studied in recent years, with the availability of SNLI, MultiNLI corpora. However, most models fail to generalize across different NLI benchmarks [15], moreover they do not perform accurately on domain-specific datasets. In this section we review textual inference models built specifically for the medical domain. Preclude [11] focuses on extracting conflicts found in health discussions posted in online forums on various health-related topics. The system follows a linguistic rule-based approach to detect inter-advice conflicts. It utilizes MetaMap for semantic clause extraction and tokenization, and then assigns polarity to extracted pairs. More recently, Zadrozny et al. suggested a conceptual framework based on the mathematical sheaf model to highlight conflicting and contradictory criteria in guidelines published by accredited medical institutes. It transforms natural language sentences to formulas with parameters, creates partial order based on common predicates and builds sheaves on these partial orders [17].

There were few scattered attempts on extracting contradictions from scientific articles available online. Sarafraz et al. [12], extracted negated molecular events from biomedical literature using a hybrid of machine learning features and semantic rules. Similarly, De Silve et al. [14], extracted inconsistencies found in miRNA research articles. The system extracts relevant triples and scores them according to an appositeness metric suggested by the authors. Alamri et al. [1], detected contradictory findings through n-grams, negation, sentiment and directionality. Our previous work combined a ranking model to find the most relevant finding per abstract and detected biomedical contradictions through semantic features and biomedical word embeddings [16].

3   Methods

3.1   Dataset

In 2016, Alamri et al. published a dataset of contradictory research claims for medical sentence classification and question answering. It is constructed out of

24 systematic reviews on 4 popular cardiovascular disease topics. Medical experts manually mapped each systematic review to a question and extracted corresponding answers from abstracts of articles referenced in the reviews. Only the most relevant sentence is chosen as answer, it is given a *YES* label if it positively answers the question or *NO* label otherwise. More details on the annotation criteria, process and the corpus statistics can be found in [1]. While the dataset is annotated by experts, its structure is not aligned with the language inference task. For that reason, we reconstruct the corpus by combining claims to build a pairwise-sentence corpus to match conventional NLI datasets. We first fetch the PubMed article ids of all 259 abstracts included in the dataset, and extract the first sentence of each abstract. The first sentence in an abstract often describe the research objective. We enrich the corpus by adding extracted sentences and assigning them with the label *NEUTRAL*. Our choice of objective sentence to fill as neutral is based on the general observation of neutral sentences across different NLI benchmarks where they are usually constructed by adding a purpose clause [7]. Given the unique set of medical questions denoted Q where each question is related to only one systematic review and multiple abstracts. For each q_i that belongs to Q, we assumed the following hypotheses while labeling the sentence pairs as entailed, contradictory or neutral:

- *claim$_2$* entails *claim$_1$* if *asr$_2$=YES* AND *asr$_1$=YES*
- *claim$_2$* contradicts *claim$_1$* if *asr$_2$=YES* AND *asr$_1$=NO*
- *claim$_2$* contradicts *claim$_1$* if *asr$_2$=NO* AND *asr$_1$=YES*
- *claim$_2$* is neutral to *claim$_1$* if *asr$_2$=YES* AND *asr$_1$=NEUTRAL*
- *claim$_2$* is neutral to *claim$_1$* if *asr$_2$=NEUTRAL* AND *asr$_1$=YES*

Where *asr* denotes the assertion value of each sentence with three possible values *YES, NO, NEUTRAL*. *Claims* refer to the question answer extracted from abstracts. It is important to note that for formulating the above guidelines, a definition of 'entailment' and 'contradiction' is needed. Therefore, we follow the original corpus interpretation of contradiction as "Two texts, T1 and T2, are said to contradict when, for a given fact F, information inferred about F from T1 is unlikely to be true at the same time as information about F inferred from T2". The final dataset consisted of 2135 sentence pairs with 1080, 608 and 447 entailment, contradiction and neutral class instances respectively.

3.2 Machine Learning

Human Engineered Features. The model has a total of 20 traditional NLP features divided into 3 main categories. The main selection criteria of features was to capture context, lexical and semantic representations of text with a limited and optimized feature set.

String-Based Features. This sub category includes *editDist, LevSim, CosSim, JacSim* to represent shortest/longest edit distance, Levenshtein similarity, Cosine similarity and jaccard similarity respectively. In addition, we calculate

4 variations of length measures between the two sentences: *LenMax, LenMin, LenAbs, LenAvg*

Contradiction-Based Features. Negation is still a robust measure of appositeness, we define 4 features to detect negation in sentences. *NegationBin* as a binary feature, *NegOverlap* as the jaccard similarity of negated words only, *AntScore* as a score between the count of antonyms found between sentences. To expand the antonyms coverage we use both WordNet and VerbOcean lexicons, and also *Mod-Overlap* as the similarity between modal words found in both input. In addition to the above set we also try to detect the outcome polarity through Subjectivity and sentiment (*SubjScore, SentLabel*) using the NLTK sentiment analyzer. Moreover, the results sentence of scientific articles are often accompanied by a "change clause" that affects the final output [9]. The key is to detect whether changes occurring in both sentences are bad, good or neutral. To measure the final pairwise polarity, we include more features such as *PolarityBin* as a binary feature set to 1 when both sentences share the same polarity and 0 otherwise, and *ChangePolarity* that scores each sentence according to a predefined list of change keywords labelled good (+ve score values) or bad (-ve score values).

Context-Based Features. To include domain knowledge we add *EntityOverlap* that computes the similarity between medical UMLS concepts identified by MetaMap[1]. We also rely on word embeddings to capture context. Our hypothesis is that models trained on domain knowledge would generate vector representation capable of learning conceptual meaning of the domain. We compute *EmbedSim* as the cosine similarity between the two embedding vectors and the *EmbedAvg* as the similarity between embedding average for each sentence pooling of all word embeddings. The word embeddings are extracted using FastText model pre-trained on the PubMed Central open access subset[2]. We add the Word Mover's Distance *WMDSim* as measure of similarity between both sentences.

Classification. We experiment with different classification algorithms available in the Scikit-learn toolkit. The experiments include Support Vector machine, Linear regression model, Random Tree, Gradient boost and Naive Bayes.

3.3 Deep Learning

Sentence Embeddings. Text embedding are considered a key element in various NLP tasks. Popular word embeddings such as Word2Vec and GloVe outperform existing models that rely on co-occurrence counts because of their ability to better represent distributional semantics. To encode sentences with one of the prior models, a simple average of their corresponding word embeddings would yield strong results. Nonetheless, during the last two years we witnessed a rise of different supervised and unsupervised approaches towards learning representations of sequences of words, such as sentences or paragraphs. They are able

---

[1] https://metamap.nlm.nih.gov/.
[2] https://github.com/lucylw/pubmed_central_fasttext_pretrained.

to identify the order of words within a sentence and hence capture more context. The developed sentence representations extend the success of earlier word vectors with interesting results and increasing potential in different tasks. We focus our research on the two of the most popular sentence encoding schemes InferSent and Universal Sentence Encoder. We argue that fine tuning these models and leveraging transfer learning could possibly lead to a good performance in a domain-specific settings. Both chosen encoders were trained partially or fully on textual inference data which fits perfectly with our task.

InferSent is a sentence encoder proposed by Facebook [6]. Its main advantage over other models is its supervised training over SNLI, a large text inference dataset manually annotated. The original model[3] is trained on 570k human-generated English sentence-pairs with a bi-directional Long Short Term Memory (BiLSTM) encoder.

Universal Sentence Encoder (USE) was developed by Google [3]. It has two variations, the first is a transformer-based encoder which yields high-accuracy at the cost of high complexity and extra computational resources. The second model uses a deep averaging network that averages word embeddings and serve as input to a deep neural network. In our model, we deploy the transformer architecture as it was proven to yield better results in several NLP tasks. The universal sentence encoder[4] training data contains supervised and unsupervised sources such as Wikipedia articles, news, discussion forms, dialogues and question/answers pairs. It is also partially augmented with instances from the SNLI corpus.

Deep Learning Network. Our DL model follows a siamese-like architecture where the first set of layers are parallel duplicates and share same weights. For merging the two inputs, we concatenate the element-wise difference and then multiply both vectors. Following that, there are multiple intermediate dense layers. The nodes are directly connected to the nodes in the next layer and use rectified linear activation (ReLU) function. Given the small dataset size, we introduce a dropout layer with a dropout rate of 0.3. Finally, the prediction layer with 3 nodes predicts the probability of each of the inference classes, and a softmax activation function. We adopt an exponentially decaying learning rate, and an l2 regularization weight of 0.01.

3.4   A Feature-Assisted Neural Network Architecture Model

With the small size of the dataset, traditional features demonstrate good performance in comparison with the neural network models. This, along with more evidence on the usefulness of combining traditional features in deep learning architecture [5,13], encouraged us to build a hybrid model. An essential dilemma for building the feature-assisted model is how to incorporate engineered features

---

[3] Pre-trained model for InferSent available at https://github.com/facebookresearch/InferSent.

[4] Pre-trained model for USE available at https://tfhub.dev/google/universal-sentence-encoder-large/3.

to sentence embeddings inputs. Directly appending the traditional ML features to the encoded representations generated from InferSent or USE would not influence the performance. In that scenario, the features' effect on the classification decision would almost be nonexistent given the large size of sentence encoding vector versus the feature vector size of 21 values. Figure 1 gives an overview of the final feature-assisted framework we propose.

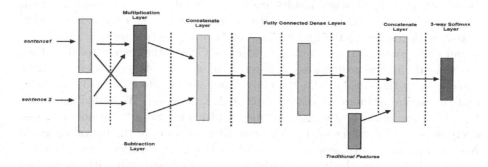

Fig. 1. The feature-assisted neural network architecture.

4   Results and Evaluation

All the following results are calculated as the average results of standard cross-validation with 10 folds. The results reported for the machine learning approach are the output of the best two classifiers: Random Forest (RF) and extreme gradient boosting (XGBoost). It is generally observed that XGBoost almost always achieves higher accuracy than RF. Table 1 shows the results details of the model, The baseline performance is 50.56% based on the majority classifier output. We note, that the ML experiments were not meant for direct comparison with the DL model. The conducted evaluations serve at choosing the best feature combination that could further boost the DL network.

Table 1. Machine learning features with Random Forest and XGBoost classifiers based on 10-fold cross validation. Reported numbers correspond to average accuracy and standard deviation

Feature set	Random Forest	XGBoost
context-based	53.26% (+/- 1.80%)	49.16 % (+/- 2.67%)
contradiction-based	67.81% (+/- 1.28%)	69.49% (+/- 1.77%)
context + string	61.01% (+/- 1.97%)	64.91% (+/- 2.98%)
all features	72.30% (+/- 2.32%)	**76.94% (+/- 1.24%)**

As for the deep learning algorithms, we ran multiple experiments while varying the number of hidden layers and the corresponding number of nodes. Adding more layers test our model capacity, in other terms, with small number of layers the model may struggle to fit the data. On the other hand, over-scaling the network size leads to great results on training data and performs poorly on the test data. Our experiments show that there was a minimal overfitting effect with increasing the number of layers, however, there was no added accuracy. Deep Learning experiments' results are shown in Table 2. In all cases, InferSent encoder outperforms USE encoder with approximately 8%. This finding is consistent with previous published findings [4]. Both encoders are considered universal and should represent sentences efficiently given the amount of data they are trained on. The performance difference between the two encoders could be attributed to the difference in the embedding vector dimension (512 vs 4096) and the nature of inference data *InferSent* is trained on. We added the traditional features to the best performing model with 3 layers and a number of nodes decreasing by 50% with each hidden layer that is deeper in the neural network. No remarkable achievement were noticed in the *USE* encoder case(only 0.6% difference). However, the hybrid model achieves the best result with an average accuracy of 96.21% and a minimum of 94.32% when combined with the *InferSent* encoder. Even with a limited dataset, the results suggest that the machine learning features and deep learning models are complementary. Their combination in an end-to-end model can enhance the learning process and improve the predictions on unseen data.

Table 2. Deep Learning performance results on 10-fold cross validation with respect to the number of hidden layers in the DNN architecture. Reported numbers corresponds to average accuracy and standard deviation

Hidden layers	Hidden units	USE *(Dim.:512)*	InferSent *(Dim:4096)*
1 layers	512	72.56% (+/- 1.14%)	89.88% (+/- 3.91%)
3 layers	512,256,128	82.27% (+/- 1.63%)	**93.95% (+/- 1.39%)**
3 layers	512,256,64	83.17% (+/- 2.20%)	93.86% (+/- 1.48%)
5 layers	512,256,256,128,128	83.68% (+/- 1.50%)	92.24% (+/- 0.79%)
3 layers	512,256,128,64,64	83.68% (+/- 1.50%)	93.18% (+/- 1.73%)

5   Conclusion

We attempt to detect medical text inference from published scientific articles. Various experiments have been applied in different scenarios including ML features and DL network built on top of sentence encoders. Our proposed hybrid architecture is the optimal configuration in terms of size and number of hidden layers. The final results are promising, however, the model must be re-evaluated on a larger corpus to generalize its effect. We could enhance the sentence encoder power by re-training them on domain-specific sources such as research articles and clinical notes. We also believe that a feature ablation analysis over a bigger range of features could potentially select a better boosting vector for assisting the neural network.

References

1. Alamri, A.: The detection of contradictory claims in biomedical abstracts. Ph.D. thesis, University of Sheffield (2016)
2. Catillon, M.: Medical Knowledge Synthesis: A Brief Overview (2017). https://www.hbs.edu/faculty/Pages/item.aspx?num=54337
3. Cer, D., et al.: Universal Sentence Encoder. arXiv preprint, March 2018
4. Chen, Q., Kim, S., Wilbur, W.J., Du, J., Lu, Z.: Combining rich features and deep learning for finding similar sentences in electronic medical records. In: Proceedings of the BioCreative/OHNLP Challenge 2018 (2018)
5. Chen, R.C., Yulianti, E., Sanderson, M., Bruce Croo, W.: On the benefit of incorporating external features in a neural architecture for answer sentence selection. ACM Ref. Format (2017). https://doi.org/10.1145/3077136.3080705
6. Conneau, A., Kiela, D., Schwenk, H., Barrault, L., Bordes, A.: Supervised Learning of Universal Sentence Representations from Natural Language Inference Data. arXiv e-prints, May 2017. http://arxiv.org/abs/1705.02364
7. Gururangan, S., Swayamdipta, S., Levy, O., Schwartz, R., Bowman, S.R., Smith, N.A.: Annotation artifacts in natural language inference data. In: Proceedings of the 2018 Conference of the North American Chapter of the Association for Computational Linguistics: Human Language Technologies (2018)
8. Ioannidis, J.P.A.: Why most published research findings are false. PLoS Med. 2(8), e124 (2005). https://doi.org/10.1371/journal.pmed.0020124
9. Niu, Y., Zhu, X., Li, J., Hirst, G.: Analysis of polarity information in medical text. In: AMIA ... Annual Symposium proceedings. AMIA Symposium 2005, pp. 570–574 (2005)
10. Prasad, V., Cifu, A., Ioannidis, J.P.A.: Reversals of established medical practices: evidence to abandon ship. Jama 307(1), 37–38 (2012)
11. Preum, S.M., Mondol, A.S., Ma, M., Wang, H., Stankovic, J.A.: Preclude2: personalized conflict detection in heterogeneous health applications. Pervasive Mob. Comput. 42, 226–247 (2017). https://doi.org/10.1016/J.PMCJ.2017.09.008
12. Sarafraz, F.: Finding conflicting statements in the biomedical literature. Ph.D. thesis, University of Manchester (2012)
13. Sequiera, R., et al.: Exploring the Effectiveness of Convolutional Neural Networks for Answer Selection in End-to-End destion Answering. arXiv e-prints (2017)
14. de Silva, N., Dou, D., Huang, J.: Discovering inconsistencies in PubMed abstracts through ontology-based information extraction. In: ACM Conference on Bioinformatics, Computational Biology, and Health Informatics (ACM BCB) (2017)
15. Talman, A., Chatzikyriakidis, S.: Testing the generalization power of neural network models across NLI benchmarks. Technical report (2018)
16. Tawfik, N.S., Spruit, M.R.: Automated contradiction detection in biomedical literature. In: Perner, P. (ed.) MLDM 2018. LNCS (LNAI), vol. 10934, pp. 138–148. Springer, Cham (2018). https://doi.org/10.1007/978-3-319-96136-1_12
17. Zadrozny, W., Garbayo, L.: A sheaf model of contradictions and disagreements. Preliminary report and discussion. In: International Symposium on Artificial Intelligence and Mathematics, Florida

Development of a Song Lyric Corpus for the English Language

Matheus Augusto Gonzaga Rodrigues[ID], Alcione de Paiva Oliveira[(✉)][ID], and Alexandra Moreira[ID]

Universidade Federal de Vicosa, Vicosa, MG 36570900, Brazil
ppgcc@ufv.br

Abstract. Web Scraping Tools are simplifying the task of creating large databases for various applications such as the construction of corpus aimed at the development of applications for natural language processing. Many of these applications require a large amount of data, and in that sense, the Web presents itself as an important data source. Among the various tasks in the NLP scope, one of the most challenging is automatic text generation. In this task the objective is to generate syntactically and semantically correct texts after a training process on a particular corpus. This article presents the elaboration of an English song lyrics Corpus, extracted from the Web, that can be used to train applications for automatic generation of lyrics, poems, or other NPL related tasks. After its normalization, an analysis of the Corpus is presented, as well as analyzes performed after the corpus vectorization (embedding) generated with the use of two current techniques.

Keywords: Text generation · Corpus linguistic · Song lyrics

1 Introduction

Corpora are linguistic resources that are difficult to create and time-consuming. Nevertheless, they are very useful resources for language studies and training of natural language processing tools. In general, these resources are constructed from texts and digitized documents, in the case of corpora based on written natural language. In the case of corpora based on spoken natural language, conversations or interviews are used. More recently, the Web has presented itself as an important source of raw material for building corpora. There is a vast amount of textual and oral material in digital form and available in several languages. There is also a growing demand for large corpora due to the emergence of modern machine learning tools. As a result, Web-based corpora propositions are emerging. Habernal et al. [3] presented a Multilingual Web-size Corpus containing over 10 billion tokens, licensed under Creative Commons license family

This study was financed in part by the Coordenação de Aperfeiçoamento de Pessoal de Nível Superior - Brasil (CAPES) - Finance Code 001, and also by the funding agencies FAPEMIG and CNPq.

E. Métais et al. (Eds.): NLDB 2019, LNCS 11608, pp. 376–383, 2019.
https://doi.org/10.1007/978-3-030-23281-8_33

in more than 50 languages. According to the authors, the texts that compose the corpus were extracted from CommonCrawl (commoncrawl.org), the largest publicly available general Web crawl to date with about 2 billion crawled URLs. The size and diversity of the Web also allows the construction of large specialized corpora. Seitner et al. [11] presented a publicly available database containing more than 400 million hypernymy relations extracted from the CommonCrawl web corpus. However, there are few corpora geared towards studies of poems and song lyrics.

In the present work we try to contribute to minimize this problem, presenting a song lyrics Corpus of music of random musical genres in English constructed with the use web scraping technique. The objective is to use the corpus as input for machine learning tools and automatic song lyrics generation. The article describes the process of extracting the song lyrics, the normalization phase and the final analysis of the created Corpus. Next section presents Corpus development work that covers issues related to what is presented in this article. In Sect. 3 is explained the web scraping method used to collect the music lyrics and how the data was processed after the collection, as well as the "cleanup" process adopted. In Sect. 4, the Corpus analysis is presented, where N-grams, noisy words and other features are presented. In Sect. 5, two different forms of embedding applied in Corpus are compared, and finally we present a brief conclusion along with the indication of future work in Sect. 6.

2   Related Works

Here we present which bear a certain resemblance to the current work. As mentioned earlier, there are not many works related to creating lyrics corpus, but it is possible to find some related datasets.

Nishina [9] investigated various features identified in the lyrics of contemporary popular songs ranked in the Billboard top 100 songs covering the 2002–2011 period. The author gathered from the site a total of 1,000 songs and, after that excluded noise characters such as leading whitespace. The resulted corpus presented an average of 502 tokens, and an average of 149 types for the 10-year period. Subsequently, the author performs a linguistic analysis on the corpus, analyzing the genre of the songs and the expressions used according to the sex of the author of the lyrics. The main difference of this work for the current work is the size of the corpus. Being such a small corpus is not suitable for quantitative analyzes and to feed machine learning tools, lending itself more for qualitative analysis.

Miethaner [6] developed BLUR (Blues Lyrics) corpus, containing blues lyrics from the early twentieth century, focusing on the study of syntactic phenomena in earlier African-American English. The corpus is composed of a computerized collection of more than 8,000 transcripts of pre-World War II blues recordings. Like the previous article, the corpus developed is too small to be handled by machine learning tools. Machine learning algorithms tend to work better on larger datasets, due to the bigger quantity of examples.

Ellis et al. [2] presented the LyricFind Corpus (www.smcnus.org), developed at the Sound & Music Computing laboratory at the National University of Singapore. The corpus consists of 275,905 distinct lyrics in bag-of-words format (67.6 million tokens). This is a corpus that is worth mentioning due to its volume, though, because it is presented in the form of bag-of-words, it is not suitable for use in training for text generation.

Kuznetsov [5] has collected 57650 songs acquired from LyricsFreak (www. lyricsfreak.com) through scraping. According to the author, he did some basic cleaning on the lyrics, removing non-English lyrics, extremely short and extremely long lyrics, and lyrics with non-ASCII symbols. Compared with current work, the number of song lyrics is less than half, although volume provided is enough to employ machine learning techniques.

3   Extraction of Lyrics and Corpus Cleaning

Following the same web scraping flow described by Milev et al. [8], the creation of the corpus began in the selection of a source where it was possible to extract a sufficiently large number of song lyrics. Initially, the site chosen was Genius.com (genius.com), due to the fact that it is one of the most used websites in the field, and contains explanatory notes in some stanze of the songs, which could later be used to enrich Corpus content. The site provides an Application Programming Interface (API) for extracting data from songs (lyrics, artists and other metadata). However, its use is limited so that a maximum daily amount of extractions of the lyrics is imposed. For this reason, it was decided not to use it. Thus, it was decided that an open site would be used, where the number of requests was unlimited. The website musica.com (www.musica.com) contained, up to October 2018, an amount of 979,972 registered song lyrics. The format of the URL and the website page layout enabled a pretty simple extraction of its information, and by applying a small script written Python it was possible to extract 120,946 lyrics of songs from the site.

The goal was to extract only lyrics in English, however, some lyrics in other languages were also downloaded. Some songs written in Spanish, German, French and Italian were detected. For this reason, the cleaning of the corpus began in the extraction process itself. The `langdetect` native language library of Python was used to identify the natural language in which the lyrics of the song were written. After extracting the song lyrics from the site, two files were generated: one with the Corpus itself, where each song was represented by a single index followed by the lyrics, and another file containing the song metadata: the artist and the title of the song.

As was said, the first phase of the cleaning process was the elimination of song lyrics in languages other than English. Still, some lyrics in English have expressions in other languages, mostly in Spanish, since many Latin artists who produce music for the American market mix the two languages. Several tracks of music were found that blended two languages, mostly Spanish and English. The song "Bailamos" by Enrique Iglesias has excerpts such as "... te quiero amor mio, bailamos, gonna live this night forever". The decision was to keep these lyrics with this characteristic due to the fact that this is a specific characteristic of certain artists, and often, of specific genres such as pop music.

The second processing phase of cleaning the Corpus involved the structure of the song lyrics. Many of the extracted song lyrics contained markers indicating repetitions of the lyrics elements and types of elements such as chorus and verses. So markers such as [Verse], [Chorus], [Repeat 2x], [Repeat 3x], among others have been removed. In this second phase, once again, a Python script was used to remove these markers. Markers with names of artists that was intended to clarify which artist was responsible for singing a certain part of the song have been removed as well. For example, in the song "Home Alone" by R. Kelly, the singer has the collaboration of another artist, Keith Murray. The piece of music in which Keith sings is demarcated by [Keith].

After finishing the cleaning, two new versions of the corpus were generated: in addition to the original corpus, a tokenized and lemmatized corpus was generated, in order to reduce the vocabulary size. The spaCy library tokenizer [4] was used and for the lemmatization, it was used the Python NLTK library [7], which has a built-in implemented lemmatizer (WordNetLemmatizer). Thus, for instance, in the song "If I Die 2Nite" by Tupac Shakur, the sentence "A coward dies a thousand deaths. A soldier dies but once" would become "A coward die a thousand death. A soldier die but once".

The second version of the corpus was generated when the stopwords were removed. Among the words present in the stopword list are "me", "I", "myself", "you", "you're", for example. Note that these are words present in almost every song, and that often contribute to the expression of some feeling in the context. For example, in the song "Forever Man" by Eric Clapton, the phrase "How many times I say I love you" after processing would look like this "many times must say love", which completely removes the sense of it. For this reason, the corpus version without stopwords was used only for N-grams analysis, described in the next section. Some other considerations were taken into account when analyzing the initial corpus. For example, the presence of onomatopoeia as "whoa", "ooooh" in its most diverse forms, as well as the emphasis on certain syllables of some words to generate musicality, as in the case of "girl", that several times was used like "Girrlll". It was decided that such structures would be maintained because they were used to promote more musicality to the song and, in a way, highlight the given word in context. In addition, by maintaining such constructs, we avoid reducing the number of Corpus types. The last consideration regarding the cleaning of Corpus was the removal of punctuation from both versions.

4   Corpus Analysis

Corpus analysis was done separately for the two versions described in the previous section. Firstly, an analysis of the frequency distribution was made in the two corpora, with the purpose of identifying the number of tokens and types in each one of them and to establish their size. In addition, an analysis of the occurrence of the unigrams in both was performed. For the lemmatized Corpus, 12,355,270 tokens and 175,412 types were counted. It is only natural that there is a much greater number of tokens than types, especially when it comes to song lyrics since there are many repetitions like what happens in choruses.

Afterwards, an analysis of the unigrams in the Corpus was carried out. In the Fig. 1, the thirty most frequent words in the corpus are presented. As expected, the most common words in Corpus were words like "the", "you", "I", "and", among others.

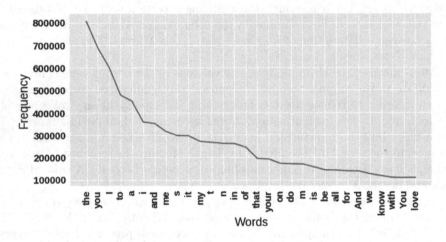

Fig. 1. The thirty most frequent unigrams in the lemmatized corpus.

In the case of the corpus without stopwords, 11,300,686 tokens and 237,786 types were counted. The Fig. 2 shows the thirty most frequent bigrams in the corpus. Notably, the pronouns "you" and "I" participate in several bigrams among the most frequent.

5   Embeddings

Two vectorization techniques were applied in the corpus and, following, an analysis was performed. First, the Word2Vec technique [7] was used through the Python Gensim package [10]. A similarity analysis was performed between words present in both versions of the corpus (lemmatized corpus and the corpus without stopwords).

Based on some of the results displayed in the Table 1, it can be stated that Word2Vec was capable of capture the context of slangs. In America, people usually refer to the word money as cheese informally, and mainly in music. On the lemmatized lyrics corpus, the word cheese can refer to the noun cheese that represents food. As an example, the song *"Summer Girls"* composed by the artist LFO has the following use of the word cheese: *"Think about that summer*

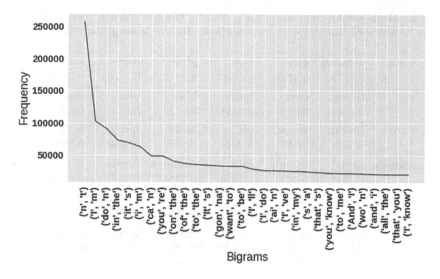

Fig. 2. The thirty most frequent bigrams in the corpus

Table 1. Words similar to "man", "car", "love", "money", "drugs" and "life", with a window of words of size 5 in the lemmatized corpus using Word2Vec.

man	woman	guy	boy	girl	person	brother
	0.8154	0.7377	0.6926	0.6897	0.6531	0.6331
car	truck	limousine	jeep	cadillac	benz	automobile
	0.7921	0.7346	0.7241	0.7208	0.7080	0.6747
love	loving	promise	life	dream	trust	hope
	0.6947	0.6529	0.6418	0.6385	0.6335	0.6326
money	cash	dollar	dough	loot	cheese	respect
	0.8595	0.8561	0.8359	0.7413	0.6123	0.6118
drugs	dope	monopoly	cocaine	auto	junky	liquor
	0.7383	0.6012	0.5849	0.5822	0.5806	0.5722
life	world	existence	destiny	lifestyle	dream	fate
	0.7143	0.6875	0.6805	0.6620	0.6604	0.6598

and I bug cuz I miss it. Like the color purple macaroni and cheese". In the other hand, the group Cypress Hill use cheese as money on the song "Superstar": "... come with me, show the sacrifice it takes to make the cheese". The same analysis can be done to the word dope. During informal conversations it can be used as "cool" or "nice". In this case, Word2Vec showed that the word dope was 0.7383 similar to drugs, which is true, due to the fact that it is used on song lyrics to refer to illegal drugs taken for recreational purposes. For the corpus without English stopwords, we have the following percentages of similarity (Table 2).

Table 2. Words similar to "man", "car", "love", "money", "drugs" and "life", with a window of words of size 5 in the corpus without stopwords using Word2Vec.

man	woman 0.7175	guy 0.67.06	boy 0.65.43	girl 0.62.07	dude 0.57.16	kid 0.56.70
car	truck 0.7261	cars 0.6689	benz 0.6254	bike 0.6232	van 0.6152	bus 0.5961
love	loving 0.6668	know 0.6131	oh 0.6058	babe 0.6004	baby 0.5986	loves 0.5851
money	cash 0.7958	dough 0.7630	loot 0.6359	moneys 0.5756	funds 0.5674	chips 0.5529
drugs	dealers 0.6477	Drug 0.6209	dope 0.6141	dealer 0.5957	cocaine 0.5790	fiend 0.5640
life	world 0.6632	lifes 0.6170	lives 0.6127	lifetime 0.5852	love 0.5551	time 0.5326

In order to perform a comparison, analyzes were also performed using another word-vectoring technique. The second vectorization technique used was fastText [1]. The great differential of fastText in relation to the representation made by Wor2Vec is that each word is represented as a bag of character n-grams in addition to the word itself. For example, be the word "money", then using the n = 4 parameter in the fastText configuration, it will generate representations for the character n-grams "<mon", "mone", "oney", and "ney>", using the characters "<" and ">" as boundary symbols. Because fastText constructs the vector for a word from n-gram vectors that constitute a word, it is able to output a vector for a word that is not in the pre-trained model. This can be quite interesting for lyrics generation since it works with phonetic similarity.

Table 3. Words similar to "man", "car", "love", "money", "drugs" and "life", with n = 3 in the corpus without stopwords using FastText.

man	moman 0.7790	catwoman 0.7782	Woman 0.7734	fellowman 0.7551	mr.man 0.7520	lawman 0.7509
car	truck 0.7957	limousine 0.76.86	cadillac 0.7617	houseboat 0.7603	driveway 0.7481	carib 0.7441
love	mylove 0.8217	trust 0.7080	loving 0.6974	promise 0.5912	babe 0.5717	dream 0.5779
money	moneys 0.8847	cash 0.8783	dough 0.8103	loot 0.7777	respect 0.6352	cheese 0.6320
drugs	drug 0.8595	drugstore 0.8075	psychiatrics 0.7593	drugdealer 0.7526	buggery 0.7525	crackheads 0.7494
life	lifeall 0.8181	lifestlye 0.7981	livelihood 0.7616	lifes 0.7585	prolife 0.7581	lifesaver 0.7388

Table 3 shows the words that are similar to the previous selected words. It is possible to see, specially in the similarity results of the words "man" and "live", that fastText is more phonetic than word2vec.

6   Conclusions

In the present article the development of a corpus containing song lyrics was presented. The corpus underwent a normalization process and two versions were created, one where the words were placed in their lemma form and the other where the stopwords were taken. No syntactic or semantic annotation process has been done, leaving these stages as future work. The corpus is, so far as it has been found in current literature, the largest, except those that are in the form of bag-of-words, but which are not suitable for text generation tools. It is expected that the availability of the corpus allows the development, testing and evaluation of tools that seek the generation of text focused on poetry and song lyrics.

References

1. Bojanowski, P., Grave, E., Joulin, A., Mikolov, T.: Enriching word vectors with subword information. Trans. Assoc. Comput. Linguist. **5**, 135–146 (2017)
2. Ellis, R.J., Xing, Z., Fang, J., Wang, Y.: Quantifying lexical novelty in song lyrics. In: ISMIR, pp. 694–700 (2015)
3. Habernal, I., Zayed, O., Gurevych, I.: C4corpus: multilingual web-size corpus with free license. In: LREC, pp. 914–922 (2016)
4. Honnibal, M., Montani, I.: spacy 2: natural language understanding with bloom embeddings, convolutional neural networks and incremental parsing (2017)
5. Kuznetsov, S.: 55000+ song lyrics. https://www.kaggle.com/mousehead/songlyrics. Accessed March 2019
6. Miethaner, U.: The blur (blues lyrics collected at the University of Regensburg) corpus: blues lyricism and the African American literary tradition. Curr. Objectives Postgrad. Am. Stud. **2** (2001). https://doi.org/10.5283/copas.64
7. Mikolov, T., Sutskever, I., Chen, K., Corrado, G.S., Dean, J.: Distributed representations of words and phrases and their compositionality. In: Advances in Neural Information Processing Systems, pp. 3111–3119 (2013)
8. Milev, P.: Conceptual approach for development of web scraping application for tracking information. Econ. Altern. (3), 475–485 (2017)
9. Nishina, Y.: A study of pop songs based on the billboard corpus. Int. J. Lang. Linguist. **4**(2), 125–134 (2017)
10. Řehůřek, R., Sojka, P.: Software framework for topic modelling with large corpora. In: Proceedings of the LREC 2010 Workshop on New Challenges for NLP Frameworks, pp. 45–50. ELRA, Valletta, Malta, May 2010. http://is.muni.cz/publication/884893/en
11. Seitner, J., et al.: A large database of hypernymy relations extracted from the web. In: LREC, pp. 360–367 (2016)

A Natural Language Interface Supporting Complex Logic Questions for Relational Databases

Ngoc Phuoc An Vo$^{(\boxtimes)}$, Octavian Popescu, Vadim Sheinin, Elahe Khorasani, and Hangu Yeo

IBM Research, Yorktown Heights, USA
ngoc.phuoc.an.vo@ibm.com,
{o.popescu,vadims,elkh,hangu}@us.ibm.com

Abstract. Natural Language Interface to Databases (NLIDB) systems accept questions in any supported natural language (i.e English) and allow users to interrogate a database. Users can access and derive information from a relational database without requirement for knowledge of database language. In this paper we introduced an NLIDB system which not only supports simple logic questions but also attempts to understand and resolve complex logic ones. The system is also equipped with a Natural Language Generation module to generate human-like description for given queries to help users to understand how a query processed and assess the correctness of the result returned. Experiment results show that our system can handle some types of complex logic questions effectively.

Keywords: Natural language interface ·
Natural Language Generation · Relational databases ·
Complex logic questions

1 Introduction

Natural language interfaces to database (NLIDB) are systems that process natural language queries into SQL queries for any specific database. The translation from natural language to a logical language is often difficult and specialized algorithms must be developed to decompose the meaning into a sequence of logical commands.

We have developed a system that is able to identify two types of queries, conveniently called, simple and complex, respectively. Simple queries are processed as described in [11]. Complex queries are decomposed into a sequence of simple queries with the information on how to interconnect them.

In this paper we report on a significant progress made in dealing with complex queries. We implemented a unified approach over what apparently are different types of complex questions. This approach, which improved the previous results

E. Métais et al. (Eds.): NLDB 2019, LNCS 11608, pp. 384–392, 2019.
https://doi.org/10.1007/978-3-030-23281-8_34

by 20%, implements a strategy of decomposing a complex query into a sequence of simple subqueries which are linked through the usage of variables.

In Sect. 2 we give a brief description of the architecture of the system, followed by Sect. 3, in which we focus on complex questions and what their automatic decomposition looks like, while the next section describes evaluation experiments. The paper ends with Related Work and Conclusion sections.

2   A Brief System Description

A deep natural language understanding module is employed to decompose the complex questions into an ordered sequence of simple questions. At the heart of the processing of simple questions resides a mechanism of transforming the query into a sequence of `data items`, i.e. terms that have a direct and specific connection to the database and SQL query. The data items are piped further to an SQL generating engine that returns an answer. However, the system does not stop here. The SQL queries and the results are interpreted and transformed into controlled English sentences by a dialog component using Natural Language Generation (NLG) module. These sentences are displayed so that users can see effectively how the initial question was processed and users have a direct way to assess whether the result returned is correct or not.

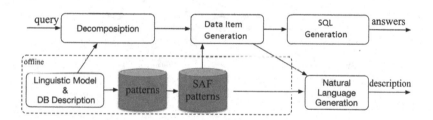

Fig. 1. System architecture

Our system is modular (see Fig. 1) and consists of the following modules:

1. Decomposition module using natural language understanding for decomposing a complex question into a set of ordered simple sub-questions,
2. Data Item Generation module to generate `data items` from simple sentences, where `data item` is a tuple (consists of table name, column name, aggregation function, and filter) which makes the connection between parts of the question and the specific database in use,
3. SQL Generation Engine that is able to produce full-fledged SQL queries and return an answer, and
4. Natural Language Generation module (NLG) which is in charge of automatically producing English sentences for each SQL query, such that the user can have a direct way to verify the correctness of the whole process.

Schema Annotation File (SAF) stores the English sentences generated by the syntactic patterns according to the database description. In our development and experiment, we used three database schemas corresponding to three SAF as below:

- Warehouse (Sales) schema stores information about sales history of products and attributes of customers, products, shops and manufacturers. Figure 2 shows the visualized SAF for Sales database schema.
- Human Resources (HR) schema contains synthetic data records and it represents a human resources data model.
- (TPOX) schema contains synthetic records of history of security transactions and holdings of customer accounts in three stock exchange markets.

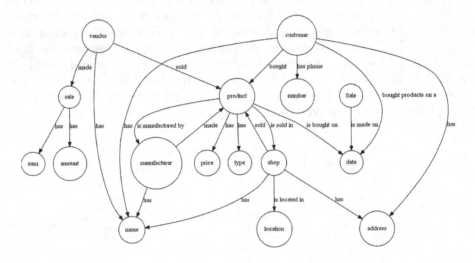

Fig. 2. Visualized SAF for sales database schema

The modules (1) and (4), Decomposition and NLG respectively, have an `off-line` component, where the deep natural language understanding models are created from the `Schema Annotation File (SAF)`. The `on-line` component applies the learned model to a specific question. Data Items extraction lies at the core of module (2), Data Item Generation. Data items are created from parts of the question's dependency parses by matching these parts against a set of phrases that describe the relationship between tables and columns in that particular database. The SQL Generation Engine built on the IBM Cognos Analytics[1] automatically generates SQL query from the data items constructed in the previous step, submits the query to the database to get the final answer and returns it to users.

---

[1] https://www.ibm.com/products/cognos-analytics.

Table 1. Examples of complex logic questions by types and database schemas.

	Complex logic questions	Type	Schema
1	what Apple product is more expensive than the average price of Microsoft products?	Q	Sales
2	What shop stocked more Iphones than iPhones in BESTBUY?	Q	Sales
3	What product was sold in shops in NY more in 2017 than in 2018	Q	Sales
4	What department has more employees than Sales?	Q	HR
5	How many managers manage more employees than Peter Gild?	Q	HR
6	What client placed more buy order in 2013 than Kozi Camps in 2012?	Q	TPOX
7	Who bought more AVEO in 2012 than they sold in 2011?	Q	TPOX
8	Which product types were sold after Willie bought Iphone?	T	Sales
9	what vendor sold Iphone to Willie before Donald bought Galaxy?	T	Sales
10	How many employees left Marketing department after Jack Smith was hired?	T	HR
11	Show me the names of employees who joined Sales after Jack London?	T	HR
12	How many clients started accounts before Kozi Camps?	T	TPOX
13	What client bought IBM shares before Kozi Camps on July 11?	T	TPOX
14	What are the types of products that were sold in Bestbuy between May and June 2014?	I	Sales
15	What product did Willie buy after Alonzo bought Iphone?	I	Sales
16	What are the employees who have birthday in July 1984?	I	HR
17	What are the departments that are managed by Jack Smith?	I	HR
18	How many agents trade IBM on accounts that are held by Kozi Camps?	I	TPOX
19	How many companies are listed on the market that lists IBM?	I	TPOX
20	What vendor sold more products in 2016 than it stocked in same year?	Q+T	Sales
21	How many shops in NY sold more Iphone than Galaxy after Donald bought Galaxy?	Q+T	Sales
22	Which departments hired more employees than Sales before Jack London was hired?	Q+T	HR
23	What managers manage more employees than Peter Guild after Jack London left Sales?	Q+T	HR
24	What client bought more ABC than IBM after ABC was listed?	Q+T	TPOX
25	What client placed more IBM buy orders than Kozi Camps after he sold ABC?	Q+T	TPOX
26	what are products that have prices greater than GALAXY?	Q+I	Sales
27	What is a name of a manufacturer that manufactured more products than Samsung?	Q+I	Sales
28	What are the departments that have salary higher than Sales?	Q+I	HR
29	what are the departments that employees joined more than Marketing?	Q+I	HR
30	When clients bought securities at a price higher than the price that ABC was sold on June 15?	Q+I	TPOX
31	Which agents manage accounts that have more IBM shares than Kozi Camps' account?	Q+I	TPOX
32	What is the addresses of shops that sold Iphones after Willie bought Galaxy?	T+I	Sales
33	Who was hired after the year in which Jack Smith was hired?	T+I	HR
34	What employee was hired after the employee in Marketing department who has the highest salary	Q+T+I	HR

3   On Query Decomposition

A complex question does not allow a direct translation to `data items`, because the true relationship between data items may be masked by a linguistic complex construct. In order to generate the correct SQL query, a deep language understanding model is created off-line, which inputs the SAF and creates a type base matching model. When this model is matched against a query, each predicate occurring in the query has its logical arguments overtly expressed according to the subcategorization frame (SCF) of each verb as specified in SAF. The model is able to infer when and how to create a sub-query that results in a value of the expected type as they are specified in SAF [8].

For example, in the question *"What company manufactured more products than Samsung"*, *Samsung* is not the logical argument of the operator *more than*, which requires a `QUANTITY` as an argument, as required by the operator *more than*. This is a quantitative complex question; we call this type, type Q. The algorithm detects the type inconsistency between *Samsung* and `quantity` argument requested by the operator, *more than*. The strategy to resolve this, is to create new predicate that connects the two types, that is, this predicate is expressed via an English sentence with a SCF matching one of the patterns in SAF.

Same decomposition algorithm is employed for temporal queries, type T queries, like *"which shop sold Iphone before BestBuy"*. The fact that the operator *before* requires a `DATE` type is just a parameter that is passed to the decomposition algorithm. The matching startegy follows the same steps as above.

For indirect references, type I queries, such as relative clauses like *"who was hired after the year in which Jack Smith was hired"*, the same algorithm first determines in a simple query which exact year is mentioned and then uses that value in a subsequent simple query.

The connection between different sub-queries is maintained through the use of variables. Each query has access to the variables defined by the previous sub-queries. The decomposition of the aforementioned relative-clause example is:

- query1 = year in which Jack Smith was hired; assignment year as Q1XA1.
- query2 = On what date employee was hired; assignment date as Q2XA1; assignment employee as Q2XA2; assignment each employee as Q2XA3.
- query3 = connection QX1 and QX2; computation Q1XA1 and Q2XA1 with operator GreaterThan; selection Q2XA3.

The variable "`Q1XA1`" retains a specific date, determined after this sub-query is executed, which later is used to have a meaningful comparison in the third sub-query.

The number of sub-queries might be variable, depending on the type of operations required to be performed in order to get to the right answer.

The system is able to cope with queries involving missing information in quantitative, temporal queries or with queries involving indirectly reference or combinations of these. In table we give examples of such queries, indicating the type of complexity by type Q (quantitative type), T (temporal type), I (relative clause type), or combination of those, like Q+T or Q+T+I, see Table 1.

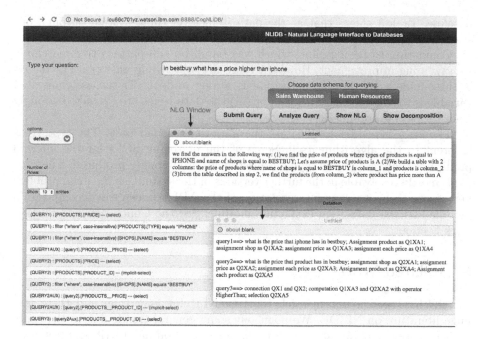

Fig. 3. Example of a complex question for sales schema.

Figure 3 shows example how complex question is handled in Sales schema. The question is entered, the database schema is selected, by default the question will be decomposed into sub-questions and displayed in Decomposition window, then these sub-questions are processed to generate `data items` as a means to produce the final SQL. The NLG text is also generated for the question as shown in the NLG window.

Natural Language Generation. The NLG model takes the decomposed SQL query as input which is a nested SQL query, and then generates the interpretation for each single query using template-based rules. Finally, our NLG model combines these interpretations by solving the cross reference and producing a human-like response to the users.

4   Experiments and Analysis

Datasets. For evaluating the whole system with embedded decomposition, we created two testing datasets: Mixed-1 (81 questions) and Mixed-2 (65 questions) in which all question types and schemas are mixed together.

Evaluation and Error Analysis. For the evaluation of performance on the training data, Table 2 shows how well the stand-alone decomposition can decipher complex questions into sets of simple logic questions. For testing the system as a whole, we implemented the system described in the literature [11] as the baseline. Table 3 shows that at the data item level (final system output), our system outperforms the baseline by a large margin for both testing datasets.

Comparing to the baseline, which was based on sub-pattern tree matching, we saw a big improvement. However, there are still wrong answers. The reason is three-folded: (i) the decomposition is wrong and/or (ii) the assignment of variable is wrong/inconsistent, and/or (iii) the data items are extracted wrongly from the simple sub-queries.

The error types (i) && (ii) are the majority, around 65%. The type (i) is inherent to the algorithm itself. Some of the matching SCFs for the inferred predicates that resolve the type inconsistencies are wrongly produced. The type (ii) is mostly resolvable in the further versions of the system, as the main problem here seems to be a wrong interpretation of the API between the decomposition module and the data item extraction module. See the Conclusion and Further work Section.

5 Related Works

NLIDB has been a hot research topic in NLP for long time and there is a huge literature on NLIDB. We point out some general trends due to the space limit of this paper. Initial systems developed based on hand written rules, were

Table 2. Training evaluation for decomposition module.

Dataset	Total	Correct	Incorrect	Accuracy(%)
Type Q				
Sales	60	45	15	75%
HR	N/A	N/A	N/A	N/A
TPOX	79	69	10	94.5%
Type T, Type Q+T				
Sales	73	51	22	69.9%
HR	26	17	9	65.4%
TPOX	50	32	18	64%
Type I, Type Q+I				
Sales	42	25	17	60%
HR	22	14	8	64%
TPOX	40	25	15	62.5%

Table 3. End-to-end testing accuracy.

Systems	Mixed-1 (81 questions)	Mixed-2 (65 questions)
Our system	70.4%	76.7%
Baseline	29.6%	51.7%

very accurate and able to resolve complicated sentences, but lacked coverage. Basically these systems can respond to a very limited (type) of natural language questions. The problem relating to the complexity of logical quantifiers and their scope was very well studied and interesting solutions, not generalizable though, were found. The literature [3] provides an excellent overview of the state of the art before the 1990's.

The focus of research in the nineties was on less NLIDB type of problems. But important steps were made towards systems that may be extremely effective tools for database access and may use some form of English (or other natural languages); however, it does not mean that the problem of completely unrestricted use of natural language queries has been solved, see [1] for a relevant survey.

The study [7] adopts a pragmatic thinking, by pointing out that the existence of very complex natural language queries does not preclude building automatic systems that are able to process simple queries. The recent developments in deep learning algorithms seem to bring back into the foreground the NLIDBs. Nevertheless, there are a lot of achievements in between. Working on related topics, such as learning inferences over constrained outputs, global inference in natural languages [9,10], open domain relation extraction [4], or more recently, resolving algebraic problems expressed in natural language [5], semantic parser [12] provide a strong ground for approaching the NLIDB by taking advantage of the power of semi-supervised and unsupervised techniques. Some papers, among which [2,6] have directly approached the NLIDB problem. However, there is no explicit reference to complex logic questions, because the evaluations were carried out in an undifferentiated manner.

6   Conclusion and Further Work

The current system implements an algorithm of decomposing of complex queries on the basis of type incompatibility as resulting from SAF files. This algorithm performs significantly better that the baseline of directly inferring the data items from dependency trees. For the next version of the system we focus on reducing firstly the type (ii) errors - that is, the API errors between decomposition and data item modules. Secondly. We focus on type (i) error - the errors due to misinterpreting the queries, and we plan to use an enhanced natural language deep understanding model that takes into account a larger class of linguistics phenomena.

References

1. Androutsopoulos, I., Ritchie, G.D., Thanisch, P.: Natural language interfaces to databases - an introduction. Nat. Lang. Eng. 1(1), 29–81 (1995)
2. Condoravdi, C., Richardson, K., Sikka, V., Suenbuel, A., Waldinger, R.: Natural language access to data: it takes common sense! In: 2015 AAAI Spring Symposium Series (2015)

3. Copestake, A., Spärck Jones, K.: Inference in a natural language front end for databases. Technical report, University of Cambridge, Computer Laboratory (1989)
4. Fader, A., Soderland, S., Etzioni, O.: Identifying relations for open information extraction. In: EMNLP (2011)
5. Koncel-Kedziorski, R., Hajishirzi, H., Sabharwal, A., Etzioni, O., Ang, S.D.: Parsing algebraic word problems into equations. TACL **3**, 585–597 (2015)
6. Li, F., Jagadish, H.: Constructing an interactive natural language interface for relational databases. Proc. VLDB Endowment **8**(1), 73–84 (2014)
7. Popescu, O.: Learning corpus patterns using finite state automata. In: Proceedings of the 10th International Conference on Computational Semantics, pp. 191–203 (2013)
8. Popescu, O., Vo, N.P.A., Sheinin, V., Khorashani, E., Yeo, H.: Tackling complex queries to relational databases. In: Proceedings of the 11th Asian Conference on Intelligent Information and Database Systems, ACIIDS 2019 (2019)
9. Punyakanok, V., Roth, D., Yih, W.t., Zimak, D.: Learning and inference over constrained output. In: IJCAI, vol. 5, pp. 1124–1129 (2005)
10. Roth, D., Yih, W.: A linear programming formulation for global inference in natural language tasks. In: HLT-NAACL, pp. 1–8 (2004)
11. Sheinin, V., Khorasani, E., Yeo, H., Xu, K., Vo, N.P.A., Popescu, O.: Quest: a natural language interface to relational databases. In: Proceedings of the Eleventh International Conference on Language Resources and Evaluation, LREC 2018 (2018)
12. Wang, Y., Berant, J., Liang, P.: Building a semantic parser overnight. In: ACL (2015)

Waste Not: Meta-Embedding of Word and Context Vectors

Selin Değirmenci$^{(\boxtimes)}$ ⓘ, Aydın Gerek ⓘ, and Murat Can Ganiz ⓘ

Marmara University, 34730 Istanbul, Turkey
selindegirmenci@marun.edu.tr, {aydin.gerek,murat.ganiz}@marmara.edu.tr

Abstract. The word2vec and fastText models train two vectors per word: a word and a context vector. Typically the context vectors are discarded after training, even though they may contain useful information for different NLP tasks. Therefore we combine word and context vectors in the framework of meta-embeddings. Our experiments show performance increases at several NLP tasks such as text classification, semantic similarity, and analogy. In conclusion, this approach can be used to increase performance at downstream tasks while requiring minimal additional computational resources.

Keywords: Meta-embedding · Word embeddings · Word2vec · FastText · Text classification · Semantic similarity · Analogy

1 Introduction and Motivation

The choice of word embedding model is an important hyperparameter for many NLP tasks, since it has been observed that different embedding models tend to provide stronger representations for different types of downstream tasks [4]. It is also known that ensembles of machine learning models tend to perform better than their individual constituents. It makes sense, then, to combine different embedding models in order to improve the performance of downstream NLP tasks.

While using ensembles of downstream models seeded with different types of word embeddings had been tried before [1], the idea of combining word embeddings directly to form meta-embeddings starts with the work of [21]. In that work the authors form meta-embeddings by concatenation, by factorization of the concatenated vectors (SVD), and a method called 1toN that learns a meta-embedding from which (also learned) projections exist to the source embeddings, with said projections minimizing the mean square error between the projected meta-embedding and source embedding of the same word for all words. A simpler but overlooked idea of averaging source embeddings is explored in [5]. In [3] autoencoders are employed to dimension reduce the concatenated (CAEME) and averaged (AAEME) meta-embeddings as well as dimension reducing source embeddings and concatenating them (DAEME).

E. Métais et al. (Eds.): NLDB 2019, LNCS 11608, pp. 393–401, 2019.
https://doi.org/10.1007/978-3-030-23281-8_35

One of the best known word embedding models is word2vec [13,14], which during its learning procedure not only learns a word vector for each word in the training corpus, but also a context vector for it. However context vectors are typically discarded after training. In [16] it is briefly mentioned that adding word and context vectors may result in a small performance boost. However it is not thoroughly investigated. In this study we investigate it in detail by forming meta-embeddings of word and context vectors in several different ways and conducting detailed experiments. We observe that combining the word and context embeddings to form a meta-embedding in several different settings yields a higher performance at the text classification, semantic similarity and analogy tasks.

In Sect. 2, we describe our novel approach and meta-embedding types. In Sect. 3, we describe our experimental setup, our implementation and NLP tasks that we perform. In Sect. 4, we present results of our meta-embedding methods on text classification, semantic similarity and word analogy tasks. In Sect. 5, we draw our conclusions based on our results and discuss possible extensions as future work.

2   Approach

Our novel approach focuses on exploiting otherwise ignored information encoded in context vectors. We formulate and experiment with seven different types of meta-embeddings. A total of nine results are given in our tables for comparison where the first two are traditional word and context embeddings which constitute the baselines.

In order to see if including context vectors help improving performance in several NLP tasks, first, we create a meta-embedding by concatenating word and context embeddings which is simply denoted by *concat*. This will result in doubling the dimensionality. Our second approach is to average word and context embeddings which is denoted by *average*. Third approach is to apply a max pooling filter to word and context embeddings to create a meta-embedding. This is donated by *maxpool*. Fourth one is a more complicated meta-embedding which is obtained by concatenation, averaging and maxpooling of word and context embeddings. This is indicated as *CAM* in our result tables. Following this, we have three additional auto-encoder based meta-embeddings [3] of word and context embeddings, namely Averaged Autoencoded Meta-Embedding (*AAEME*), Concatenated Autoencoded Meta-Embedding (*CAEME*), and Decoupled Autoencoded Meta-Embedding (*DAEME*). Please note that different meta-embedding approaches we have taken result in different dimensional vectors. This can be seen in Table 1.

In order to obtain word and context embeddings we use two of the most popular word embedding models; the word2vec and fastText [2,8]. For both models we use the skip-gram negative sampling architecture as it is more popular. In the case of fastText models while character n-grams are trained alongside word and context embeddings, we've choose not to include those in our meta-embeddings for comparability reasons.

Table 1. Embedding and dimension

Embedding	word	context	concat	average	maxpool	CAM	AAEME	CAEME	DAEME
Dimension	200	200	400	200	200	400	300	400	400

3   Experiments

3.1   Datasets

We trained our embeddings on Text8 [11], which is a corpus based on the first 10^9 bytes of the Wikipedia dump of March 3, 2006. For comparison we also trained our embeddings on a large Wikipedia dump. This corpus contains 19,251,790 articles and occupies approximately 16 GB of disk space.

For the text classification task we use the following datasets: AG's News Corpus [22] consisting of 120,000 documents in 4 classes, WEBKB which is a highly imbalanced dataset of 8,282 documents in 7 classes [12,17], Yelp Reviews Polarity [20,22] consisting of 560,000 documents in 2 classes, and DBPedia [9,22] also consisting of 560,000 documents but in 14 classes.

For the semantic similarity test we use the following datasets: WS [6] (353 word pairs), RG [19] (65 word pairs), RW [10] (2034 word pairs), SL [7] (999 word pairs).

For the analogy test we use the GL [14] dataset (19,557 analogy questions).

3.2   Experimental Setup

We use the gensim library [18] implementations of word2vec and fastText. For both we train vectors dimension of 200, and use default hyperparameters otherwise. We also use the word similarity and analogy tests implemented in the gensim library. We report the Spearman Correlation between the cosine similarity of word vectors and human assigned similarity scores.

For the text classification experiments we use Support Vector Machines (SVM) algorithm, more specifically Linear Support Vector Classifier (LinearSVC) which is commonly used in this domain. We use the one implemented in the scikit-learn library [15] with the default hyper parameters. Documents to be classified are represented as averages of their words' vectors.

The text classification experiments were run with 10-fold cross validation. We report the average accuracy and the standard deviations for the classification experiments.

In text classification experiments, in order to see if the performance improvement of meta-embeddings such as *concat* is due to the increased (actually doubled) number of dimensions or not, we conduct two sets of experiments. First, we compare meta-embeddings of size 200 (100 word + 100 context) with a baseline word embedding vectors of 200. In the second set of similar experiments we double the vector sizes.

4   Results and Discussion

4.1   Text Classification

According to our results, as seen in Tables 2 and 3, we see that for text classification the concatenation approach has a distinct advantage over all other approaches. The auto-encoder meta-embeddings appear to perform better than the average and the baseline meta-embeddings. However for the WEBKB dataset, which is a highly class imbalanced one, we observe a different pattern. In this dataset autoencoder based meta-embeddings perform poorly compared to others.

The concatenation meta-embeddings of both word2vec and fastText models exceed the classification performance of the other meta-embeddings in all datasets.

Table 2. Performance of word2vec meta-embeddings trained on text8 for text classification task

	AG News	WEBKB	Yelp Polarity	DBpedia
word	85.41 +/- 1.11	67.61 +/- 2.07	79.51 +/- 0.78	94.78 +/- 1.49
context	86.26 +/- 1.12	67.11 +/-2.45	81.29 +/- 0.79	94.75 +/- 1.52
concat	**87.42** +/- 1.07	69.56 +/- 2.27	**83.31** +/- 0.80	96.02 +/- 1.21
average	85.68 +/- 1.14	67.40 +/- 2.14	79.79 +/- 0.74	94.87 +/- 1.47
maxpool	85.51 +/- 1.07	66.97 +/- 2.17	79.33 +/- 1.00	94.63 +/- 1.49
CAM	86.79 +/- 1.06	**69.79** +/- 1.88	82.44 +/- 0.89	**96.05** +/- 1.20
AAEME	86.71 +/- 1.21	58.53 +/- 2.72	82.21 +/- 0.73	94.75 +/- 1.58
CAEME	86.90 +/- 1.15	59.14 +/- 2.72	82.96 +/- 0.75	94.99 +/- 1.54
DAEME	86.33 +/- 1.12	56.79 +/- 2.92	82.19 +/- 0.77	94.56 +/- 1.59

The improvement is most obvious in the Yelp Reviews Polarity dataset with an increase of 3.8% points over word embeddings for word2vec, and 4.27% points for fastText.

As the second step of experiments we run the same text classification tasks using our meta-embedding models trained using Wikipedia. According to results, as seen in Tables 4 and 5, again concatenation meta-embeddings of both word2vec and fastText models exceed the classification performance of the other meta-embeddings in all datasets.

For text classification, as seen in Table 6, we also compare the performance of same size meta-embedding and baseline embedding vectors. We observe that concatenation of word and context vectors still shows higher accuracy than word vectors by themselves, even though their dimensionalities are equal.

Table 3. Performance of fastText meta-embeddings trained on text8 for text classification task

	AG News	WEBKB	Yelp Polarity	DBpedia
word	85.32 +/- 1.16	68.08 +/- 2.37	79.78 +/- 0.74	94.29 +/- 1.75
context	86.61 +/- 1.23	67.51 +/-2.32	81.26 +/- 0.84	94.79 +/- 1.54
concat	**87.49 +/- 1.17**	70.13 +/- 2.30	**84.05 +/- 0.88**	**95.98 +/- 1.25**
average	85.77 +/- 1.16	68.23 +/- 2.23	80.16 +/- 0.79	94.51 +/- 1.66
maxpool	85.21 +/-1.28	67.21 +/- 2.69	79.43 +/- 0.67	94.79 +/- 1.86
CAM	86.86 +/- 1.08	**70.85 +/-2.60**	83.05+/- 0.80	95.72 +/- 1.31
AAEME	86.75 +/- 1.17	59.72 +/- 2.70	83.05 +/- 0.88	94.75 +/- 1.72
CAEME	87.07 +/- 1.20	60.31 +/- 2.74	83.86 +/- 0.83	95.15 +/- 1.59
DAEME	86.61 +/- 1.26	58.25 +/- 2.49	83.08 +/- 0.86	94.63 +/- 1.73

Table 4. Performance of word2vec meta-embeddings trained on Wikipedia for text classification task

	AG News	WEBKB	Yelp Polarity	DBpedia
word	88.82 +/- 1.11	68.31 +/- 3.02	84.06 +/- 0.73	96.53 +/- 1.20
context	88.98 +/- 1.21	67.96 +/- 3.30	84.71 +/-0.70	96.56 +/- 1.20
concat	**89.42 +/- 1.10**	**69.89 +/- 3.36**	86.07 +/- 0.67	96.90 +/- 1.11
average	88.98 +/- 1.16	68.13 +/- 3.19	84.50 +/- 0.69	96.53 +/-1.18
maxpool	88.87 +/- 1.21	68.23 +/- 3.10	84.09 +/- 0.73	96.40 +/- 1.22
CAM	89.38 +/- 1.14	69.04 +/-3.64	**86.23 +/- 0.68**	**96.97 +/- 1.11**

Table 5. Performance of fastText meta-embeddings trained on Wikipedia for text classification task

	AG News	WEBKB	Yelp Polarity	DBpedia
word	88.04 +/-1.35	67.30 +/- 1.97	82.12 +/- 1.47	96.32 +/-1.18
context	88.56 +/-1.13	66.48+/ -2.18	84.77 +/ -0.74	96.35 +/- 1.26
concat	**88.76 +/-1.10**	67.97 +/- 2.11	**85.34 +/- 1.01**	**96.83 +/- 1.07**
average	88.23 +/- 1.29	68.17 +/- 1.84	83.37+/- 0.78	96.38 +/- 1.19
maxpool	87.61 +/- 1.54	67.05 +/-4.01	81.79 +/-1.11	95.98 +/- 1.39
CAM	88.33 +/-1.21	**69.47 +/- 2.24**	83.17 +/- 2.79	96.69 +/- 1.20

Table 6. Performance comparison of word2vec and fastText meta-embedding *concat* with word embeddings of the same dimensionality on the text classification task

Model	Type	Dim	AG News	WEBKB	Yelp Polarity	DBpedia
word2vec	concat	400	**87.42 +/- 1.07**	**69.56 +/- 2.27**	**83.31 +/- 0.80**	**96.02 +/- 1.21**
word2vec	word	400	86.26 +/- 1.03	68.46 +/- 1.82	81.32 +/- 0.73	95.67 +/-1.24
word2vec	concat	200	**86.07 +/- 1.24**	67.50 +/- 2.55	**81.17 +/- 0.71**	**94.79 +/- 1.51**
word2vec	word	200	85.41 +/- 1.11	**67.61 +/- 2.07**	79.51 +/- 0.78	94.78 +/- 1.49
fastText	concat	400	**87.49 +/- 1.17**	**70.13 +/- 2.30**	**84.05 +/- 0.88**	**95.98 +/- 1.25**
fastText	word	400	86.30 +/- 1.07	69.59 +/-2.66	82.42 +/- 0.87	95.46 +/- 1.39
fastText	concat	200	**86.54 +/- 1.22**	**68.25 +/- 2.29**	**80.64 +/- 0.73**	**94.49 +/- 1.72**
fastText	word	200	85.32 +/- 1.16	68.08 +/- 2.37	79.78 +/- 0.74	94.29 +/- 1.75

4.2   Semantic Similarity and Word Analogy

Word2vec meta-embedding semantic similarity and analogy results that can be seen in Table 7, for three of the five datasets the average meta-embedding performs better. For the RW dataset auto-encoder based meta-embedding DAEME slightly outperforms the average meta-embedding. Interestingly other auto-encoder based meta-embeddings under perform the average meta-embedding in the same dataset. One outlier in the semantic similarity task is the SL dataset. In this one concatenation outperforms all other methods by a large margin.

In the fastText meta-embedding semantic similarity and analogy results which can be seen in Table 7, we observe a pattern differing from its word2vec counterparts. We see a much better picture for auto-encoder based meta-embedding methods. For four of the five datasets the auto-encoder base meta-embeddings outperform all others visibly. The only dataset where they do not is the RG dataset which is the smallest dataset used in the semantic similarity task. In this dataset the average meta-embeddings perform better.

Of note is the fact that for every dataset except the SL dataset, the average meta-embeddings outperform the concatenation meta-embeddings at the semantic similarity task, and at the analogy task as well.

We also see a difference in the performance of fastText word and context embeddings. For instance in the analogy task the context embeddings only solve 13.8% of the analogy questions, whereas the word vectors manage 40.6%. In the SL dataset for the semantic similarity task the context vectors significantly outperform word vectors (Spearman correlation of 0.3 versus 0.242). This should be due to the difference in the training of fastText and word2vec vectors. Namely in the word2vec model the similarity score is calculated as a function of the dot product between word and context vectors, whereas in the fastText model word and character n-gram embeddings are summed before computing a dot product with the context vectors.

Thus while word and context vectors are symmetric in the word2vec model, they are not in the fastText model. We suspect the differences in performance are due to this fundamental asymmetry.

Table 7. Performance of word2vec/fastText meta-embeddings trained on text8 for semantic similarity and analogy tasks

	WS	RG	RW	SL	GL
word2vec					
word	0.622	0.504	0.328	0.261	24.4
context	0.445	0.320	0.336	0.258	18.0
concat	0.625	0.487	0.347	**0.336**	24.4
average	**0.642**	**0.525**	0.367	0.269	**27.1**
maxpool	0.608	0.441	0.373	0.262	20.5
CAM	0.629	0.480	0.378	0.264	24.2
AAEME	0.593	0.441	0.348	0.285	25.4
CAEME	0.576	0.409	0.353	0.279	25.8
DAEME	0.598	0.444	**0.375**	0.270	24.9
	WS	RG	RW	SL	GL
fastText					
word	0.435	0.377	0.305	0.242	40.6
context	0.393	0.352	0.304	0.300	13.8
concat	0.437	0.365	0.309	0.251	41.1
average	0.473	**0.414**	0.328	0.254	42.1
maxpool	0.425	0.418	0.320	0.220	36.1
CAM	0.451	0.430	0.328	0.239	39.7
AAEME	**0.475**	0.398	0.345	0.316	**44.5**
CAEME	0.471	0.389	0.345	**0.320**	44.1
DAEME	0.427	0.397	**0.349**	0.317	39.2

5   Conclusions and Future Work

By combining word and context vectors of word2vec and fastText models using several different meta-embedding approaches we evaluate how much improvement context vectors can provide to word vectors' performances in downstream NLP tasks such as text classification, semantic similarity, and analogy. Furthermore we investigate which meta-embedding approaches are better at these tasks.

We show that even when we use a much larger training corpus for embedding models, resulting meta-embeddings show similar behavior, the concatenation of word and context embeddings usually leads to higher accuracy in text classification task.

It is interesting to note that just as the performances of word embedding models differ according to task, so do those of meta-embeddings of word and context vectors. In particular concatenation meta-embeddings perform better at text classification tasks, and average meta-embeddings tend to perform better at semantic similarity and analogy tasks.

We plan to combine word and context embeddings using a greater variety of meta-embedding methods. Namely, we think that the averaging method will perform better if the word and context embeddings are aligned via an orthogonal transformation first. We would also like to evaluate the 1toN [21] in this context. Another interesting approach will be inclusion of character n-gram embeddings of fastText in the various combinations.

In the future we would like to shed some light onto performance differences of auto-encoder based meta-embeddings by throughout analysis.

Acknowledgements. This work is supported in part by The Scientific and Technological Research Council of Turkey (TÜBİTAK) grant number 116E047. Points of view in this document are those of the authors and do not necessarily represent the official position or policies of the TÜBİTAK.

References

1. Bansal, M., Gimpel, K., Livescu, K.: Tailoring continuous word representations for dependency parsing. In: Proceedings of the 52nd Annual Meeting of the Association for Computational Linguistics (Volume 2: Short Papers), vol. 2, pp. 809–815 (2014)
2. Bojanowski, P., Grave, E., Joulin, A., Mikolov, T.: Enriching word vectors with subword information. Trans. Assoc. Comput. Linguist. **5**, 135–146 (2017). http://aclweb.org/anthology/Q17-1010
3. Bollegala, D., Bao, C.: Learning word meta-embeddings by autoencoding. In: Proceedings of the 27th International Conference on Computational Linguistics, pp. 1650–1661 (2018)
4. Chen, Y., Perozzi, B., Al-Rfou, R., Skiena, S.: The expressive power of word embeddings. In: ICML 2013 Workshop on Deep Learning for Audio, Speech, and Language Processing, Atlanta, GA, USA, July 2013. https://sites.google.com/site/deeplearningicml2013/TheExpressive-PowerOfWordEmbeddings.pdf
5. Coates, J., Bollegala, D.: Frustratingly easy meta-embedding-computing meta-embeddings by averaging source word embeddings. In: Proceedings of the 2018 Conference of the North American Chapter of the Association for Computational Linguistics: Human Language Technologies, Volume 2 (Short Papers), vol. 2, pp. 194–198 (2018)
6. Finkelstein, L., Gabrilovich, E., Matias, Y., Rivlin, E., Solan, Z., Wolfman, G., Ruppin, E.: Placing search in context: the concept revisited. ACM Trans. Inf. Syst. **20**(1), 116–131 (2002)
7. Hill, F., Reichart, R., Korhonen, A.: SimLex-999: evaluating semantic models with (genuine) similarity estimation. Comput. Linguist. **41**(4), 665–695 (2015)
8. Joulin, A., Grave, E., Bojanowski, P., Mikolov, T.: Bag of tricks for efficient text classification. In: Proceedings of the 15th Conference of the European Chapter of the Association for Computational Linguistics: Volume 2, Short Papers, pp. 427–431. Association for Computational Linguistics (2017). http://aclweb.org/anthology/E17-2068
9. Lehmann, J., Isele, R., Jakob, M., Jentzsch, A., Kontokostas, D., Mendes, P.N., Hellmann, S., Morsey, M., Van Kleef, P., Auer, S., et al.: Dbpedia–a large-scale, multilingual knowledge base extracted from Wikipedia. Semant. Web **6**(2), 167–195 (2015)

10. Luong, T., Socher, R., Manning, C.: Better word representations with recursive neural networks for morphology. In: Proceedings of the Seventeenth Conference on Computational Natural Language Learning, pp. 104–113 (2013)
11. Mahoney, M.: About the test data (2011). http://mattmahoney.net/dc/textdata
12. McCallum, A., Nigam, K., et al.: A comparison of event models for Naive Bayes text classification. In: AAAI-98 Workshop on Learning for Text Categorization, vol. 752, pp. 41–48. Citeseer (1998)
13. Mikolov, T., Chen, K., Corrado, G., Dean, J.: Efficient estimation of word representations in vector space. CoRR abs/1301.3781 (2013). arxiv:1301.3781
14. Mikolov, T., Sutskever, I., Chen, K., Corrado, G., Dean, J.: Distributed representations of words and phrases and their compositionality. In: Proceedings of the 26th International Conference on Neural Information Processing Systems - Volume 2, NIPS 2013, pp. 3111–3119. Curran Associates Inc., USA (2013). http://dl.acm.org/citation.cfm?id=2999792.2999959
15. Pedregosa, F., et al.: Scikit-learn: machine learning in Python. J. Mach. Learn. Res. 12, 2825–2830 (2011)
16. Pennington, J., Socher, R., Manning, C.: Glove: Global vectors for word representation. In: Proceedings of the 2014 Conference on empirical methods in Natural Language Processing (EMNLP), pp. 1532–1543 (2014)
17. Poyraz, M., Kilimci, Z.H., Ganiz, M.C.: Higher-order smoothing: a novel semantic smoothing method for text classification. J. Comput. Sci. Technol. 29(3), 376–391 (2014)
18. Řehůřek, R., Sojka, P.: Software framework for topic modelling with large corpora. In: Proceedings of the LREC 2010 Workshop on New Challenges for NLP Frameworks, Valletta, Malta, pp. 45–50. ELRA, May 2010. http://is.muni.cz/publication/884893/en
19. Rubenstein, H., Goodenough, J.B.: Contextual correlates of synonymy. Commun. ACM 8(10), 627–633 (1965)
20. Yelp: Yelp reviews dataset challenge (2015). https://www.yelp.com/dataset/challenge
21. Yin, W., Schütze, H.: Learning word meta-embeddings. In: Proceedings of the 54th Annual Meeting of the Association for Computational Linguistics (Volume 1: Long Papers), vol. 1, pp. 1351–1360 (2016)
22. Zhang, X., Zhao, J., LeCun, Y.: Character-level convolutional networks for text classification. In: Advances in Neural Information Processing Systems, pp. 649–657 (2015)

Extracting Statistical Mentions from Textual Claims to Provide Trusted Content

Tien Duc Cao[1,2(✉)], Ioana Manolescu[1,2(✉)], and Xavier Tannier[3(✉)]

[1] Inria Saclay Île-de-France, Palaiseau, France
{tien-duc.cao,ioana.manolescu}@inria.fr
[2] LIX (UMR 7161, CNRS and École Polytechnique), Palaiseau, France
[3] Sorbonne Université, Inserm, LIMICS (UMRS 1142), Paris, France
xavier.tannier@sorbonne-universite.fr

Abstract. Claims on statistic (numerical) data, e.g., immigrant populations, are often fact-checked. We present a novel approach to extract from text documents, e.g., online media articles, mentions of statistic entities from a reference source. A claim states that an entity has certain value, at a certain time. This completes a fact-checking pipeline from text, to the reference data closest to the claim. We evaluated our method on the INSEE dataset and show that it is efficient and effective.

1 Introduction

With the increase of disinformation in online media, social networks and the Web in general, we witness a strong interest in computational fact-checking, defined as a set of computer-assisted techniques capable of assessing the truthfulness of a given statement [4]. In this context, computational fact-checking is a many-stage pipeline, whereas (*i*) claims are extracted from text, (*ii*) possible sources of reference are identified, (*iii*) a check is made combining automated and manual means; (*iv*) an interpretation is produced.

In this paper, we focus on steps (*i*) and (*ii*). We use data from French national institute for statistics and economic studies (INSEE) as an example as high-quality, trustful reference database. In previous work, we have **extracted tens of thousands of RDF graphs** out of INSEE statistic tables [2][1]. We also developed a novel **keyword search algorithm** which, given a set of search terms, e.g. *"unemployment"*, *"Île-de-France"*, *"2018"* locates the RDF nodes corresponding to the most relevant table cells [3][2].

In this work, we describe the last missing step of our system: the **extraction of claims referring to statistical mentions from text sources**. This step allows to automatically formulate the search queries which our system [3] can solve against the RDF corpus we gathered [2]. Our whole system can help fact-checking journalists to find checkable claims in massive text sources, as well as

---

[1] https://gitlab.inria.fr/tcao/insee-search/blob/master/insee-rdf.ttl.gz.
[2] The search algorithm is deployed online at http://statsearch.inria.fr.

© Springer Nature Switzerland AG 2019
E. Métais et al. (Eds.): NLDB 2019, LNCS 11608, pp. 402–408, 2019.
https://doi.org/10.1007/978-3-030-23281-8_36

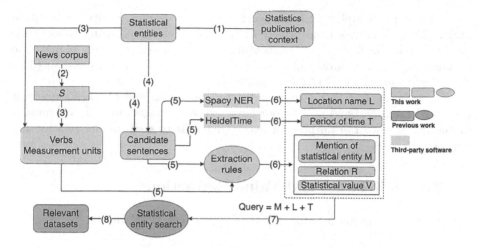

Fig. 1. Main processing steps of our statistical claim extraction method.

the closest reference datasource value for the given claim. Based on these, the journalists can choose the truth label which seems most appropriate.

The architecture of the system is presented by Fig. 1. From the publication context of statistic data (the text in header of statistics tables) we extract a set of **statistical entities** (step (1) in the figure), those whose reference values are known in the statistic dataset for some time periods and/or geographical area, such as *"unemployment"*, *"youth unemployment"*, *"unemployment in Aquitaine in 2015"*, *"gross domestic product"*. From 111,145 tables published by INSEE, we have obtained a total of 1,397 statistic entities, as we detail in Sect. 2.1.

We have built a **text corpus** which we selected with an interest in topics that INSEE studies. We focused on news articles from three French newspapers, and because most INSEE metrics refer to the economy domain, we looked only for articles on such topics, by using URL keywords or an LDA [1] topic selection[3]. From these articles, we have extracted (step (2)) 322,873 sentences containing at least one numerical value. From now on, we will refer to these sentences as S. From S, we extract (step (3)) all the verbs which state a numerical value, e.g., "amounts to", "is worth", "decreases" etc., as well as all the measurement units, e.g., "people", "euros", "percentage" etc.

Next, we identify among S sentences the **candidate sentences** which could claim a relationship between a statistical entity and a value. This is done (step (4)) by selecting those S sentences which mention statistic entities. From each candidate sentence, e.g., *"France's public debt fell slightly, by 11.4 billion euros, between the second and third quarters of 2013"*, we extract: **(1)** a mention of statistical entity M, e.g., *public debt*; **(2)** optionally, a location L, e.g., *France*,

---

[3] All topics and their keywords are available at https://gitlab.inria.fr/tcao/news-scraper/blob/master/lesechos_topics_all.txt. In this work, we use topics 1, 2, 3, 7.

by extracting geographical places using the spaCy Named Entity Recognition tool[4]; **(3)** optionally, a time period T, e.g., *2013*, extracted using HeidelTime [9]; **(4)** a relation R, e.g., *fell*, connecting M to V in the sentence. R may also be missing, e.g., in a phrase such as *"France's 60 million inhabitants..."*; **(5)** a statistical value V, e.g., *"11.4 billion euros"*.

For each (M, L, T, R, V) tuple extracted as above, the (M, L, T) query is generated (step (7)) and sent to our keyword search algorithm [3]. We omit R in the query since the purpose of extracting R is to confirm the relationship between M and V.

2   Entity, Relation and Value Extraction

We present our approach to extract the components M, R and V.

2.1   Statistical Entities

We made a hypothesis of the existence of statistical entities in the headers of statistic tables. For example, one header of table[5] is *"Taux de chômage au T1 2015"* ("Unemployment rate in the first quarter of 2015"). We keep only headers that contain a measurement unit such as euro, %, etc. These headers are usually noun phrases in format *Entity + (Unit)* such as "Unemployment rate in 2015 (in %)". We prefer to rely on table headers and not on table titles and comments, since the latter are longer sentences that could (or could not) contain the entities, and customarily do contain much more irrelevant information. We also filter out possible date time values and their associated prepositions. In the above example,

Table 1. Sample extracted statistical entities

Extracted statistical entities	Frequency
intensité de la pauvreté (intensity of poverty)	190
nombre d'entreprises (number of companies)	176
taux de pauvreté au seuil de 60% (poverty rate at 60% median wages)	130
chômeurs (unemployed people)	104
excédent brut d'exploitation (Earnings before Interest, Taxes and Amortization)	68
PIB (gross domestic product, GDP)	54
taux de population en sous-emploi (share of people working less than they would like)	54
solde migratoire (net migration)	44
taux de marge (margin rate)	28
taux de pauvreté (poverty rate)	21

---

[4] https://spacy.io/models/fr#fr_core_news_md.
[5] https://www.insee.fr/fr/statistiques/1288156#tableau-Figure_2.

this leads to the snippet "Unemployment rate". A final manual filtering allowed us to weed out some text snippets which do not in fact comprise relevant entities.

We thus obtained 1,397 statistical entities, some of which are presented, with their frequencies from the statistics publication context, in Table 1.

2.2   Relevant Verbs and Measurement Units

We use the annotation S_I to refer to the candidate sentences that contain the word "insee". These sentences are likely to feature a relationship between a mention of statistical entity M as a noun phrase (e.g. "unemployment rate") and a statistical value V as a numerical value, optionally followed by a measurement unit (e.g. "5%").

We used spaCy [7] to collect the syntactic dependency paths connecting M, R and V. For each NOUN node, we located the paths that connect it to a NUM node. Many paths start with (NOUN, nsubj, VERB) (a noun is subject of a verb); we refer to them as $Paths_I$. As the relation R of M and V is generally introduced by specific verbs, we collected all the verbs associated with VERB nodes from $Paths_I$. To make sure of the quality of the collected verbs, we filtered manually from the original list to retain 129 relevant ones; in the sequel, we denote them by L_verbs. Based on $Paths_I$, we also gathered a set of measurement units by collecting all the NOUN nodes connected to a NUM node via a nmod edge (nominal modifiers of nouns or noun phrases). We call this list L_units.

2.3   Extraction Rules

Given the input sentence i and a statistical entity e, we extract the mention of statistical entity M, the statistical value V and their relation R. If there is no relationship between e and the statistical value, or there is no statistical value in i, we return the value $M = None$. We identify from the dependency tree the statistical entity e and the numerical value(s), as follows.

1. We filter out the year values (e.g. 2018) since we only want to search for the relationship of statistical entity and statistical value.
2. We define the distance $d(n_1, n_2)$ of two nodes n_1 and n_2 in $t(i)$ as the absolute value of n_1's position - n_2's position. For instance, $d(inflation, établie) = 3$.
3. The distance $D(e, v)$ from e to a numerical value v is the minimum value of $d(e$'s first word, $v)$ and $d(e$'s last word, $v)$. In case there are more than one numerical values, we select the one that has the smallest $D(e, v)$ as the statistical value of e.
4. We identify the dependency path $p(i)$ that connects **the first word of** e (let's call it s) and e**'s statistical value** (if available), let's call it n. With our sample dependency tree, $p(i) = $ (NOUN, nsubj:pass, VERB, obl, NOUN, nummod, NUM)

5. We look for the node u directly connected to n (the last one before n) in $p(i)$. If u is a noun and there is a nmod edge (nominal modifiers of nouns or noun phrases) between u and n, we return $M = None$ in the following cases:
   - u does not appear in L_units.
   - u appears in L_units and in the input sentence, there is an article or an adposition between s and u.

   On the contrary, we extract the relevant nodes from:
   (a) the first NOUN node s: we identify the nodes that connect to s via nmod and amod (adjectival modifier) edges, and we collect their subtrees.
   (b) the VERB node $verb$: the subtree of nodes that connect to $verb$ via obl edge (a nominal dependent of a verb), the leftmost node of subtree must be a preposition among en, $à$, $dans$ and $verb$ has to appear in L_verbs.

   If the nodes from these subtrees appear in $p(i)$, we do not include them.

All the extracted nodes form the mention of statistical entity M. The statistical value V is composed of n and u. The relation R is composed the nodes from $p(i)$ which do not belong to M and V.

3 Evaluation

Evaluation of the Extraction Rules. We select some statistical entities[6] from the list of statistical entities in Sect. 2.1. For each entity e we pick randomly 50 sentences that contain e then we split randomly 25 sentences for development set and 25 sentences for test set. Finally there are 200 sentences for each set. If there is no relationship between e and the statistical value, or there is no statistical value in the given sentence, we assign a label *NoStats*. Otherwise we annotate each sentence with e and the relevant phrases (we call these phrases *contexts* of e) to form a mention of statistical entity[7]. For a given sentence, if the extraction rules return $M = None$ and we have the *NoStats* label from the annotated sentence then the extraction is an accurate one. On the contrary, we verify if the extracted M contains e and one of its contexts. In that case, the extraction is also accurate. The accuracy of our extraction rules in the development, resp. test set and obtain is 71.35%, respectively and 69.63%.

Evaluation of the End-to-End System. We selected randomly 38 sentences for the test set (from which 26 were considered as extracted correctly at previous step – Sect. 3). We gave the corresponding generated queries $q = M + L + T$ as input to the INSEE-Search system [3]. We evaluated the accuracy of the system using a modified version of the mean average precision metric, (MAP) widely used for evaluating ranked lists of results. MAP is traditionally defined based on a binary relevance judgment (relevant or irrelevant in our case). We experimented with the two possibilities:

---

[6] *"taux de chômage"*, *"nombre de demandeurs d'emploi"*, *"niveau de vie"*, *"consommation des ménages"*, *"PIB"*, *"inflation"*, *"SMIC"*, *"taux d'emploi"*.

[7] The annotated data is available at https://gitlab.inria.fr/tcao/text2insee/.

- MAP_h is the mean average precision where only highly relevant datasets are considered as relevant ($MAP_h(10)$ is computed on the top 10 search results).

- MAP_p is the mean average precision where both partially and highly relevant datasets are considered relevant.

Note that there is no guarantee that any "highly relevant" element at all exists in the dataset for each query.

The results (Table 2) show that, given an arbitrary claim (related to statistic entities), fine-grained and relevant information can be returned in the vast majority of the cases. They also show that, as in all keyword-based search systems, building a perfect query is neither necessary or sufficient for obtaining good results. Even if a good entity extraction improves the results, we can still find highly or partially relevant information even if the entity extraction is not perfectly achieved. Our findings should be confirmed by an evaluation on more claims, more databases and in a real-user study. We also showed in [3] that the performance of our query system was similar to a document-level search engine such as Google, but with a much better granularity (data cell instead of page).

Table 2. Evaluation of INSEE-Search

		$MAP_h(10)$	$MAP_p(10)$
Overall performance (38 sentences)		0.672	0.789
among which	M extracted correctly (26)	0.725	0.829
	M extracted incorrectly (12)	0.559	0.703

4 Related Work and Perspectives

BONIE [8] claims to be the first open numerical relation extractor. The system is based on high precision patterns to extract seed facts from input sentences and on bootstrapping to increase the number of seed facts and to learn patterns. We tried their approach, but found that the learnedt patterns were either too generic or too specific and failed to capture the correct dependency path in the new texts. ClausIE [5] is an open information extraction system. It first detects clauses in a sentence and then apply specific rules for each type of clause in order to extract the entity of interest. ClausIE also makes use of a hand-crafted dictionary of verbs to identify the existence of relation in sentence. Compare to their approach, we have a "semi-automated" solution to identify the list of verbs. ClaimBuster [6] was the first work on check-worthiness. They used annotated sentences from US election debates to train a SVM classifier in order to determine whether or not a sentence is a check-worthy claim. This is the common approach when having a large amount of training data, which is not the case in French.

In this article we have presented an end-to-end system for identifying statistic claims and finding in a statistic database the relevant statistic data for checking this claim. A classic defect of these pipeline approaches in NLP systems is that errors accumulate at each step. Nevertheless, our results show that we often manage to find useful information for the user, which will make the human work of fact-checking easier and faster. To make the RDF graph up-to-date, our crawler works on a daily basis to collect the latest statistic tables. We also leave journalists state whether the claim is "true", "mostly true", "mostly false" etc.

References

1. Blei, D.M., Ng, A.Y., Jordan, M.I.: Latent dirichlet allocation. J. Mach. Learn. Res. **3**, 993–1022 (2003)
2. Cao, T., Manolescu, I., Tannier, X.: Extracting linked data from statistic spreadsheets. In: International Workshop on Semantic Big Data (2017)
3. Cao, T.D., Manolescu, I., Tannier, X.: Searching for truth in a database of statistics. In: WebDB (2018)
4. Cazalens, S., Lamarre, P., Leblay, J., Manolescu, I., Tannier, X.: A content management perspective on fact-checking. In: WWW (2018)
5. Corro, L.D., Gemulla, R.: ClausIE : Clause-based open information extraction. In: WWW (2013)
6. Hassan, N., et al.: Claimbuster: the first-ever end-to-end fact-checking system. In: PVLDB (2017)
7. Honnibal, M., Johnson, M.: An improved non-monotonic transition system for dependency parsing. In: EMNLP (2015)
8. Saha, S., Pal, H.: Mausam: bootstrapping for numerical open IE. In: ACL (2017)
9. Strötgen, J., Gertz, M.: HeidelTime: High quality rule-based extraction and normalization of temporal expressions. In: International Workshop on Semantic Evaluation (2010)

Aspect Extraction from Reviews Using Convolutional Neural Networks and Embeddings

Peiman Barnaghi, Georgios Kontonatsios, Nik Bessis, and Yannis Korkontzelos[✉]

Edge Hill University, Liverpool, UK
{barnaghp,Georgios.Kontonatsios,Nik.Bessis,
Yannis.Korkontzelos}@edgehill.ac.uk

Abstract. Aspect-based sentiment analysis is an important natural language processing task that allows to extract the sentiment expressed in a review for parts or aspects of a product or service. Extracting all aspects for a domain without manual rules or annotations is a major challenge. In this paper, we propose a method for this task based on a Convolutional Neural Network (CNN) and two embedding layers. We address shortcomings of state-of-the-art methods by combining a CNN with an embedding layer trained on the general domain and one trained the specific domain of the reviews to be analysed. We evaluated our system on two SemEval datasets and compared against state-of-the-art methods that have been evaluated on the same data. The results indicate that our system performs comparably well or better than more complex systems that may take longer to train.

Keywords: Aspect-based sentiment analysis · Aspect extraction · Convolutional Neural Networks · Deep learning · NLP

1 Introduction

Currently immense volumes of text-based reviews are available, in a great variety of diverse domains. Consumers can share their experience on services and products. Natural Language Processing (NLP) methods can be used to extract meaningful information from this data. Quantifying sentiment expressed for various aspects of a product or service can help producers and consumers to monitor, assess and make decisions. Significant volume of research has focused on Extensive research has focussed on analysing online reviews for a variety of topics or products, e.g. movies, restaurants, mobile applications and software projects.

Aspect-based sentiment analysis is a variation of sentiment analysis that considers different aspects of the object of a text-based review and classifies the comments for each aspect as positive, negative or neutral. For example, in the comment "the food is great but expensive and service is slow" three aspects are mentioned, i.e. quality, price and service. Lately, neural networks

© Springer Nature Switzerland AG 2019
E. Métais et al. (Eds.): NLDB 2019, LNCS 11608, pp. 409–415, 2019.
https://doi.org/10.1007/978-3-030-23281-8_37

have been shown to perform very well in sentiment analysis when combined with word embeddings. Word embeddings are vector representations of textual vocabularies, useful for finding similar words. Each word is mapped to a vector that captures its context in different sentences. Embeddings retain syntactic and semantic similarities and relations among words. Most neural network based systems for text analysis have employed Convolutional Neural Networks (CNN) and Recurrent Neural Networks (RNN).

Given the success of CNN [7] on aspect extraction, we propose a CNN-based system for extracting aspects from reviews and we combine it with different embedding layers. CNN models are less complex than RNN models. Experiments in the literature show that CNN models train faster than RNN models and tuning their hyper-parameters is simpler. State-of-the-art aspect extraction systems combine neural networks with word embeddings. The contribution or this paper is the combination of a CNN with two word embeddings concurrently: one trained on the general domain and one trained on the domain of the reviews. The model performs comparably or better than methods that integrate more complex architectures, such as RNNs. It outperforms a CNN-based model that uses either general domain embeddings or domain-specific ones, only.

2   Related Work

Aspect based sentiment analysis identifies sentiment expressed for each aspect of a product or service. It was introduced for summarising customer reviews and was addressed by a rule-based model [5]. Since then, a variety of systems have been proposed and several competition tasks have been organised in the SemEval (Semantic Evaluation) series. Task 4 A in SemEval 2014 focussed on the extraction of aspects in reviews. Liu [9] discussed four approaches for aspect identification: frequent terms, opinion and target relations, supervised classification and topic modelling algorithms. Conditional Random Fields (CRF) have been employed to consider long term dependencies when extracting aspects [6], and performed better than other supervised models for feature extraction [23]. Toh and Wang [22] used a tagging model with linguistic features that consider resources, such as WordNet, for aspect extraction and polarity classification. Brun et al. [1] combined word features, parsing and a sentiment lexicon to train a Support Vector Machine (SVM) for aspect-based sentiment classification.

In SemEval 2016, the best performing system used CRFs for sequential labelling, i.e. aspect extraction, and a single-layer feed-forward neural network for classification [18]. A CNN-based aspect extraction method tagged each word in subjective text [16]. The CNN tags each word as aspect or not, in different layers. The model performed better than state-of-the-art approaches. CNNs, as non-linear models, fit the data better that linear models, such CRFs.

In summary, latest research uses deep learning to improve aspect extraction and aspect-based sentiment classification performance, as it has been very successful in supervised and unsupervised settings. Shortcomings of these models

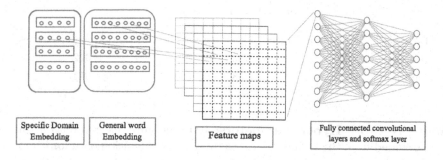

Fig. 1. Proposed CNN model with domain and general word embedding.

for extracting hidden aspects concern long distance dependencies and domain-specific expressions. In this paper, we address the latter shortcoming, by combining general and domain-specific embeddings.

3   Method

To extract aspects of reviews, we use a CNN [7] with fully connected layers combined with two independent embedding layers, as shown in Fig. 1. The input is a sentence of any size that mentions zero or more aspects. Each word of the sentence is looked up in both embeddings and the two resulting vectors are concatenated together. The general embedding is a pre-trained Global Vectors for word representation (GloVe) model [13], trained on 840 billion tokens. The vocabulary size is 2.2 million vectors of dimension 300. We selected this model due to the size of the data it was trained on and its popularity for aspect extraction. The domain-specific embeddings are trained either on *Yelp* [11], a restaurant review dataset, or on *Amazon reviews* for laptops [4]. Reviews in both datasets come labeled with aspect terms.

The joined vector is the input of the multi-layer CNN. Each layer uses a convolutional filter of fixed window width and kernel size. For example, with kernel size $k = 5$, two words on the left and right of the current one are kept. Each filter represents each word and its nearby words. An activation function is used to choose the maximum value of each features node, and a dropout is applied to prevent overfitting during training. Finally, a Softmax layer is applied to a fully connected layer to select the sequence with the highest position weight and assign a label to each word, accordingly.

4   Experiments

Datasets: For evaluation, we use two benchmark SemEval datasets: the laptop review dataset in SemEval-2014 Task 4 and the restaurant review dataset in SemEval-2016 Task 5. Table 1 shows statistics of the two datasets.

Table 1. Dataset statistics

Dataset	Training instances	Aspect terms	Test instances	Aspect terms
Restaurant	2000	1743	676	622
Laptop	3,045	2358	800	654

Tuning Network Hyper-Parameters: 100 randomly-selected data instances were excluded from the training data, to be used as validation data for tuning. A popular technique for avoiding underfitting, is to evaluate the model for various layer sizes, parameters and learning rates. If validation accuracy is higher than training accuracy then the model is underfitting, or otherwise it is overfitting. Each CNN layer consists of 256 filters of kernel size 3. Processing continues to the end of the vector and feature weights are computed. We used common parameter values for the dropout and learning rates: 0.5 and 10^{-4}, respectively.

Evaluation: Following common practice, we use F-score (F1), the geometric mean of precision and recall. For evaluation, we used the SemEval script. We compare our proposed model with all methods, for which results on the SemEval 2014 and 2016 datasets have been made available [8].

IHS_RD [2] was the best system on laptop reviews in task 4 of SemEval 2014. It used conditional random fields for cross-domain feature extraction [15].

NLANGP [18] was the best system in aspect extraction on restaurant reviews in task 5 of SemEval 2016. It is also based on neural networks [14].

AUEB [20] a CRF-based method for sequence labeling that uses hand-crafted features and embeddings. It was ranked among the top systems in SemEval 2016.

CRF [12] a CRF-based method using general embeddings and basic features.

Semi-Markov CRF (Semi-CRF) [17] uses features in Cuong et al. [3].

DLIREC [22] a CRF-based classifier that uses semantic features and clustering on unlabeled data. It was ranked second in both SemEval tasks.

WDEmb [21] a CRF that uses linear and dependency context information.

RNCRF [19] a CRF and RNN combination for aspect extraction.

LSTM [10] uses an RNN and general pre-trained word embeddings.

MIN [8] uses two LSTM models for aspect extraction and one for sentiment classification.

We used two baselines: (1) the proposed multilayer CNN model with general word embeddings, only; and (2) the proposed model with in-domain word embeddings, only. The performance gap between the baselines and the proposed system shall highlight the impact of combining the two word embeddings.

5   Discussion

The experimental results in Table 2 show that the proposed method, (PM-G&D), performs better than state-of-the-art systems for aspect extraction. It outperforms the two baselines that use the same model with either general (PM-G) or in-domain embeddings (PM-D), only. This result stresses the contribution of the combination of the two embeddings, since all other settings are kept the same.

The performance difference between the two datasets indicates that the in-domain embedding is more effective in laptop reviews. This is probably because this domain has more keywords than restaurant reviews, which mainly contains general words possibly available in general embeddings. Our model performs better than CRF-based models that specialise on label dependencies, since both datasets mostly contain single-word aspects.

Table 2. F-score results on the restaurant (R) and laptop (L) review dataset. PM, G and D stand for our proposed model, general and in-domain embedding, respectively.

Data	IHS_RD	NLANGP	AUEB	CRF	Semi-CRF	DLIREC	WDEmb	RNCRF	LSTM	MIN	PM-G	PM-D	PM-G&D
R	-	72.34	70.44	69.56	66.35	-	-	69.74	71.26	73.44	69.80	68.23	**73.81**
L	74.55	-	-	74.01	68.75	73.78	75.16	77.26	75.25	77.58	73.19	73.37	**78.26**

As all systems in Table 2 are using general embeddings, the results show that combining large general and small in-domain word embeddings can improve aspect extraction performance. Figure 2 shows the effect of increasing the training data size on performance. Results improve mildly as the size of the training data for the in-domain embeddings increases.

Embeddings can be combined to improve the aspect extraction performance in other domains and probably other languages. Although recent work shows that RNN models are state-of-the-art, we have achieved comparable results using a much simpler model, which is faster to train, by adding an extra learning layer.

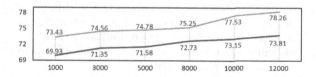

Fig. 2. Performance for increasing size of the domain-specific embedding layer for restaurants (blue line) and laptops (red line) (color figure online)

6 Conclusion and Future Work

We proposed a new model for aspect extraction from text-based reviews. It uses a convolutional neural network and two word embedding layers: a general and a domain-specific one, trained on data of the specific domain of the reviews. Evaluation on two benchmark SemEval datasets, containing restaurant and laptop reviews, shows that the model performs comparably or better than more complex neural network architectures, that take longer to train. In the future, we plan to comparatively evaluate more aspect extraction methods, deep learning architectures and embedding types on diverse domains.

Acknowledgment. This research work is part of the TYPHON Project, which has received funding from the European Union's Horizon 2020 Research and Innovation Programme under grant agreement No. 780251.

References

1. Brun, C., Popa, D.N., Roux, C.: XRCE: hybrid classification for aspect-based sentiment analysis. In: Proceedings of SemEval, pp. 838–842 (2014)
2. Chernyshevich, M.: IHS R&D belarus: cross-domain extraction of product features using CRF. In: Proceedings of SemEval, pp. 309–313 (2014)
3. Cuong, N.V., Ye, N., Lee, W.S., et al.: Conditional random field with high-order dependencies for sequence labeling and segmentation. J. Mach. Learn. Res. **15**(1), 981–1009 (2014)
4. He, R., McAuley, J.: Ups and downs: modeling the visual evolution of fashion trends with one-class collaborative filtering. In: Proceedings of WWW, pp. 507–517 (2016)
5. Hu, M., Liu, B.: Mining and summarizing customer reviews. In: Proceedings of SIGKDD, pp. 168–177. ACM (2004)
6. Jakob, N., Gurevych, I.: Extracting opinion targets in a single-and cross-domain setting with conditional random fields. In: Proceedings of EMLNP, pp. 1035–1045 (2010)
7. Kim, Y.: Convolutional neural networks for sentence classification. arXiv preprint arXiv:1408.5882 (2014)
8. Li, X., Lam, W.: Deep multi-task learning for aspect term extraction with memory interaction. In: Proceedings of EMNLP, pp. 2886–2892 (2017)
9. Liu, B.: Sentiment analysis and opinion mining. Synth. Lect. Hum. Lang. Technol. **5**(1), 1–167 (2012)
10. Liu, P., Joty, S., Meng, H.: Fine-grained opinion mining with recurrent neural networks and word embeddings. In: Proceedings of EMNLP, pp. 1433–1443 (2015)
11. Mikolov, T., Chen, K., Corrado, G., et al.: Efficient estimation of word representations in vector space. arXiv preprint arXiv:1301.3781 (2013)
12. Okazaki, N.: CRFsuite: a fast implementation of conditional random fields (CRFs) (2007). www.chokkan.org/software/crfsuite
13. Pennington, J., Socher, R., Manning, C.: Glove: global vectors for word representation. In: Proceedings of EMNLP, pp. 1532–1543 (2014)
14. Pontiki, M., Galanis, D., Papageorgiou, H., et al.: SemEval-2016 task 5: aspect based sentiment analysis. In: Proceedings of SemEval, pp. 19–30 (2016)
15. Pontiki, M., Galanis, D., Pavlopoulos, J., et al.: SemEval-2014 task 4: aspect based sentiment analysis. In: Proceedings of SemEval, pp. 27–35 (2014)
16. Poria, S., Cambria, E., Gelbukh, A.: Aspect extraction for opinion mining with a deep convolutional neural network. Knowl. Based Syst. **108**, 42–49 (2016)
17. Sarawagi, S., Cohen, W.: Semi-markov conditional random fields for information extraction. In: Proceedings of NIPS, pp. 1185–1192 (2005)
18. Toh, Z., Su, J.: Nlangp at SemEval-2016 task 5: Improving aspect based sentiment analysis using neural network features. In: Proceedings of SemEval, pp. 282–288 (2016)
19. Wang, W., Pan, S.J., Dahlmeier, D., et al.: Recursive neural conditional random fields for aspect-based sentiment analysis. arXiv preprint arXiv:1603.06679 (2016)

20. Xenos, D., Theodorakakos, P., Pavlopoulos, J., et al.: Aueb-absa at SemEval-2016 task 5: ensembles of classifiers and embeddings for aspect based sentiment analysis. In: Proceedings of SemEval, pp. 312–317 (2016)
21. Yin, Y., Wei, F., Dong, L., et al.: Unsupervised word and dependency path embeddings for aspect term extraction. arXiv preprint: arXiv:1605.07843 (2016)
22. Zhiqiang, T., Wenting, W.: DLIREC: aspect term extraction and term polarity classification system. In: Proceedings of the 8th International Workshop on Semantic Evaluation (SemEval 2014), pp. 235–240 (2014)
23. Zhuang, L., Jing, F., Zhu, X.Y.: Movie review mining and summarization. In: Proceedings of CIKM, pp. 43–50. ACM (2006)

Author Index

Printed in the United States
By Bookmasters